Vaughan Public Libraries
2191 Major Mackenzie Dr.
Maple, Ontario
905-653-READ (7323)

Patrick Moore's
Yearbook of Astronomy
2015

Patrick Moore's Yearbook of Astronomy 2015

EDITED BY
John Mason

MACMILLAN

First published 2014 by Macmillan
an imprint of Pan Macmillan, a division of Macmillan Publishers Limited
Pan Macmillan, 20 New Wharf Road, London N1 9RR
Basingstoke and Oxford
Associated companies throughout the world
www.panmacmillan.com

ISBN 978-1-4472-8517-5

Copyright © Macmillan Publishers 2014

The right of John Mason to be identified as the
author of this work has been asserted by him in accordance
with the Copyright, Designs and Patents Act 1988.

All rights reserved. No part of this publication may be reproduced,
stored in a retrieval system, or transmitted, in any form, or by any means
(electronic, mechanical, photocopying, recording or otherwise)
without the prior written permission of the publisher.

1 3 5 7 9 8 6 4 2

A CIP catalogue record for this book is available from the British Library.

Typeset by Ellipsis Digital Limited
Printed and bound by CPI Group (UK) Ltd, Croydon, CR0 4YY

This book is sold subject to the condition that it shall not, by way of
trade or otherwise, be lent, hired out, or otherwise circulated without
the publisher's prior consent in any form of binding or cover other than
that in which it is published and without a similar condition including
this condition being imposed on the subsequent purchaser.

Visit **www.panmacmillan.com** to read more about all our books
and to buy them. You will also find features, author interviews and
news of any author events, and you can sign up for e-newsletters
so that you're always first to hear about our new releases.

Contents

Part Two

Article Section

Part Three
Miscellaneous

Editor's Foreword

I am delighted to be writing this Foreword to the 2015 *Yearbook of Astronomy*. My only sadness is that my good friend Sir Patrick Moore, who guided the publication of this Yearbook for over fifty years, is no longer with us. I have missed him greatly while putting this edition together – not only his encyclopaedic astronomical knowledge which was at all times so invaluable, but also the great fun that we always had working on it together.

When Patrick invited me to join him in editing the *Yearbook* back in 2001 it soon became very clear to me that he wanted its publication to continue after his death, and that it should go on for as long as it fulfilled a need. I am indebted both to Patrick's executors and to the team at Pan Macmillan for their great support in securing the future of the *Yearbook*.

I have done my best to continue the fine tradition maintained by Patrick over more than half a century and have followed his long-established pattern. In this edition we have the usual mix of contributions both for those who want to know exactly what is going on in the night sky, month-by-month, and for those who enjoy reading the longer articles. As usual, we have done our best to give you a wide choice, both of subject and of technical level.

Of all the exciting and important advances in astronomy and astrophysics in recent years, one of the most significant has been the announcement by the BICEP2 team of evidence of a slight twist in the light from the cosmic microwave background radiation – a twist that had its origins in the beginning of the universe, moments after the Big Bang – the so-called B-mode polarization. As Stephen Webb carefully explains, if the result is verified by other scientists, it will be one of the most important discoveries of recent years – and very exciting times will lie ahead. The coming year will also see a landmark event in solar-system exploration – the first ever spacecraft flyby of tiny Pluto. David Harland examines the background to the New Horizons mission to Pluto and looks ahead to the exciting days of the flyby in July 2015, an event that has been eagerly awaited for so many years – and one that I know Patrick would have loved to see.

For the amateur observer, Pete Lawrence describes the techniques needed to safely observe and image the Sun, both in white light and with narrow-band filters. Keeping with the solar theme, we also provide a detailed overview of the solar eclipse on 20 March 2015, which will be total in the Arctic and visible as a very large partial eclipse all over the British Isles – the best since August 1999.

Martin Mobberley gives an absorbing account of the life and work of the legendary Horace Dall, one of the UK's best lunar and planetary photographers and the most knowledgeable authority on amateur astronomical telescopes. We also have Richard Baum's fascinating story of the ninth-magnitude star that was at first thought to have disappeared but eventually, after much speculation, was shown to have had its position misrecorded due to a simple observational error – and was, in fact, the star Lalande 36613. Finally, 300 years after the path of the total solar eclipse on 22 April 1715 crossed the British Isles, historian Allan Chapman provides a captivating biography of the life of Dr Edmond Halley, astronomer, geophysicist, meteorologist and Royal Navy captain, and the man who did more than any other to promote this 'people's eclipse'. He not only alerted the English and Irish to the forthcoming spectacle, but was the first to invite the general population to take part in the 'mass observation' of an astronomical event

I am, as always, most grateful to all of the regular contributors for their support in completing this edition. Martin Mobberley has supplied the notes on eclipses, comets and minor planets, and Nick James has produced the data for the phases of the Moon, longitudes of the Sun, Moon and planets, and details of lunar occultations. As usual, John Isles and Bob Argyle have provided the information on variable stars and double stars, respectively. Wil Tirion, who produced our stars maps for the Northern and Southern Hemispheres, has again drawn all of the line diagrams showing the positions and movements of the planets to accompany the Monthly Notes. These provide a detailed guide to what's happening in the night sky throughout the year, on a month-by-month basis.

I very much hope that you will enjoy reading the range of articles that have been prepared for you this year.

John Mason
Barnham, August 2014

Preface

New readers will find that all the information in this *Yearbook* is given in diagrammatic or descriptive form; the positions of the planets may easily be found from the specially designed star charts, while the Monthly Notes describe the movements of the planets and give details of other astronomical phenomena visible in both in the Northern and Southern Hemispheres. Two sets of star charts are provided. The **Northern Star Charts** (pp. 7 to 31) are designed for use at latitude 52°N, but may be used without alteration throughout the British Isles, and (except in the case of eclipses and occultations) in other countries of similar northerly latitude. The **Southern Star Charts** (pp. 33 to 57) are drawn for latitude 35°S, and are suitable for use in South Africa, Australia and New Zealand, and other locations in approximately the same southerly latitude. The reader who needs more detailed information will find *Norton's Star Atlas* an invaluable guide, while more precise positions of the planets and their satellites, together with predictions of occultations, meteor showers and periodic comets, may be found in the *Handbook of the British Astronomical Association*. Readers will also find details of forthcoming events given in the American monthly magazine *Sky & Telescope* and the British periodicals *The Sky at Night*, *Astronomy Now* and *Astronomy and Space*.

Important note

The times given on the star charts and in the Monthly Notes are generally given as local times, using the twenty-four-hour clock, the day beginning at midnight. All the dates, and the times of a few events (e.g. eclipses) are given in Greenwich Mean Time (GMT), which is related to local time by the formula:

Local Mean Time = GMT – west longitude

In practice, small differences in longitude are ignored, and the observer will use local clock time, which will be the appropriate Standard (or

Zone) Time. As the formula indicates, places in west longitude will have a Standard Time slow on GMT, while places in east longitude will have a Standard Time fast on GMT. As examples we have:

Standard Time in	
New Zealand	GMT + 12 hours
Victoria, NSW	GMT + 10 hours
Western Australia	GMT + 8 hours
South Africa	GMT + 2 hours
British Isles	GMT
Eastern ST	GMT − 5 hours
Central ST	GMT − 6 hours, etc.

If Summer Time is in use, the clocks will have been advanced by one hour, and this hour must be subtracted from the clock time to give Standard Time.

Part One

Monthly Charts and Astronomical Phenomena

Notes on the Star Charts

The stars, together with the Sun, Moon and planets, seem to be set on the surface of the celestial sphere, which appears to rotate about the Earth from east to west. Since it is impossible to represent a curved surface accurately on a plane, any kind of star map is bound to contain some form of distortion.

Most of the monthly star charts which appear in the various journals and some national newspapers are drawn in circular form. This is perfectly accurate, but it can make the charts awkward to use. For the star charts in this volume, we have preferred to give two hemispherical maps for each month of the year, one showing the northern aspect of the sky and the other showing the southern aspect. Two sets of monthly charts are provided, one for observers in the Northern Hemisphere and one for those in the Southern Hemisphere.

Unfortunately, the constellations near the overhead point (the zenith) on these hemispherical charts can be rather distorted. This would be a serious drawback for precision charts, but what we have done is to give maps which are best suited to star recognition. We have also refrained from putting in too many stars, so that the main patterns stand out clearly. To help observers with any distortions near the zenith, and the lack of overlap between the charts of each pair, we have also included two circular maps, one showing all the constellations in the northern half of the sky, and one showing those in the southern half. Incidentally, there is a curious illusion that stars at an altitude of 60° or more are actually overhead, and beginners may often feel that they are leaning over backwards in trying to see them.

The charts show all stars down to the fourth magnitude, together with a number of fainter stars which are necessary to define the shapes of constellations. There is no standard system for representing the outlines of the constellations, and triangles and other simple figures have been used to give outlines which are easy to trace with the naked eye. The names of the constellations are given, together with the proper names of the brighter stars. The apparent magnitudes of the stars are indicated roughly by using different sizes of dot, the larger dots representing the brighter stars.

The two sets of star charts – one each for Northern and Southern Hemisphere observers – are similar in design. At each opening there is a single circular chart which shows all the constellations in that hemisphere of the sky. (These two charts are centred on the North and South Celestial Poles, respectively.) Then there are twelve double-page spreads, showing the northern and southern aspects for each month of the year for observers in that hemisphere. In the **Northern Star Charts** (drawn for latitude 52°N) the left-hand chart of each spread shows the northern half of the sky (lettered 1N, 2N, 3N . . . 12N), and the corresponding right-hand chart shows the southern half of the sky (lettered 1S, 2S, 3S . . . 12S). The arrangement and lettering of the charts is exactly the same for the **Southern Star Charts** (drawn for latitude 35°S).

Because the sidereal day is shorter than the solar day, the stars appear to rise and set about four minutes earlier each day, and this amounts to two hours in a month. Hence the twelve pairs of charts in each set are sufficient to give the appearance of the sky throughout the day at intervals of two hours, or at the same time of night at monthly intervals throughout the year. For example, charts 1N and 1S here are drawn for 23 hours on 6 January. The view will also be the same on 6 October at 05 hours; 6 November at 03 hours; 6 December at 01 hours and 6 February at 21 hours. The actual range of dates and times when the stars on the charts are visible is indicated on each page. Each pair of charts is numbered in bold type, and the number to be used for any given month and time may be found from the following table:

Local Time	**18h**	**20h**	**22h**	**0h**	**2h**	**4h**	**6h**
January	11	12	1	2	3	4	5
February	12	1	2	3	4	5	6
March	1	2	3	4	5	6	7
April	2	3	4	5	6	7	8
May	3	4	5	6	7	8	9
June	4	5	6	7	8	9	10
July	5	6	7	8	9	10	11
August	6	7	8	9	10	11	12
September	7	8	9	10	11	12	1

October	8	9	10	11	12	1	2
November	9	10	11	12	1	2	3
December	10	11	12	1	2	3	4

On these charts, the ecliptic is drawn as a broken line on which longitude is marked every 10°. The positions of the planets are then easily found by reference to the table on page 64. It will be noticed that on the **Southern Star Charts** the ecliptic may reach an altitude in excess of 62.5° on the star charts showing the northern aspect (5N to 9N). The continuations of the broken line will be found on the corresponding charts for the southern aspect (5S, 6S, 8S and 9S).

Northern Star Charts

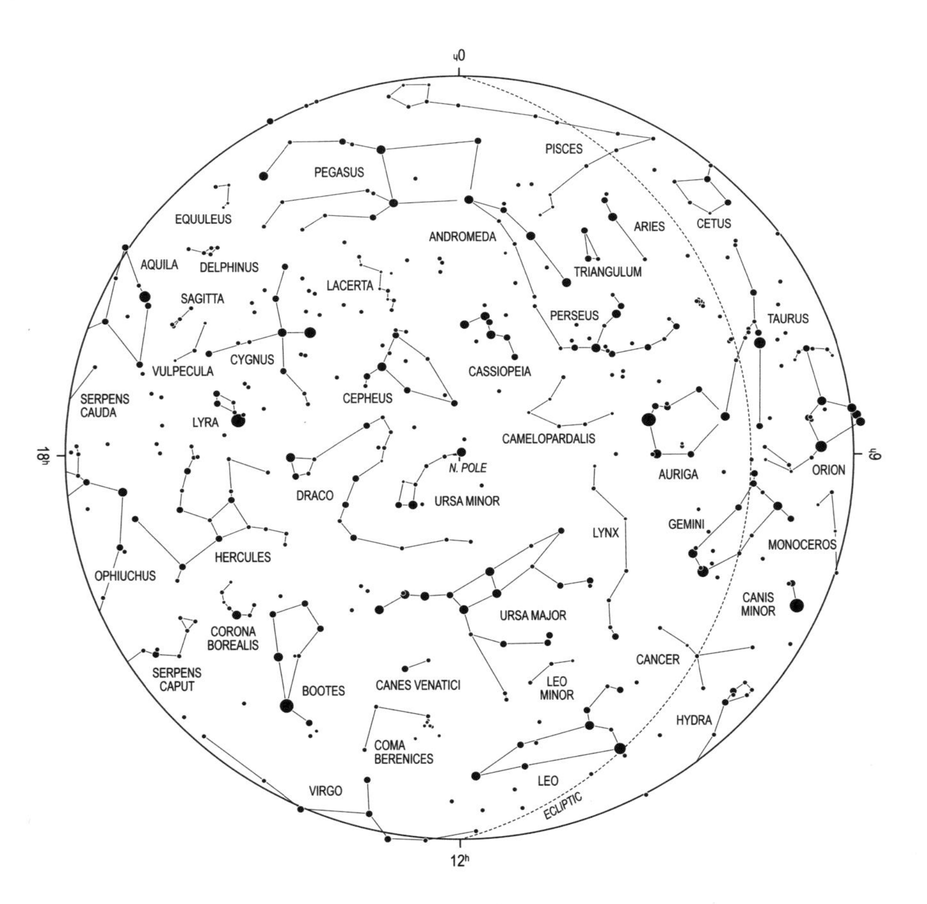

Northern Hemisphere

Note that the markers at 0h, 6h, 12h and 18h indicate hours of Right Ascension.

1N

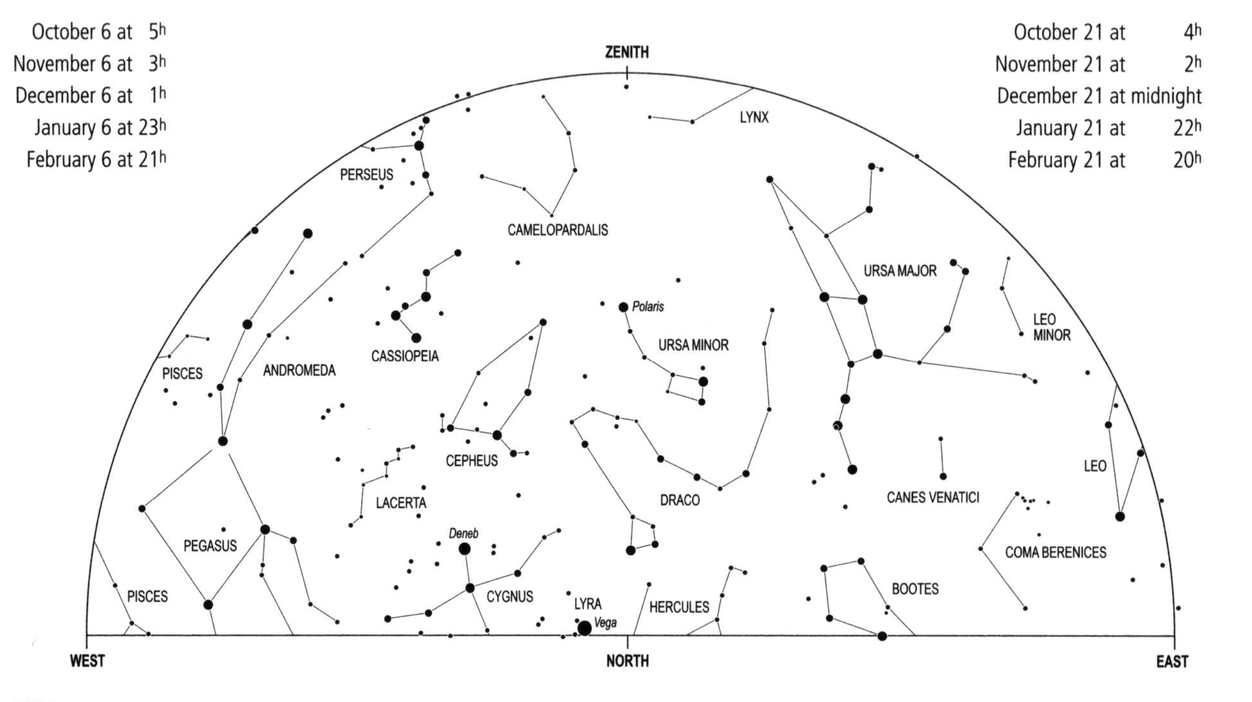
October 6 at 5h
November 6 at 3h
December 6 at 1h
January 6 at 23h
February 6 at 21h
October 21 at 4h
November 21 at 2h
December 21 at midnight
January 21 at 22h
February 21 at 20h
ZENITH
LYNX
PERSEUS
CAMELOPARDALIS
URSA MAJOR
Polaris
URSA MINOR
LEO
MINOR
PISCES
ANDROMEDA
CASSIOPEIA
CEPHEUS
LEO
LACERTA
DRACO
CANES VENATICI
Deneb
PEGASUS
COMA BERENICES
PISCES
CYGNUS
LYRA
HERCULES
BOOTES
Vega
WEST
NORTH
EAST

1S

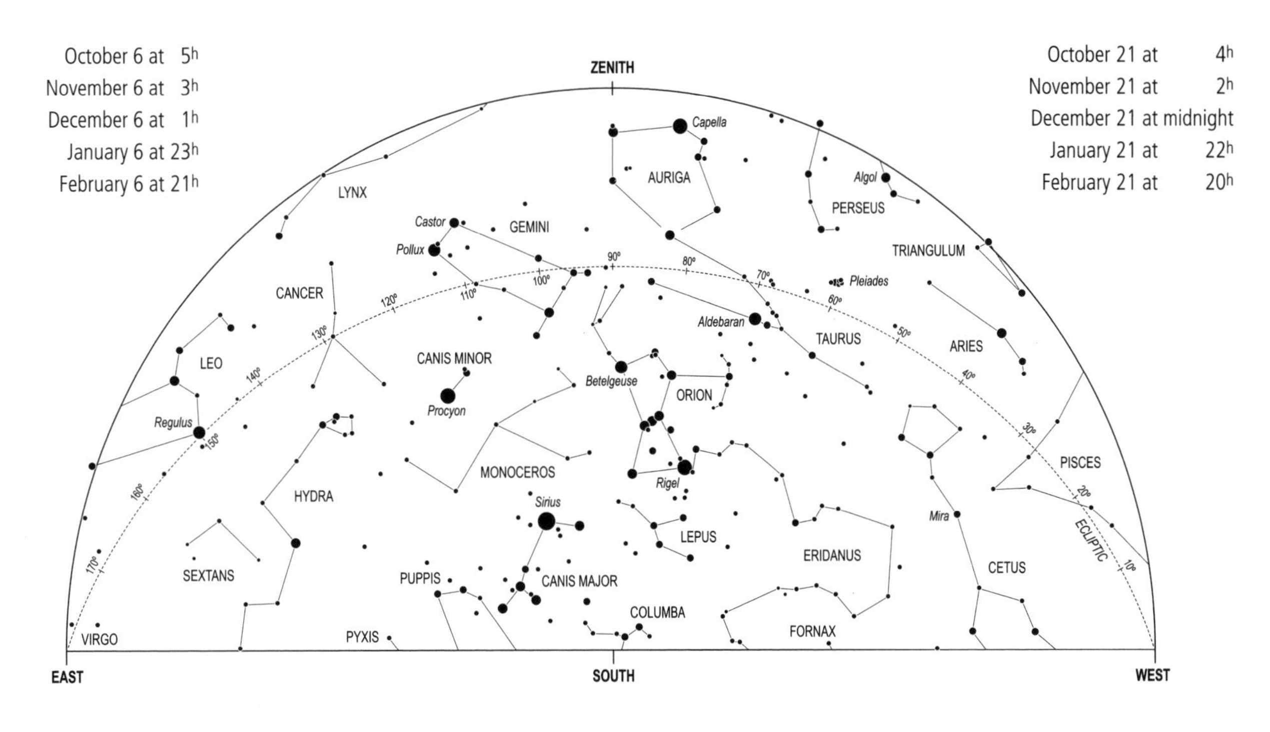
October 6 at 5h
November 6 at 3h
December 6 at 1h
January 6 at 23h
February 6 at 21h
October 21 at 4h
November 21 at 2h
December 21 at midnight
January 21 at 22h
February 21 at 20h
ZENITH
EAST
SOUTH
WEST
LYNX
Castor
Pollux
GEMINI
Capella
AURIGA
Algol
PERSEUS
TRIANGULUM
Pleiades
CANCER
Aldebaran
TAURUS
ARIES
LEO
CANIS MINOR
Procyon
Betelgeuse
ORION
Regulus
MONOCEROS
Rigel
PISCES
HYDRA
Sirius
Mira
ECLIPTIC
LEPUS
ERIDANUS
CETUS
SEXTANS
PUPPIS
CANIS MAJOR
COLUMBA
FORNAX
VIRGO
PYXIS

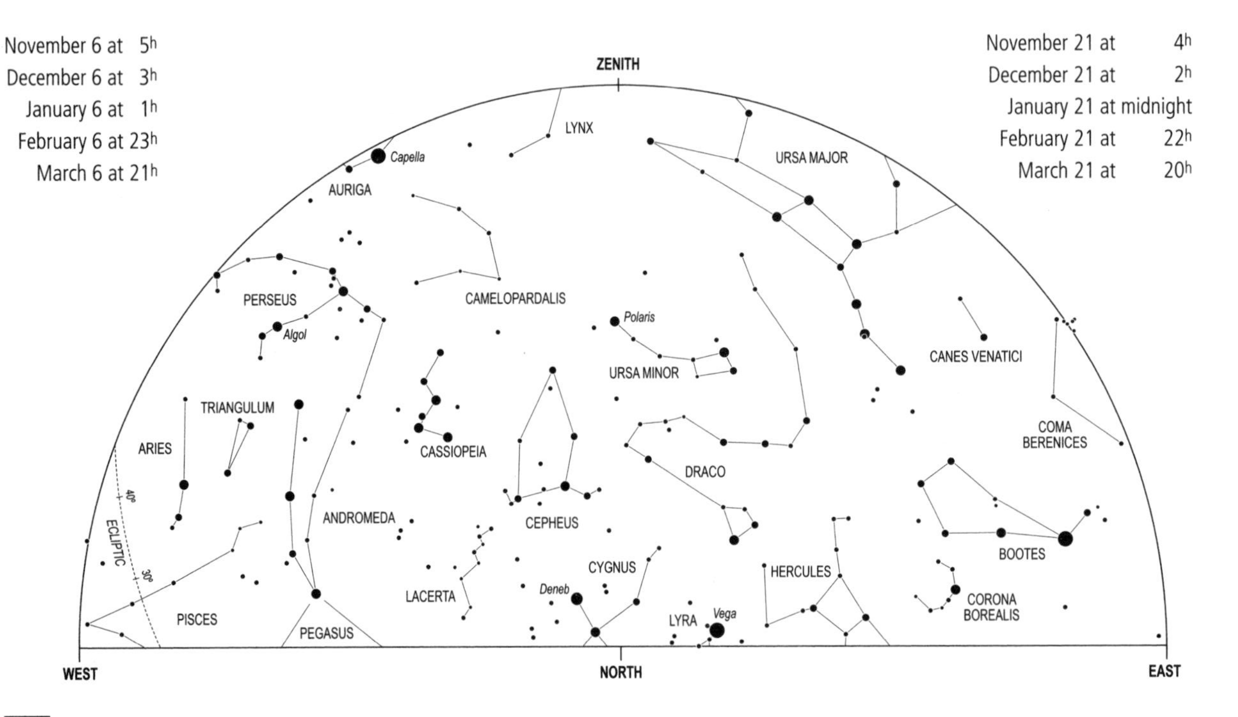
November 6 at 5h
December 6 at 3h
January 6 at 1h
February 6 at 23h
March 6 at 21h
ZENITH
November 21 at 4h
December 21 at 2h
January 21 at midnight
February 21 at 22h
March 21 at 20h
Capella
AURIGA
LYNX
URSA MAJOR
PERSEUS
Algol
CAMELOPARDALIS
Polaris
CANES VENATICI
URSA MINOR
TRIANGULUM
ARIES
CASSIOPEIA
COMA
BERENICES
DRACO
40°
ECLIPTIC
ANDROMEDA
CEPHEUS
BOOTES
30°
CYGNUS
HERCULES
LACERTA
Deneb
CORONA
BOREALIS
PISCES
LYRA
Vega
PEGASUS
WEST
NORTH
EAST

2N

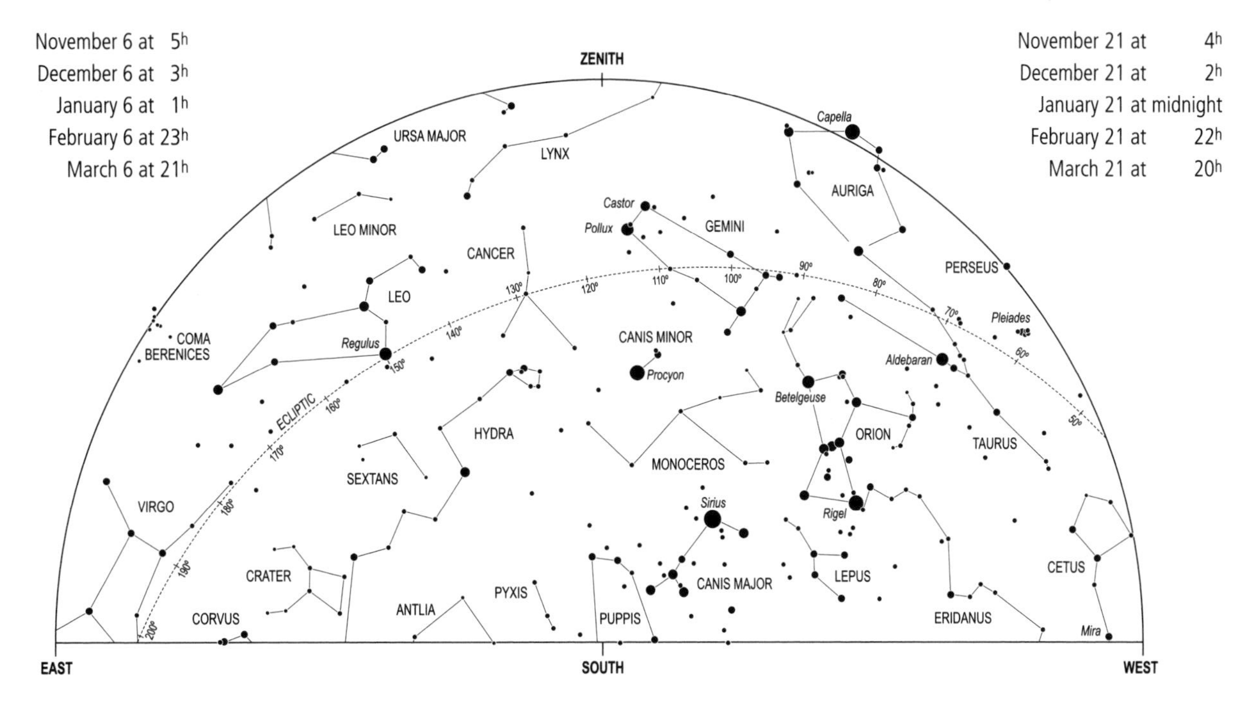
November 6 at 5h
December 6 at 3h
January 6 at 1h
February 6 at 23h
March 6 at 21h
November 21 at 4h
December 21 at 2h
January 21 at midnight
February 21 at 22h
March 21 at 20h
ZENITH
EAST
SOUTH
WEST
URSA MAJOR
LYNX
Capella
AURIGA
LEO MINOR
Castor
Pollux
GEMINI
CANCER
PERSEUS
LEO
COMA
BERENICES
Regulus
CANIS MINOR
Pleiades
Aldebaran
Procyon
Betelgeuse
ECLIPTIC
HYDRA
ORION
TAURUS
SEXTANS
MONOCEROS
Sirius
Rigel
VIRGO
CRATER
CANIS MAJOR
LEPUS
CETUS
PYXIS
CORVUS
ANTLIA
PUPPIS
ERIDANUS
Mira
50°
60°
70°
80°
90°
100°
110°
120°
130°
140°
150°
160°
170°
180°
190°
200°

2S

3N

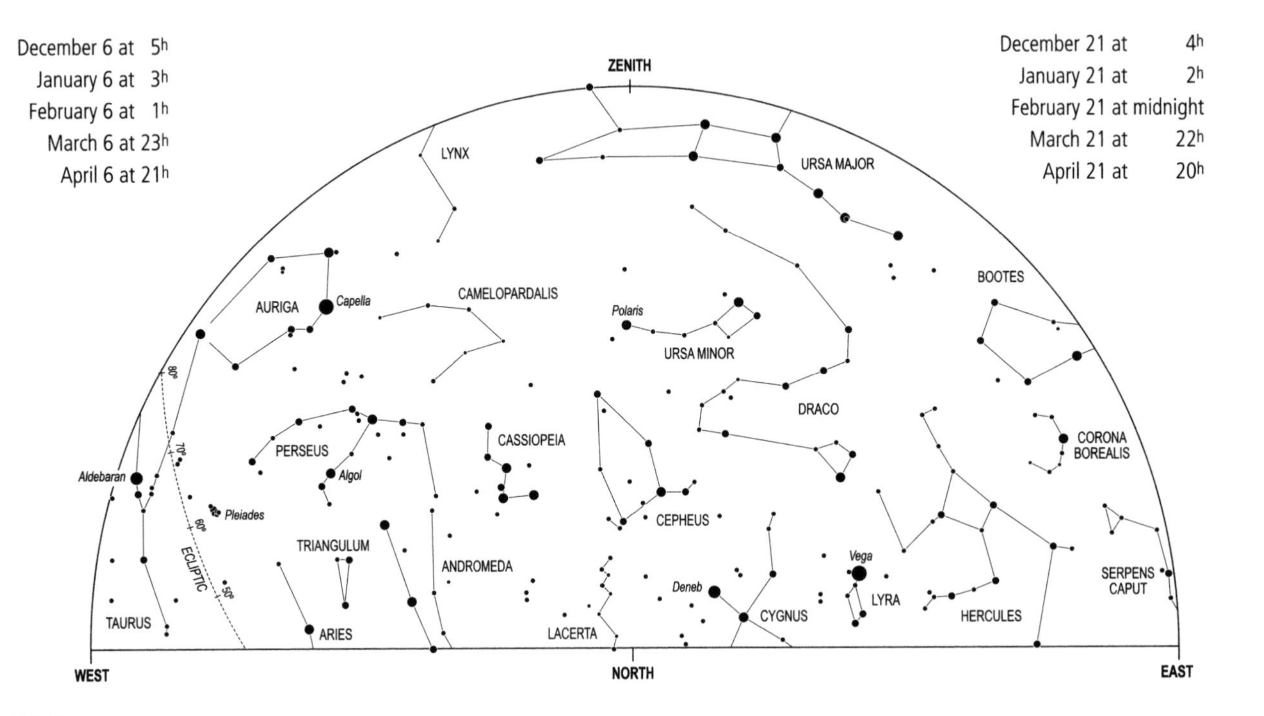
December 6 at 5h
January 6 at 3h
February 6 at 1h
March 6 at 23h
April 6 at 21h
December 21 at 4h
January 21 at 2h
February 21 at midnight
March 21 at 22h
April 21 at 20h
ZENITH
WEST
NORTH
EAST
ECLIPTIC
50°
60°
70°
80°
TAURUS
Aldebaran
Pleiades
AURIGA
Capella
LYNX
CAMELOPARDALIS
PERSEUS
Algol
TRIANGULUM
ARIES
ANDROMEDA
CASSIOPEIA
LACERTA
CEPHEUS
Polaris
URSA MINOR
URSA MAJOR
DRACO
Deneb
CYGNUS
Vega
LYRA
HERCULES
BOOTES
CORONA BOREALIS
SERPENS CAPUT

3S

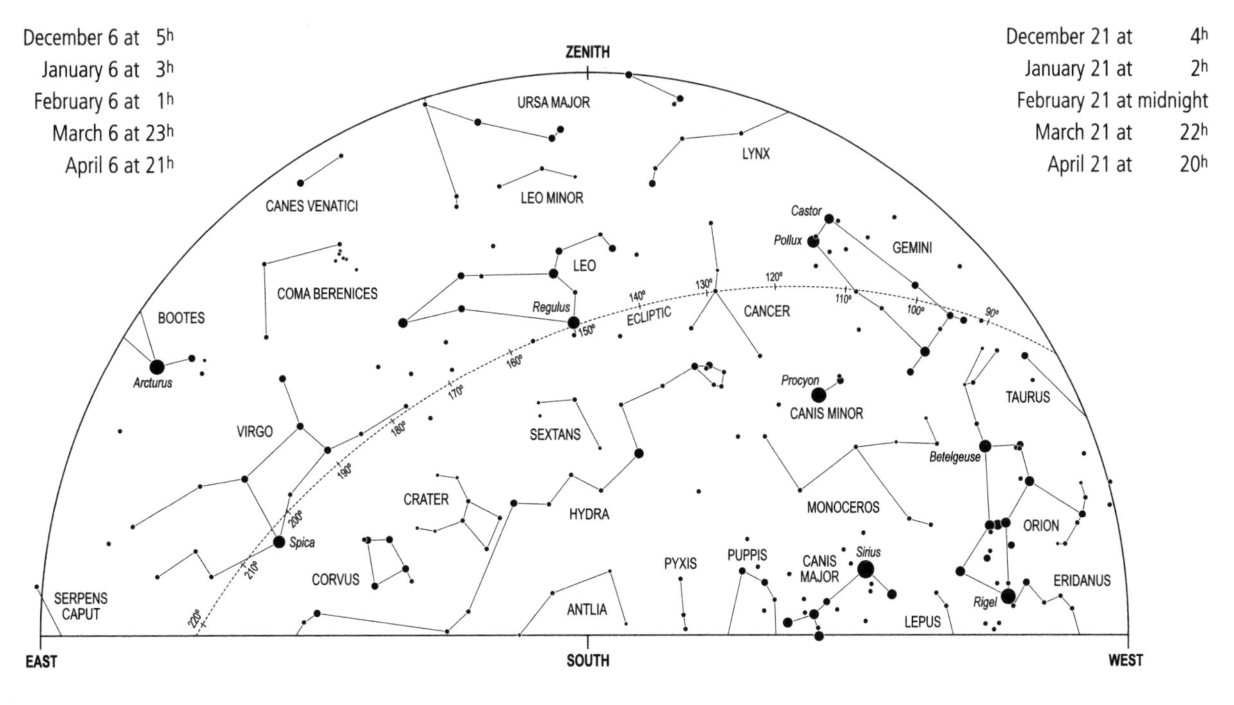
December 6 at 5h
January 6 at 3h
February 6 at 1h
March 6 at 23h
April 6 at 21h
December 21 at 4h
January 21 at 2h
February 21 at midnight
March 21 at 22h
April 21 at 20h
ZENITH
URSA MAJOR
LYNX
LEO MINOR
CANES VENATICI
Castor
Pollux
GEMINI
LEO
COMA BERENICES
Regulus
ECLIPTIC
CANCER
BOOTES
Arcturus
Procyon
CANIS MINOR
TAURUS
VIRGO
SEXTANS
Betelgeuse
CRATER
HYDRA
MONOCEROS
ORION
Spica
CORVUS
PYXIS
PUPPIS
CANIS MAJOR
Sirius
ERIDANUS
SERPENS CAPUT
ANTLIA
LEPUS
Rigel
EAST
SOUTH
WEST
90°
100°
110°
120°
130°
140°
150°
160°
170°
180°
190°
200°
210°
220°

4N

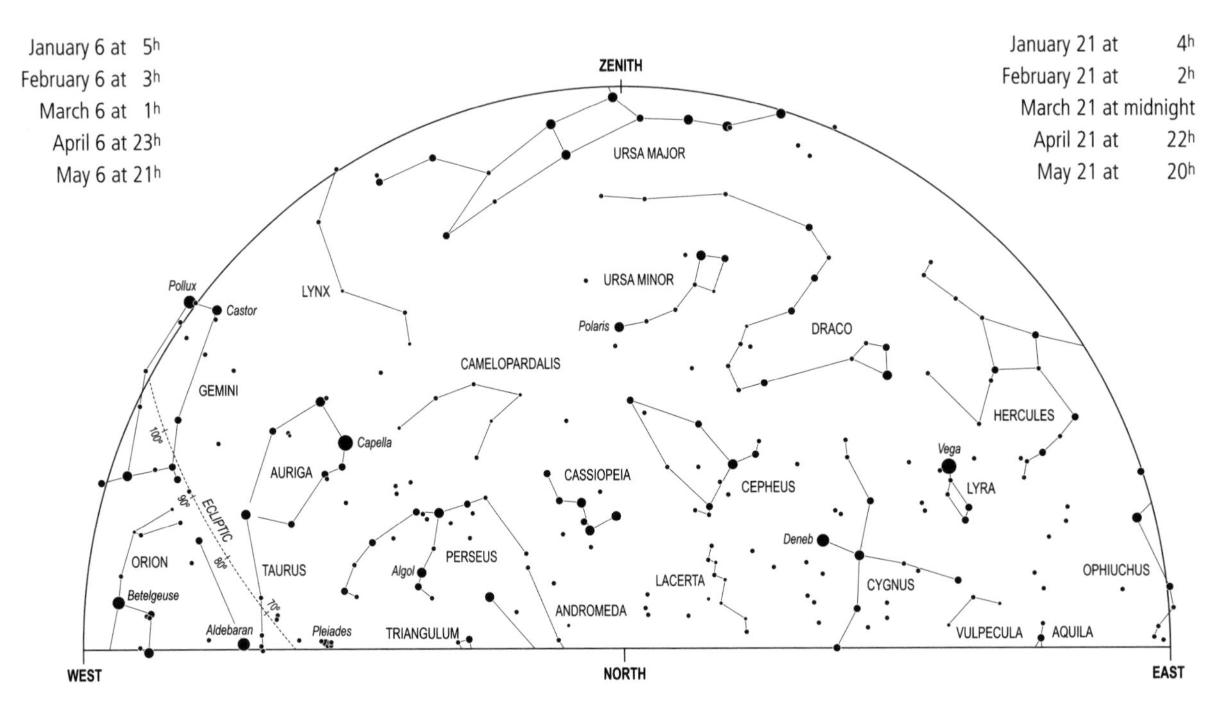
January 6 at 5h
February 6 at 3h
March 6 at 1h
April 6 at 23h
May 6 at 21h
January 21 at 4h
February 21 at 2h
March 21 at midnight
April 21 at 22h
May 21 at 20h
ZENITH
WEST
NORTH
EAST
ECLIPTIC
100°
90°
80°
70°
URSA MAJOR
URSA MINOR
Polaris
LYNX
Pollux
Castor
GEMINI
CAMELOPARDALIS
DRACO
HERCULES
Capella
AURIGA
CASSIOPEIA
CEPHEUS
Vega
LYRA
ORION
TAURUS
Betelgeuse
Aldebaran
Pleiades
Algol
PERSEUS
TRIANGULUM
ANDROMEDA
LACERTA
Deneb
CYGNUS
OPHIUCHUS
VULPECULA
AQUILA

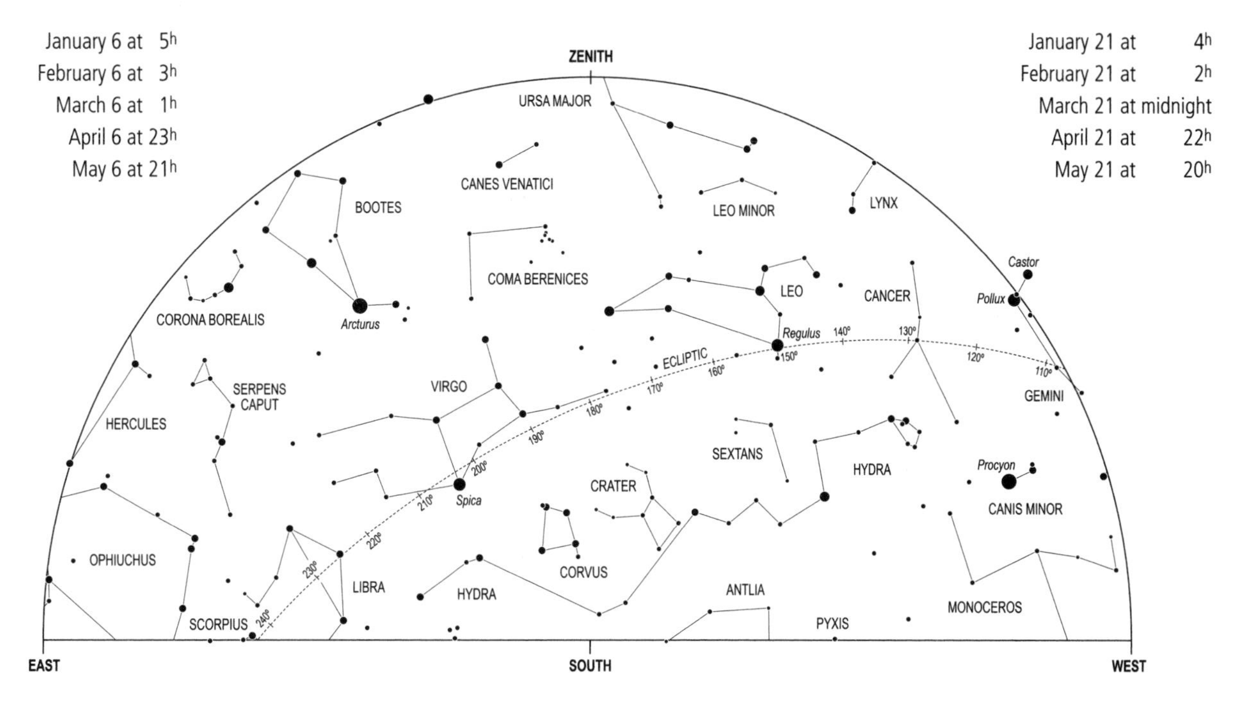
January 6 at 5h
February 6 at 3h
March 6 at 1h
April 6 at 23h
May 6 at 21h
January 21 at 4h
February 21 at 2h
March 21 at midnight
April 21 at 22h
May 21 at 20h
ZENITH
URSA MAJOR
CANES VENATICI
BOOTES
LEO MINOR
LYNX
COMA BERENICES
CORONA BOREALIS
Arcturus
LEO
CANCER
Castor
Pollux
Regulus
ECLIPTIC
140°
130°
150°
160°
170°
120°
110°
GEMINI
SERPENS CAPUT
VIRGO
HERCULES
180°
190°
SEXTANS
HYDRA
Procyon
200°
Spica
210°
CRATER
CANIS MINOR
220°
OPHIUCHUS
230°
CORVUS
LIBRA
HYDRA
ANTLIA
MONOCEROS
SCORPIUS
240°
PYXIS
EAST
SOUTH
WEST

4S

5N

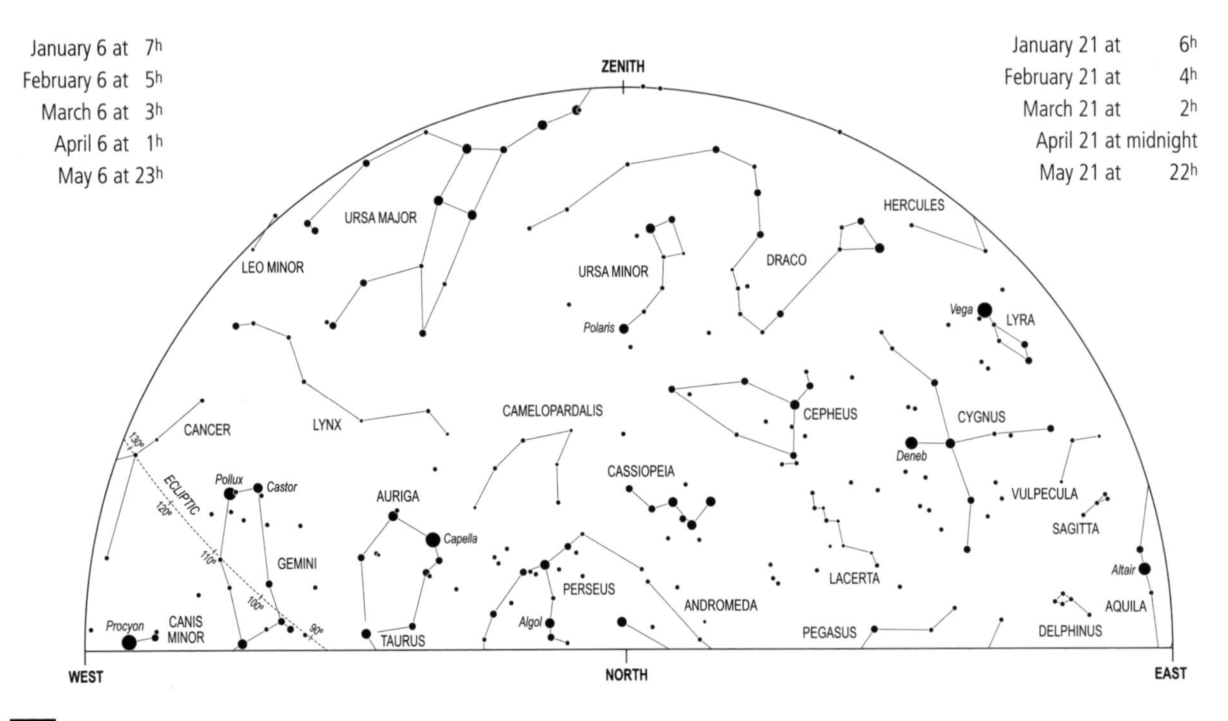
January 6 at 7h
February 6 at 5h
March 6 at 3h
April 6 at 1h
May 6 at 23h
January 21 at 6h
February 21 at 4h
March 21 at 2h
April 21 at midnight
May 21 at 22h
ZENITH
WEST
NORTH
EAST
ECLIPTIC
130°
120°
110°
100°
90°
URSA MAJOR
LEO MINOR
URSA MINOR
Polaris
DRACO
HERCULES
Vega
LYRA
CANCER
LYNX
CAMELOPARDALIS
CEPHEUS
CYGNUS
Deneb
Pollux
Castor
GEMINI
AURIGA
Capella
CASSIOPEIA
VULPECULA
SAGITTA
Procyon
CANIS MINOR
TAURUS
Algol
PERSEUS
ANDROMEDA
LACERTA
PEGASUS
Altair
AQUILA
DELPHINUS

5S

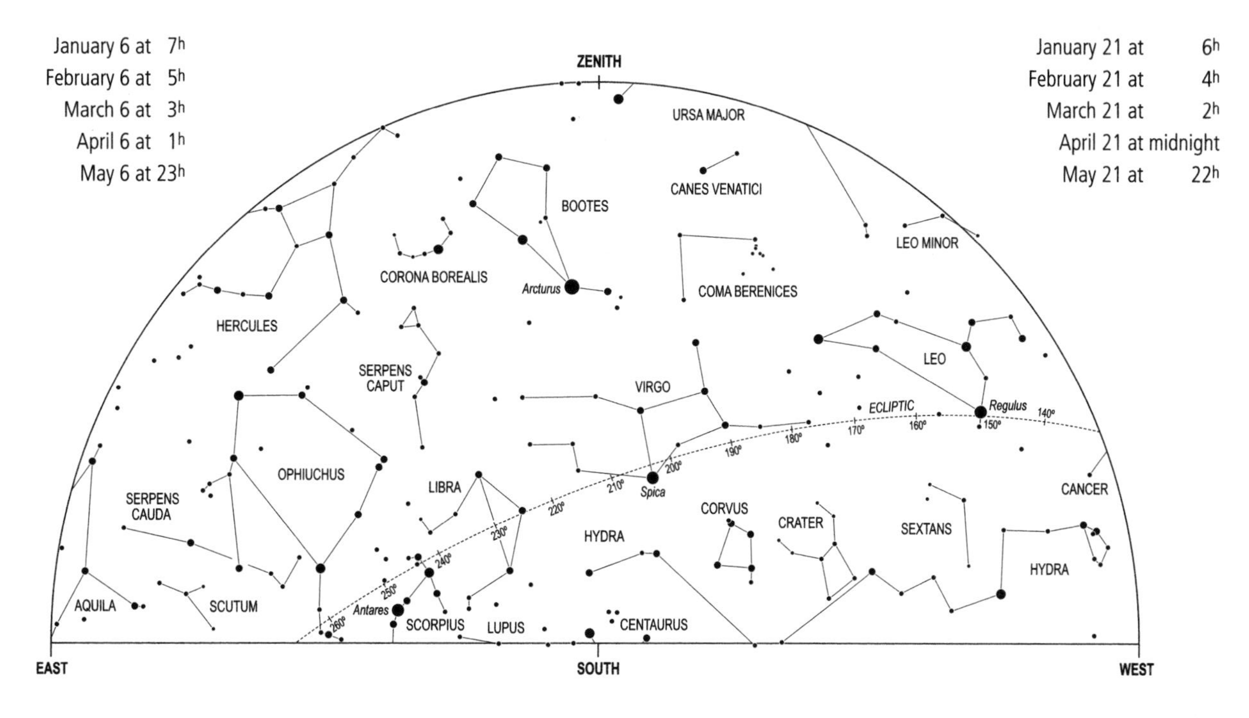
January 6 at 7h
February 6 at 5h
March 6 at 3h
April 6 at 1h
May 6 at 23h
January 21 at 6h
February 21 at 4h
March 21 at 2h
April 21 at midnight
May 21 at 22h
ZENITH
URSA MAJOR
CANES VENATICI
BOOTES
LEO MINOR
CORONA BOREALIS
Arcturus
COMA BERENICES
HERCULES
LEO
SERPENS CAPUT
VIRGO
ECLIPTIC
Regulus
140°
150°
160°
170°
180°
190°
200°
210°
220°
230°
240°
250°
260°
OPHIUCHUS
LIBRA
Spica
CANCER
SERPENS CAUDA
CORVUS
CRATER
SEXTANS
HYDRA
HYDRA
AQUILA
SCUTUM
Antares
SCORPIUS
LUPUS
CENTAURUS
EAST
SOUTH
WEST

6N

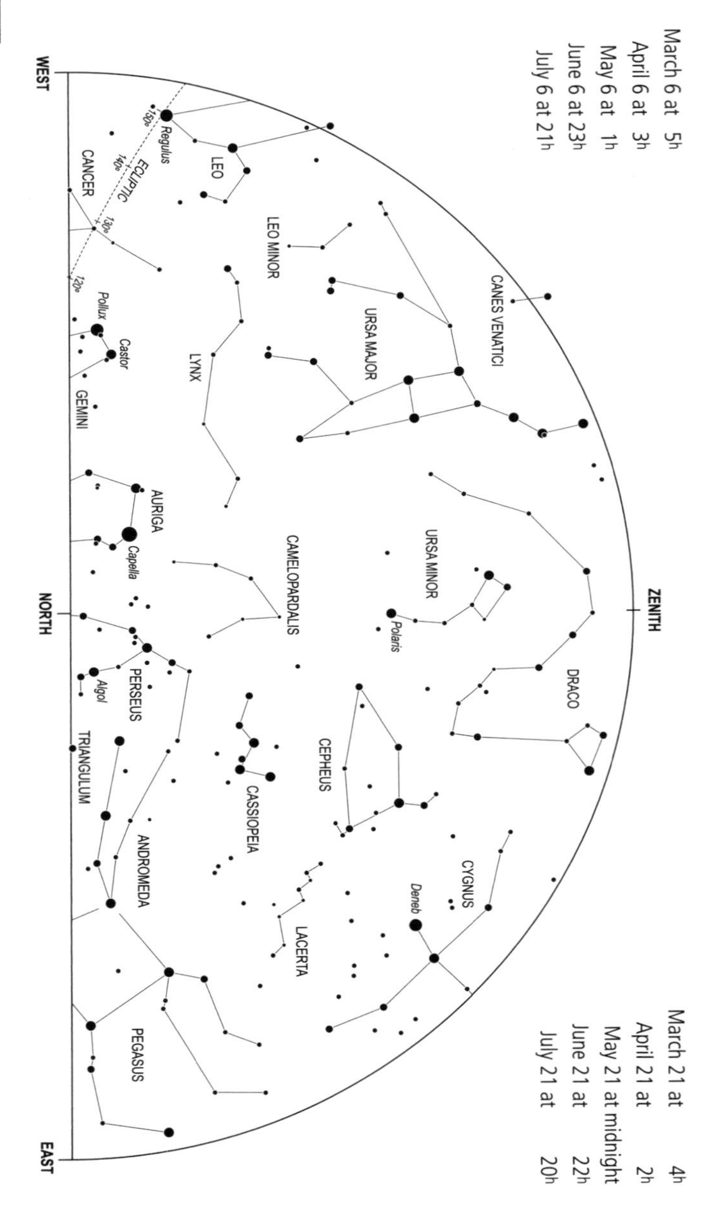
March 6 at 5h
April 6 at 3h
May 6 at 1h
June 6 at 23h
July 6 at 21h
March 21 at 4h
April 21 at 2h
May 21 at midnight
June 21 at 22h
July 21 at 20h
WEST
NORTH
EAST
ZENITH
ECLIPTIC
150°
140°
130°
120°
Regulus
LEO
CANCER
LEO MINOR
CANES VENATICI
URSA MAJOR
LYNX
Pollux
Castor
GEMINI
AURIGA
Capella
CAMELOPARDALIS
URSA MINOR
Polaris
DRACO
PERSEUS
Algol
TRIANGULUM
CASSIOPEIA
CEPHEUS
CYGNUS
Deneb
ANDROMEDA
LACERTA
PEGASUS

6S

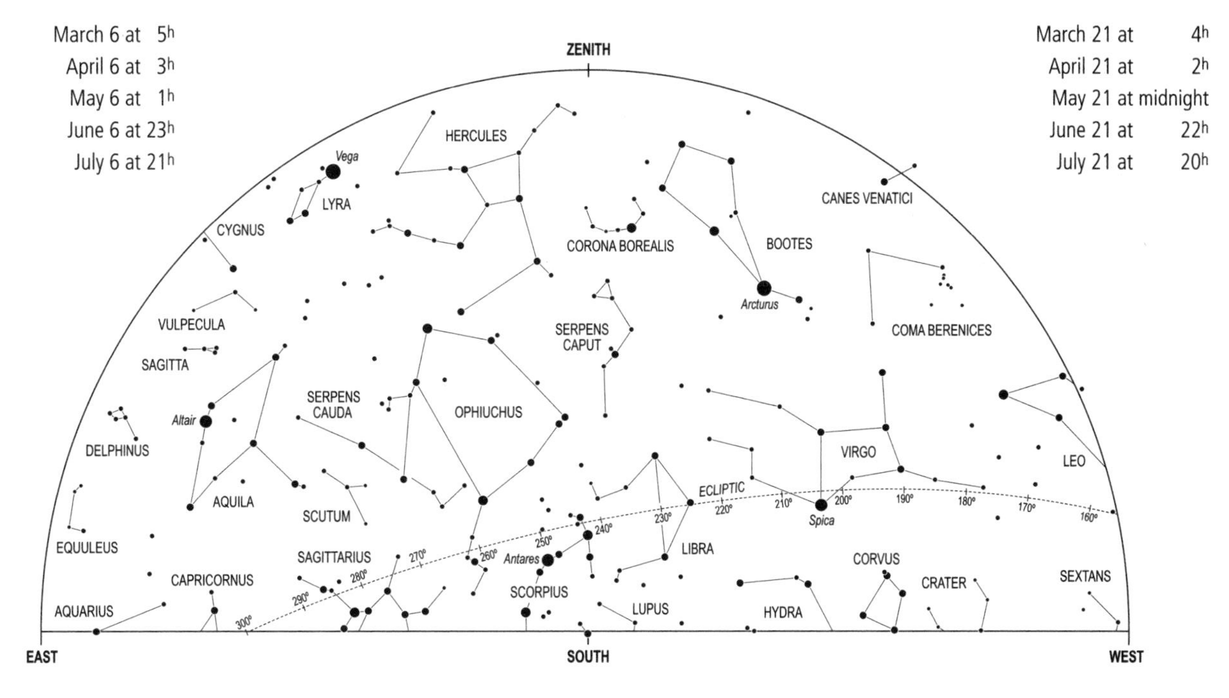
March 6 at 5h
April 6 at 3h
May 6 at 1h
June 6 at 23h
July 6 at 21h
March 21 at 4h
April 21 at 2h
May 21 at midnight
June 21 at 22h
July 21 at 20h
ZENITH
HERCULES
Vega
LYRA
CYGNUS
CORONA BOREALIS
BOOTES
Arcturus
CANES VENATICI
COMA BERENICES
VULPECULA
SAGITTA
SERPENS CAPUT
SERPENS CAUDA
OPHIUCHUS
Altair
DELPHINUS
AQUILA
SCUTUM
VIRGO
LEO
ECLIPTIC
Spica
EQUULEUS
SAGITTARIUS
Antares
LIBRA
CORVUS
SEXTANS
CAPRICORNUS
CRATER
SCORPIUS
AQUARIUS
LUPUS
HYDRA
160°
170°
180°
190°
200°
210°
220°
230°
240°
250°
260°
270°
280°
290°
300°
EAST
SOUTH
WEST

7N

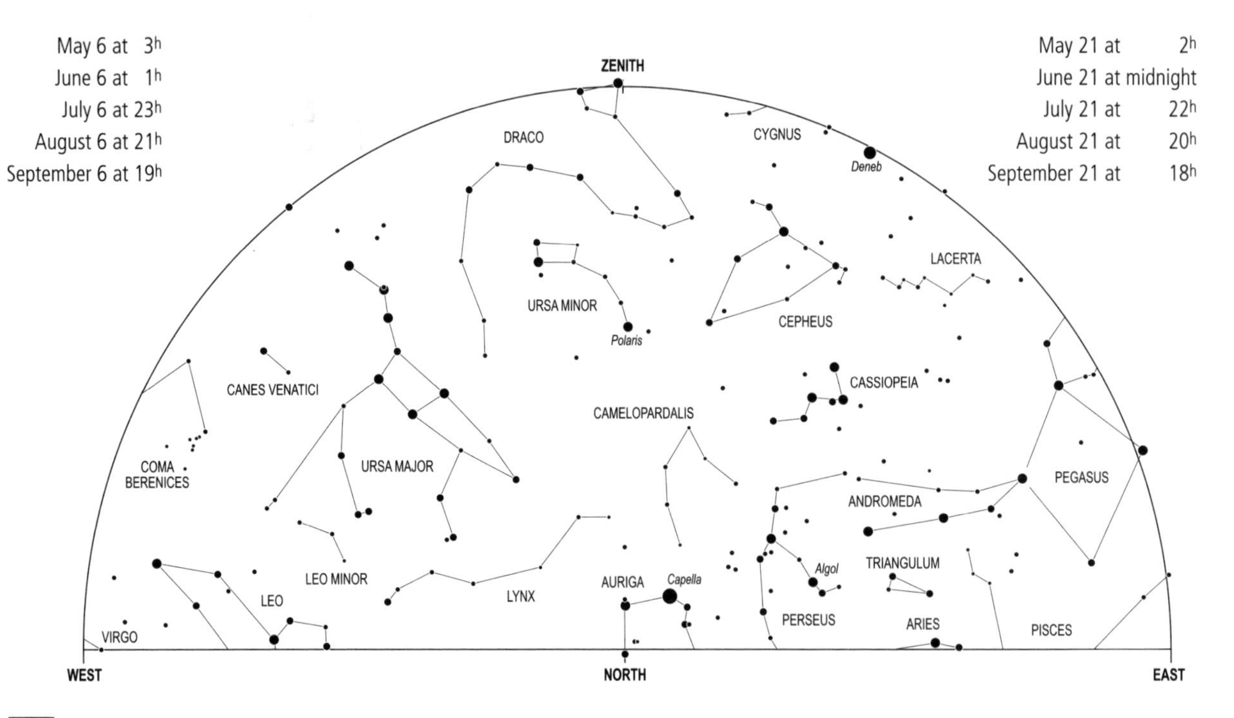
May 6 at 3h
June 6 at 1h
July 6 at 23h
August 6 at 21h
September 6 at 19h
May 21 at 2h
June 21 at midnight
July 21 at 22h
August 21 at 20h
September 21 at 18h
ZENITH
DRACO
CYGNUS
Deneb
LACERTA
URSA MINOR
Polaris
CEPHEUS
CANES VENATICI
CASSIOPEIA
CAMELOPARDALIS
COMA
BERENICES
URSA MAJOR
PEGASUS
ANDROMEDA
TRIANGULUM
Algol
LEO MINOR
LYNX
AURIGA
Capella
LEO
PERSEUS
ARIES
PISCES
VIRGO
WEST
NORTH
EAST

7S

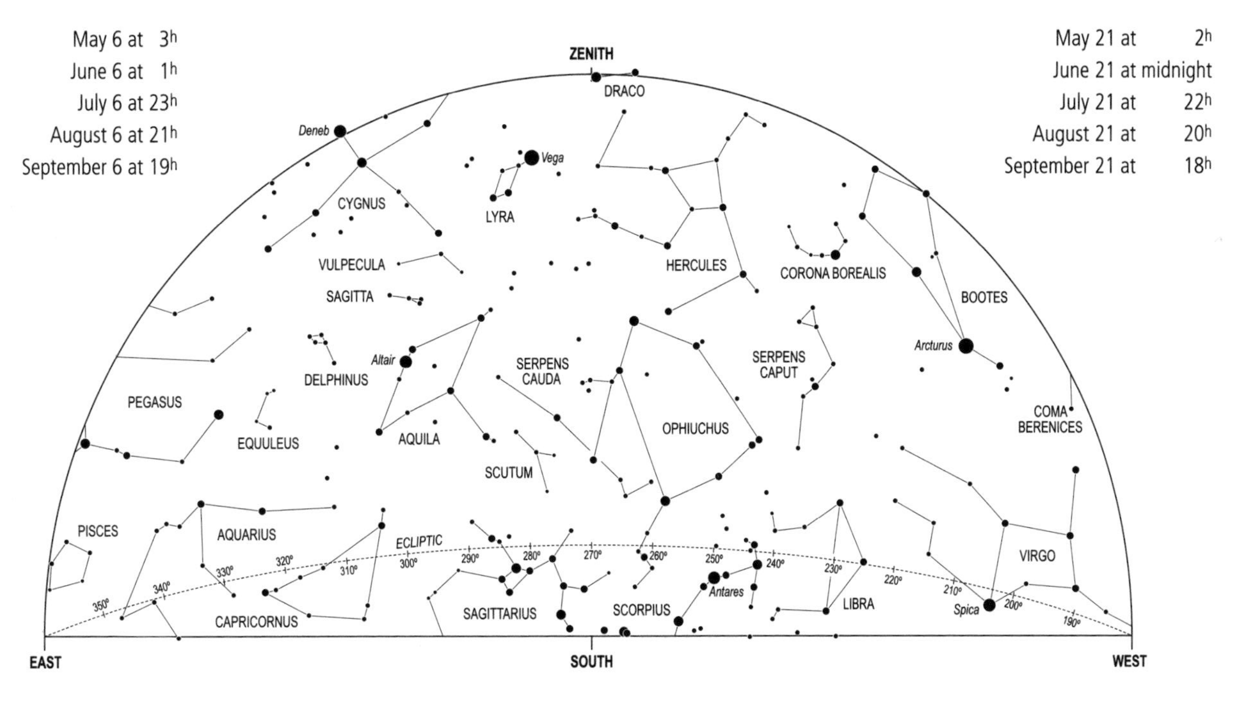

May 6 at 3h
June 6 at 1h
July 6 at 23h
August 6 at 21h
September 6 at 19h
May 21 at 2h
June 21 at midnight
July 21 at 22h
August 21 at 20h
September 21 at 18h
ZENITH
DRACO
Deneb
Vega
CYGNUS
LYRA
VULPECULA
SAGITTA
HERCULES
CORONA BOREALIS
BOOTES
Arcturus
Altair
DELPHINUS
SERPENS CAUDA
SERPENS CAPUT
PEGASUS
EQUULEUS
AQUILA
OPHIUCHUS
COMA BERENICES
SCUTUM
PISCES
AQUARIUS
ECLIPTIC
VIRGO
CAPRICORNUS
SAGITTARIUS
SCORPIUS
Antares
LIBRA
Spica
350°
340°
330°
320°
310°
300°
290°
280°
270°
260°
250°
240°
230°
220°
210°
200°
190°
EAST
SOUTH
WEST

8N

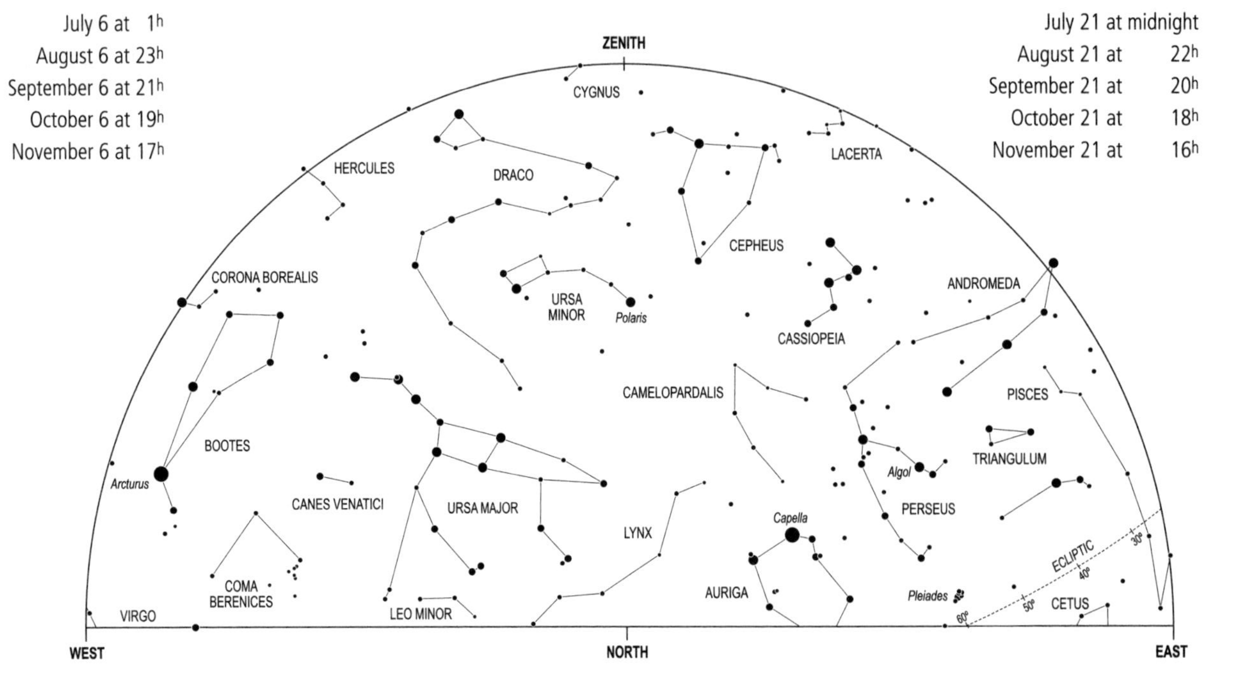

July 6 at 1h
August 6 at 23h
September 6 at 21h
October 6 at 19h
November 6 at 17h
July 21 at midnight
August 21 at 22h
September 21 at 20h
October 21 at 18h
November 21 at 16h
ZENITH
CYGNUS
HERCULES
DRACO
LACERTA
CEPHEUS
CORONA BOREALIS
URSA MINOR
Polaris
ANDROMEDA
CASSIOPEIA
CAMELOPARDALIS
PISCES
BOOTES
TRIANGULUM
Algol
Arcturus
CANES VENATICI
URSA MAJOR
PERSEUS
Capella
LYNX
ECLIPTIC
30°
40°
COMA BERENICES
AURIGA
Pleiades
50°
CETUS
VIRGO
LEO MINOR
60°
WEST
NORTH
EAST

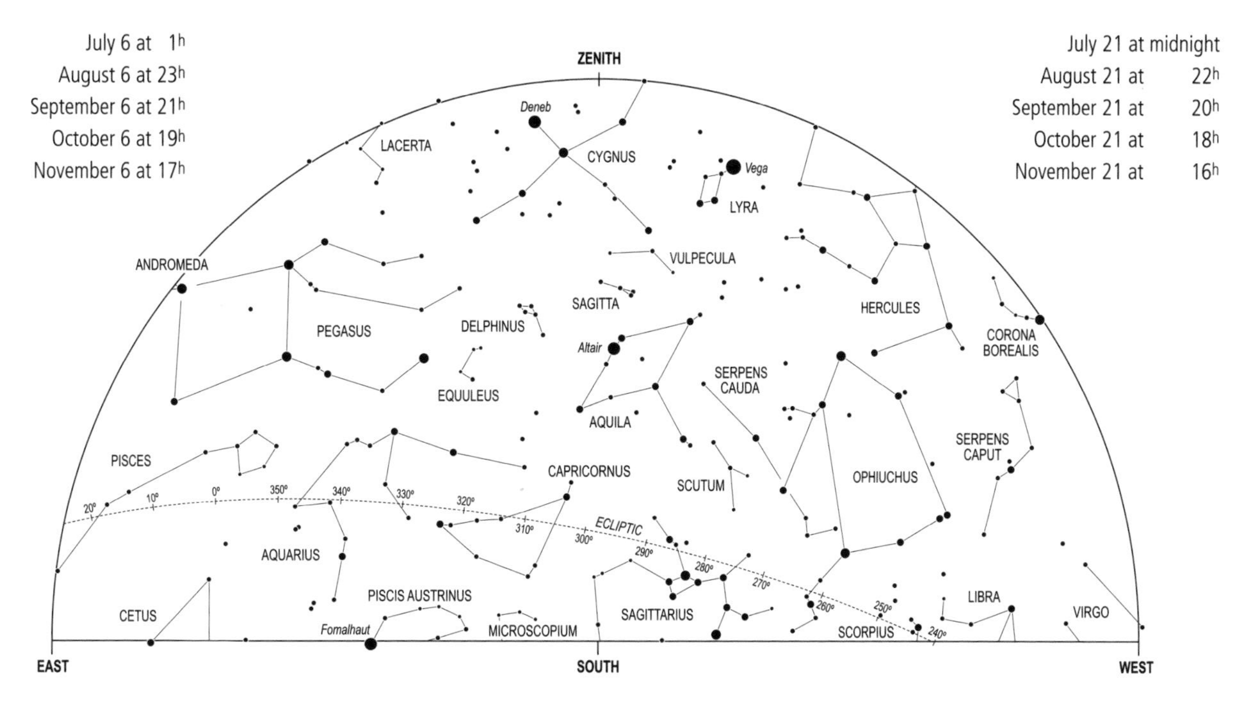

July 6 at 1h
August 6 at 23h
September 6 at 21h
October 6 at 19h
November 6 at 17h
July 21 at midnight
August 21 at 22h
September 21 at 20h
October 21 at 18h
November 21 at 16h
8S
ZENITH
Deneb
LACERTA
CYGNUS
Vega
LYRA
ANDROMEDA
VULPECULA
SAGITTA
HERCULES
PEGASUS
DELPHINUS
Altair
CORONA BOREALIS
SERPENS CAUDA
EQUULEUS
AQUILA
SERPENS CAPUT
PISCES
CAPRICORNUS
SCUTUM
OPHIUCHUS
20°
10°
0°
350°
340°
330°
320°
310°
300°
ECLIPTIC
290°
280°
270°
260°
250°
240°
AQUARIUS
PISCIS AUSTRINUS
LIBRA
CETUS
SAGITTARIUS
VIRGO
Fomalhaut
MICROSCOPIUM
SCORPIUS
EAST
SOUTH
WEST

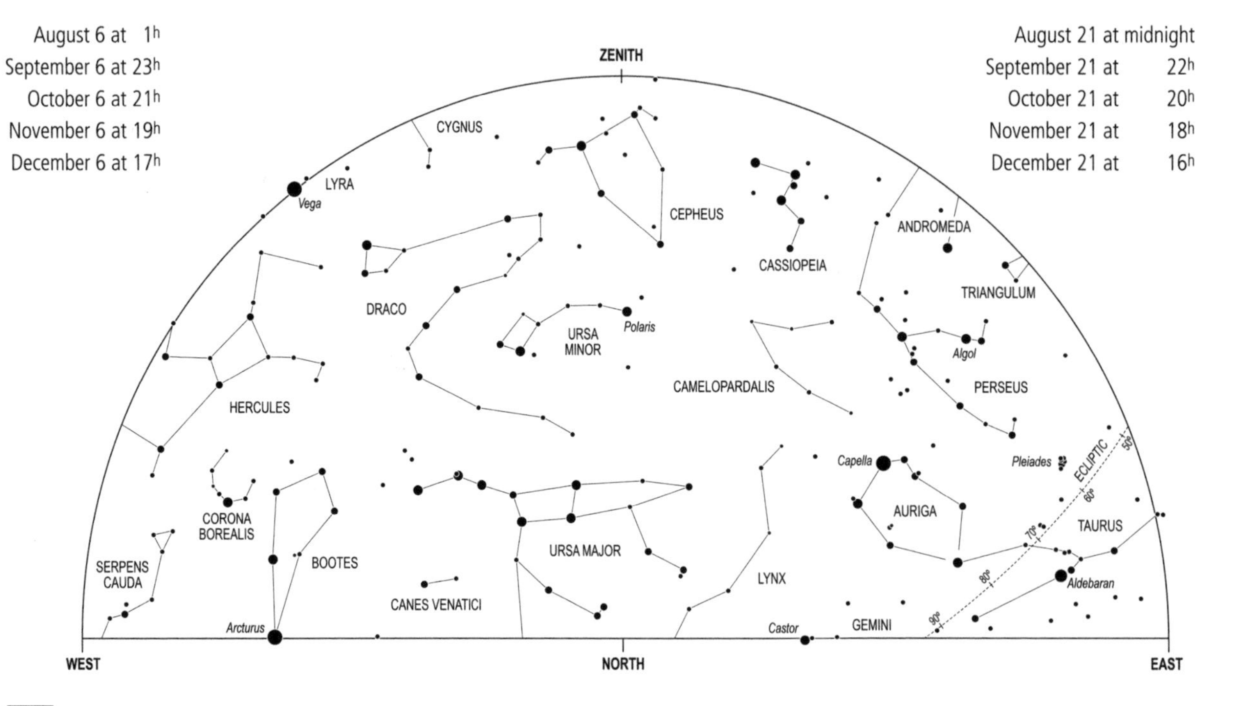
9N
August 6 at 1h
September 6 at 23h
October 6 at 21h
November 6 at 19h
December 6 at 17h
August 21 at midnight
September 21 at 22h
October 21 at 20h
November 21 at 18h
December 21 at 16h
ZENITH
CYGNUS
LYRA
Vega
CEPHEUS
CASSIOPEIA
ANDROMEDA
TRIANGULUM
DRACO
URSA MINOR
Polaris
HERCULES
CAMELOPARDALIS
Algol
PERSEUS
Capella
Pleiades
ECLIPTIC
50°
60°
70°
80°
90°
CORONA BOREALIS
AURIGA
TAURUS
BOOTES
URSA MAJOR
SERPENS CAUDA
LYNX
Aldebaran
CANES VENATICI
GEMINI
Arcturus
Castor
WEST
NORTH
EAST

9S

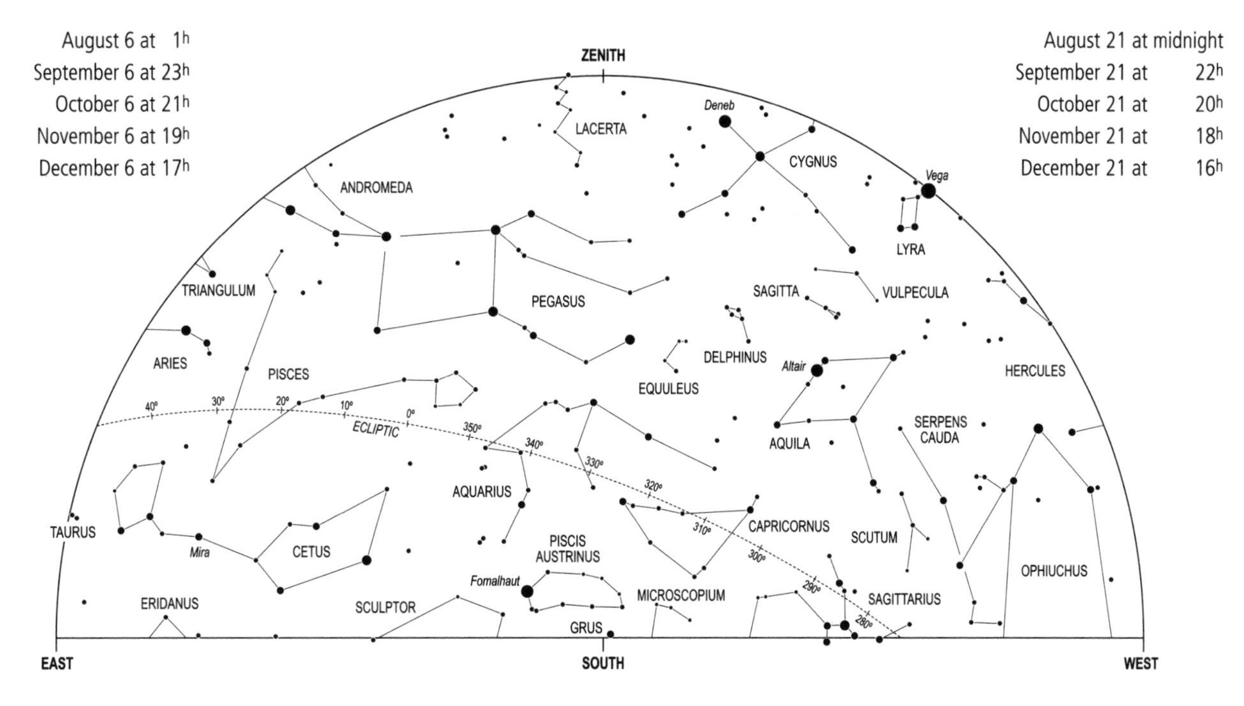
August 6 at 1h
September 6 at 23h
October 6 at 21h
November 6 at 19h
December 6 at 17h
August 21 at midnight
September 21 at 22h
October 21 at 20h
November 21 at 18h
December 21 at 16h
ZENITH
LACERTA
Deneb
CYGNUS
Vega
LYRA
ANDROMEDA
TRIANGULUM
PEGASUS
SAGITTA
VULPECULA
ARIES
PISCES
DELPHINUS
Altair
EQUULEUS
HERCULES
40°
30°
20°
10°
0°
350°
340°
330°
320°
310°
300°
290°
280°
ECLIPTIC
AQUILA
SERPENS
CAUDA
AQUARIUS
TAURUS
Mira
CETUS
PISCIS
AUSTRINUS
CAPRICORNUS
SCUTUM
OPHIUCHUS
Fomalhaut
ERIDANUS
SCULPTOR
MICROSCOPIUM
SAGITTARIUS
GRUS
EAST
SOUTH
WEST

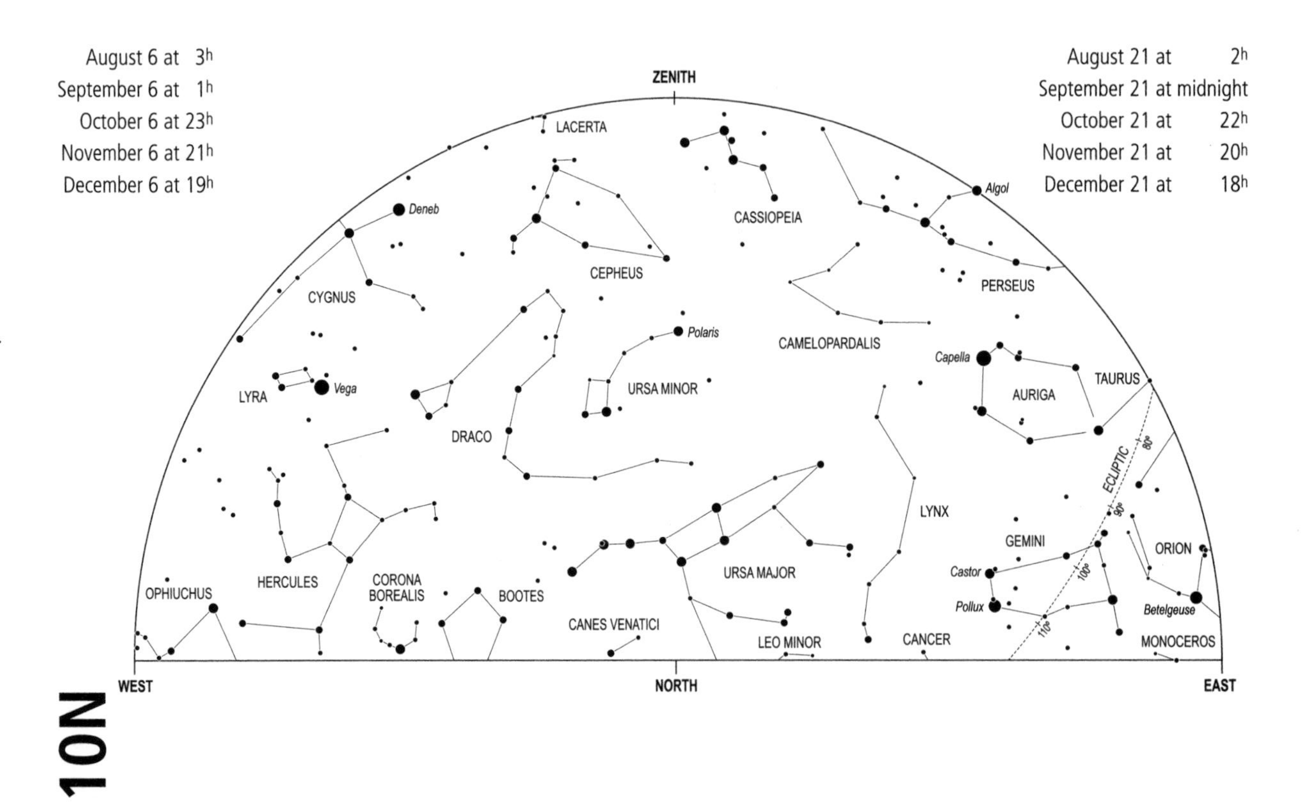
10N
August 6 at 3h
September 6 at 1h
October 6 at 23h
November 6 at 21h
December 6 at 19h
August 21 at 2h
September 21 at midnight
October 21 at 22h
November 21 at 20h
December 21 at 18h
ZENITH
LACERTA
CASSIOPEIA
Deneb
Algol
CEPHEUS
CYGNUS
PERSEUS
Polaris
CAMELOPARDALIS
Capella
LYRA
Vega
URSA MINOR
TAURUS
AURIGA
DRACO
ECLIPTIC
80°
90°
100°
110°
LYNX
GEMINI
ORION
Castor
Pollux
Betelgeuse
URSA MAJOR
OPHIUCHUS
HERCULES
CORONA BOREALIS
BOOTES
CANES VENATICI
LEO MINOR
CANCER
MONOCEROS
WEST
NORTH
EAST

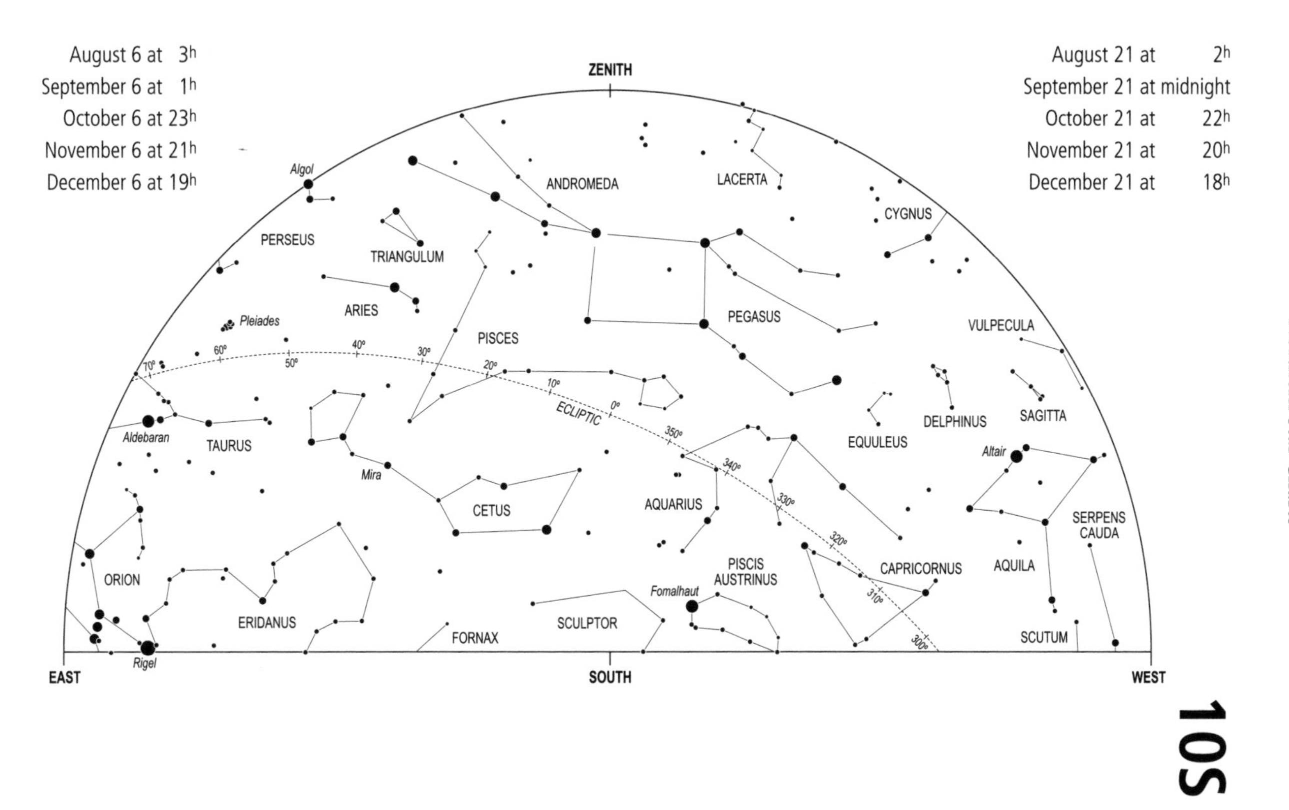
August 6 at 3h
September 6 at 1h
October 6 at 23h
November 6 at 21h
December 6 at 19h
August 21 at 2h
September 21 at midnight
October 21 at 22h
November 21 at 20h
December 21 at 18h
ZENITH
Algol
ANDROMEDA
LACERTA
CYGNUS
PERSEUS
TRIANGULUM
ARIES
Pleiades
PEGASUS
VULPECULA
PISCES
60°
40°
30°
20°
10°
70°
50°
0°
ECLIPTIC
350°
340°
330°
320°
310°
300°
Aldebaran
TAURUS
Mira
CETUS
AQUARIUS
EQUULEUS
DELPHINUS
SAGITTA
Altair
SERPENS CAUDA
ORION
ERIDANUS
FORNAX
SCULPTOR
Fomalhaut
PISCIS AUSTRINUS
CAPRICORNUS
AQUILA
SCUTUM
Rigel
EAST
SOUTH
WEST
10S

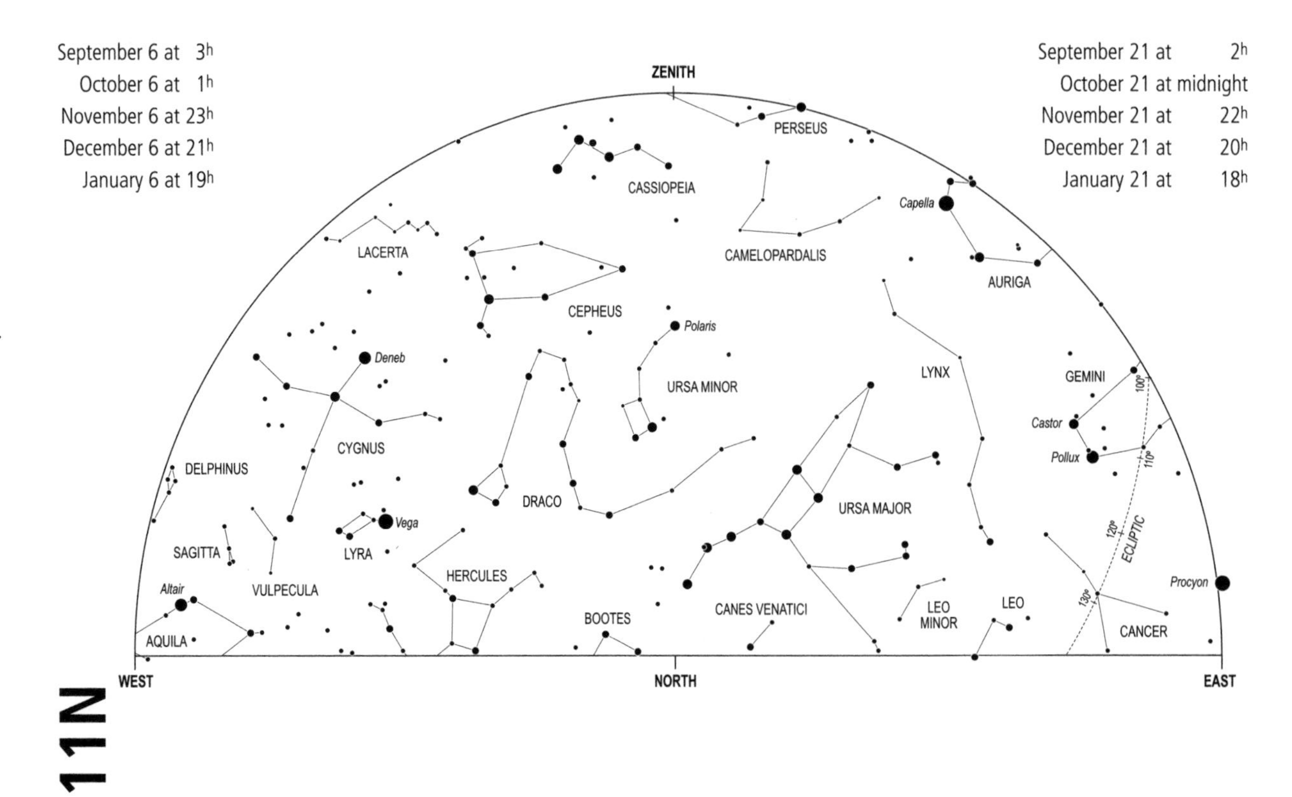
11N
September 6 at 3h
October 6 at 1h
November 6 at 23h
December 6 at 21h
January 6 at 19h
September 21 at 2h
October 21 at midnight
November 21 at 22h
December 21 at 20h
January 21 at 18h
ZENITH
PERSEUS
CASSIOPEIA
CAMELOPARDALIS
Capella
AURIGA
LACERTA
CEPHEUS
Polaris
Deneb
URSA MINOR
LYNX
GEMINI
Castor
Pollux
CYGNUS
DELPHINUS
DRACO
URSA MAJOR
Vega
LYRA
SAGITTA
HERCULES
Altair
VULPECULA
BOOTES
CANES VENATICI
LEO MINOR
LEO
CANCER
Procyon
ECLIPTIC
100°
110°
120°
130°
AQUILA
WEST
NORTH
EAST

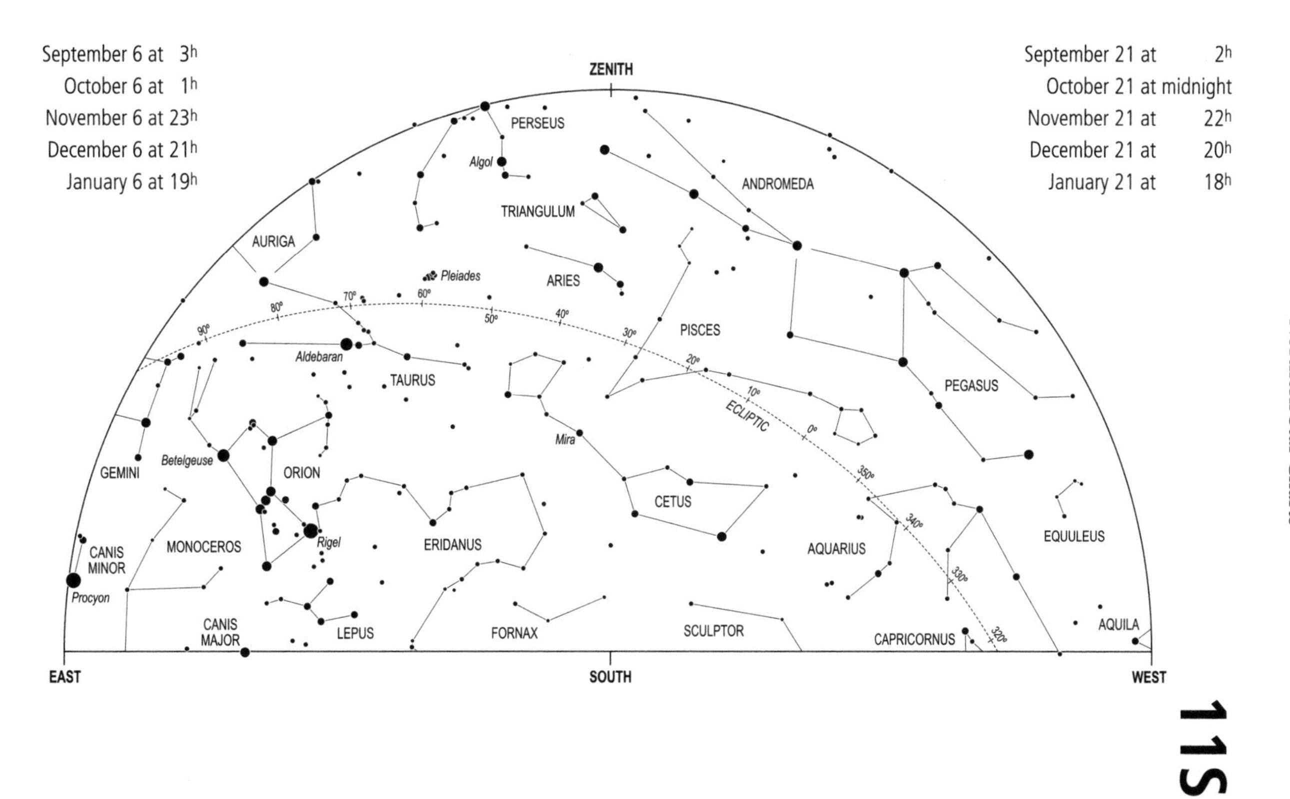
11S
September 6 at 3h
October 6 at 1h
November 6 at 23h
December 6 at 21h
January 6 at 19h
September 21 at 2h
October 21 at midnight
November 21 at 22h
December 21 at 20h
January 21 at 18h
ZENITH
EAST
SOUTH
WEST
PERSEUS
Algol
ANDROMEDA
TRIANGULUM
AURIGA
Pleiades
ARIES
PISCES
PEGASUS
Aldebaran
TAURUS
ECLIPTIC
Mira
GEMINI
Betelgeuse
ORION
CETUS
EQUULEUS
CANIS MINOR
Procyon
MONOCEROS
Rigel
ERIDANUS
AQUARIUS
CANIS MAJOR
LEPUS
FORNAX
SCULPTOR
CAPRICORNUS
AQUILA

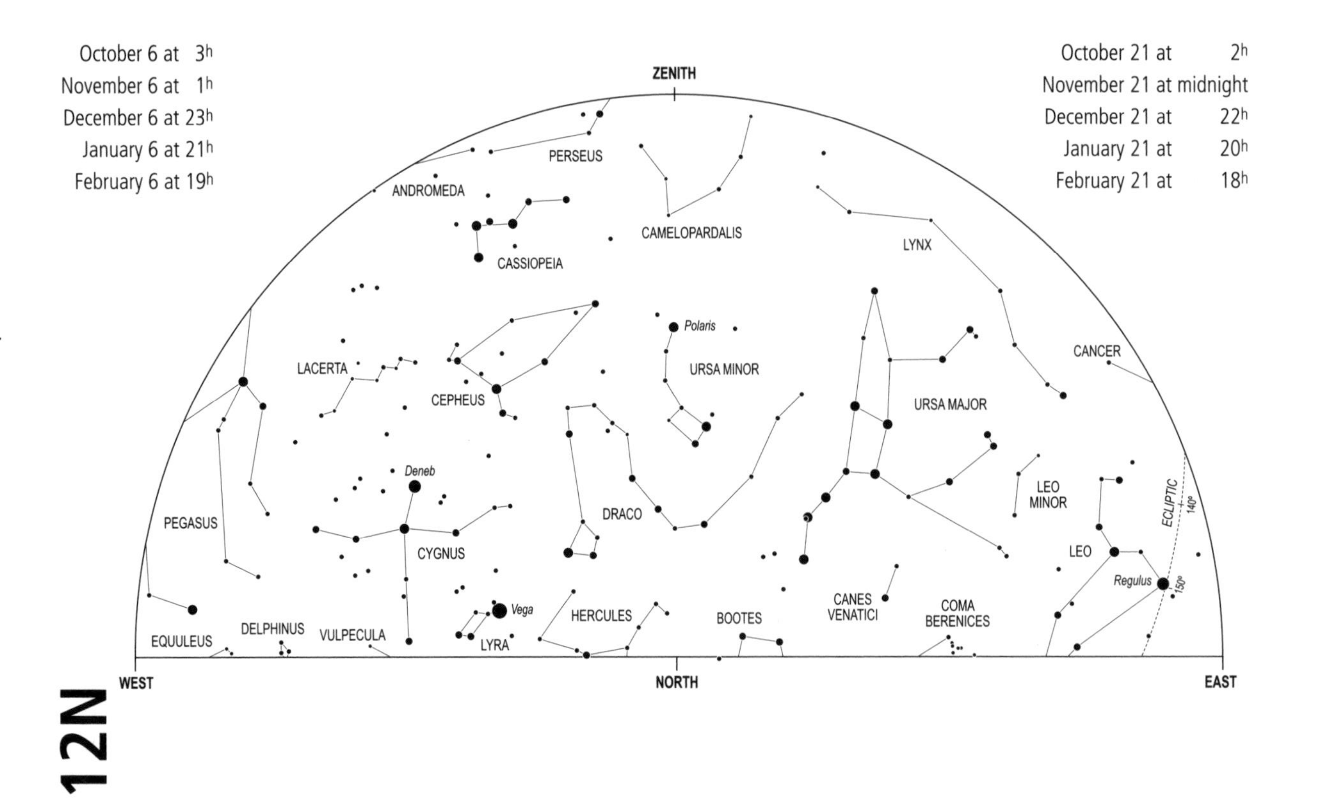
October 6 at 3h
November 6 at 1h
December 6 at 23h
January 6 at 21h
February 6 at 19h
October 21 at 2h
November 21 at midnight
December 21 at 22h
January 21 at 20h
February 21 at 18h
ZENITH
PERSEUS
ANDROMEDA
CASSIOPEIA
CAMELOPARDALIS
LYNX
Polaris
LACERTA
CEPHEUS
URSA MINOR
CANCER
URSA MAJOR
Deneb
LEO MINOR
PEGASUS
DRACO
ECLIPTIC
140°
CYGNUS
LEO
Regulus
150°
Vega
HERCULES
CANES VENATICI
COMA BERENICES
BOOTES
EQUULEUS
DELPHINUS
VULPECULA
LYRA
WEST
NORTH
EAST
12N

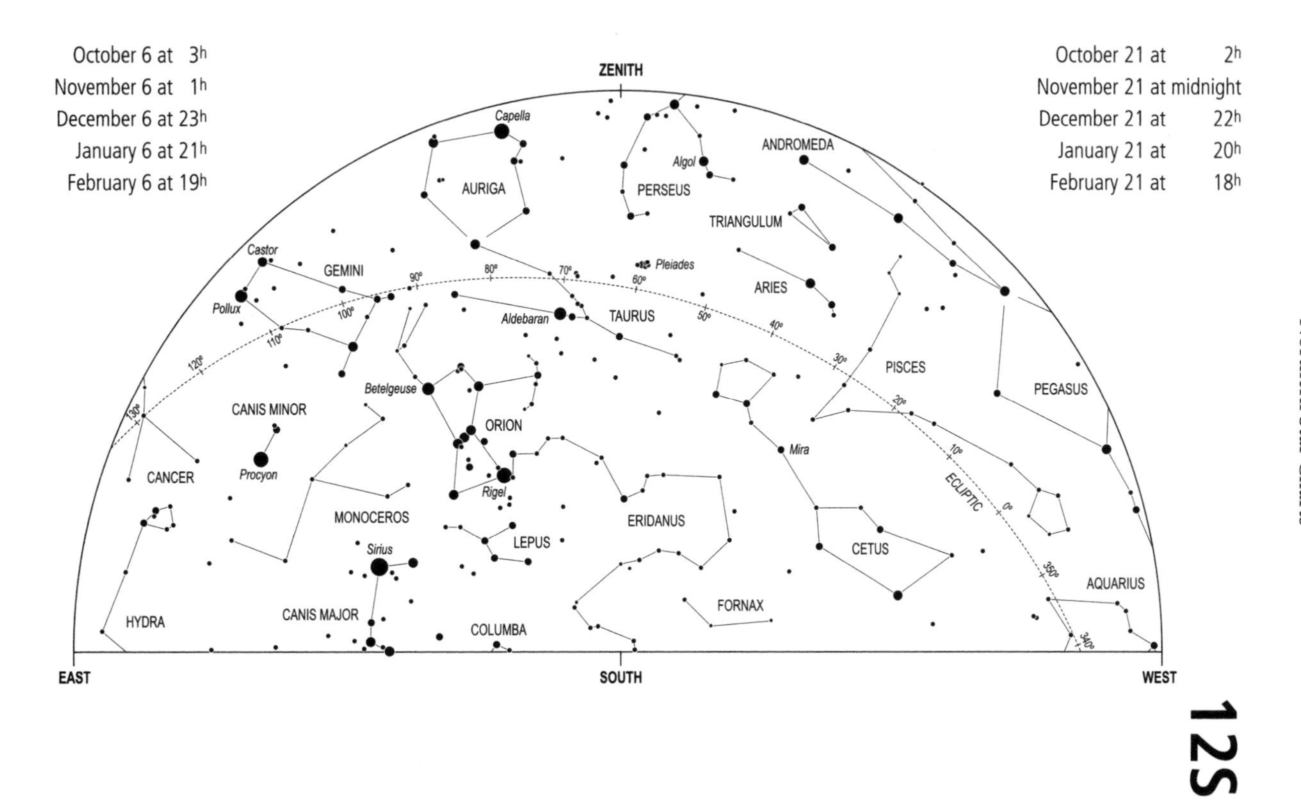
12S
October 6 at 3h
November 6 at 1h
December 6 at 23h
January 6 at 21h
February 6 at 19h
October 21 at 2h
November 21 at midnight
December 21 at 22h
January 21 at 20h
February 21 at 18h
ZENITH
EAST
SOUTH
WEST
ECLIPTIC
AURIGA
Capella
PERSEUS
Algol
ANDROMEDA
TRIANGULUM
Pleiades
ARIES
TAURUS
Aldebaran
GEMINI
Castor
Pollux
PISCES
PEGASUS
Betelgeuse
ORION
Rigel
CANIS MINOR
Procyon
CANCER
MONOCEROS
LEPUS
ERIDANUS
Mira
CETUS
FORNAX
AQUARIUS
Sirius
CANIS MAJOR
COLUMBA
HYDRA
130°
120°
110°
100°
90°
80°
70°
60°
50°
40°
30°
20°
10°
0°
350°
340°

Southern Star Charts

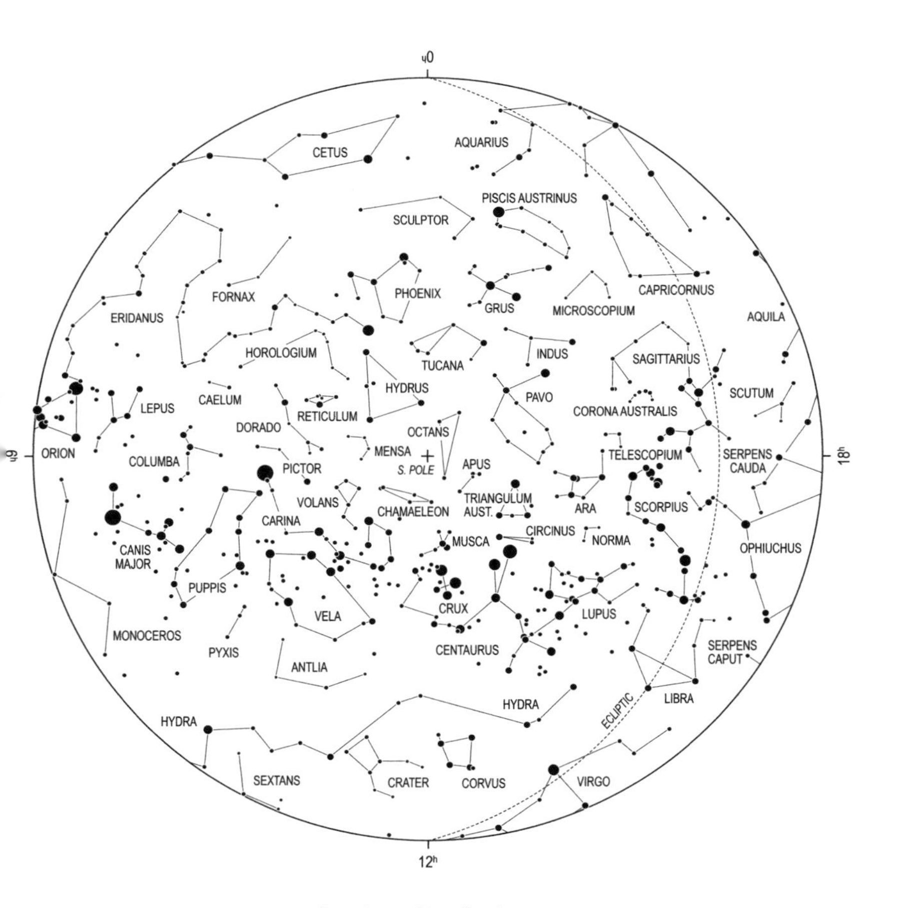

Southern Hemisphere

Note that the markers at 0h, 6h, 12h and 18h indicate hours of Right Ascension.

1N

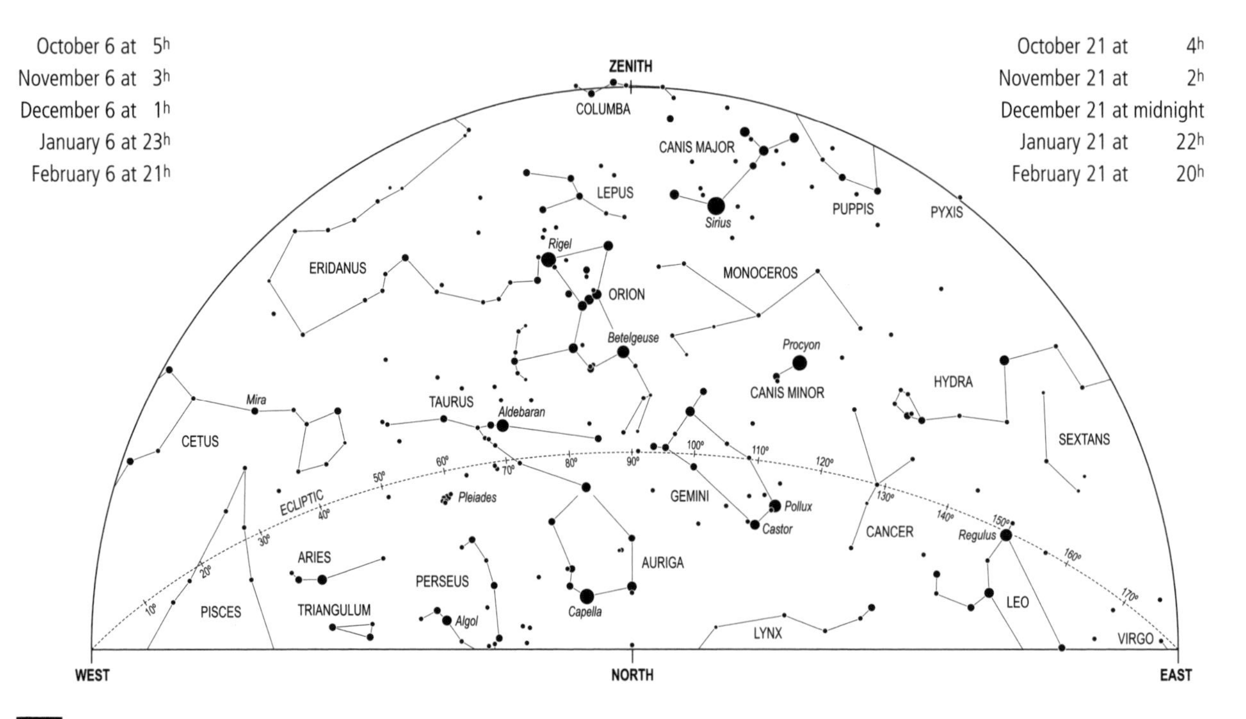
October 6 at 5h
November 6 at 3h
December 6 at 1h
January 6 at 23h
February 6 at 21h
October 21 at 4h
November 21 at 2h
December 21 at midnight
January 21 at 22h
February 21 at 20h
ZENITH
COLUMBA
CANIS MAJOR
LEPUS
Sirius
PUPPIS
PYXIS
Rigel
ERIDANUS
MONOCEROS
ORION
Betelgeuse
Procyon
CANIS MINOR
HYDRA
Mira
TAURUS
Aldebaran
CETUS
SEXTANS
ECLIPTIC
Pleiades
GEMINI
Pollux
Castor
CANCER
Regulus
ARIES
PERSEUS
AURIGA
PISCES
TRIANGULUM
Algol
Capella
LEO
LYNX
VIRGO
WEST
NORTH
EAST

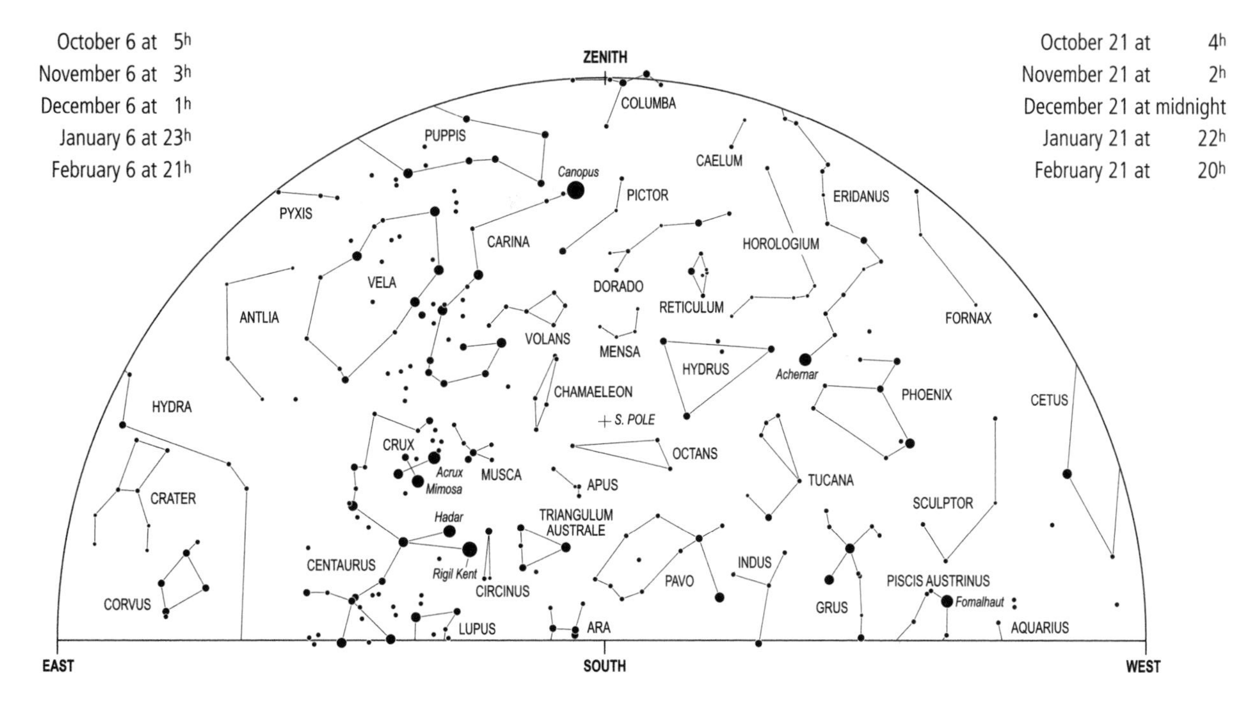
October 6 at 5h
November 6 at 3h
December 6 at 1h
January 6 at 23h
February 6 at 21h
October 21 at 4h
November 21 at 2h
December 21 at midnight
January 21 at 22h
February 21 at 20h
ZENITH
COLUMBA
PUPPIS
Canopus
PICTOR
CAELUM
ERIDANUS
PYXIS
CARINA
HOROLOGIUM
VELA
DORADO
RETICULUM
FORNAX
ANTLIA
VOLANS
MENSA
HYDRUS
Achernar
HYDRA
CHAMAELEON
PHOENIX
CETUS
S. POLE
CRUX
OCTANS
Acrux
MUSCA
Mimosa
APUS
TUCANA
CRATER
TRIANGULUM AUSTRALE
SCULPTOR
Hadar
CENTAURUS
Rigil Kent
INDUS
PAVO
CIRCINUS
PISCIS AUSTRINUS
Fomalhaut
CORVUS
GRUS
LUPUS
ARA
AQUARIUS
EAST
SOUTH
WEST
1S

2N

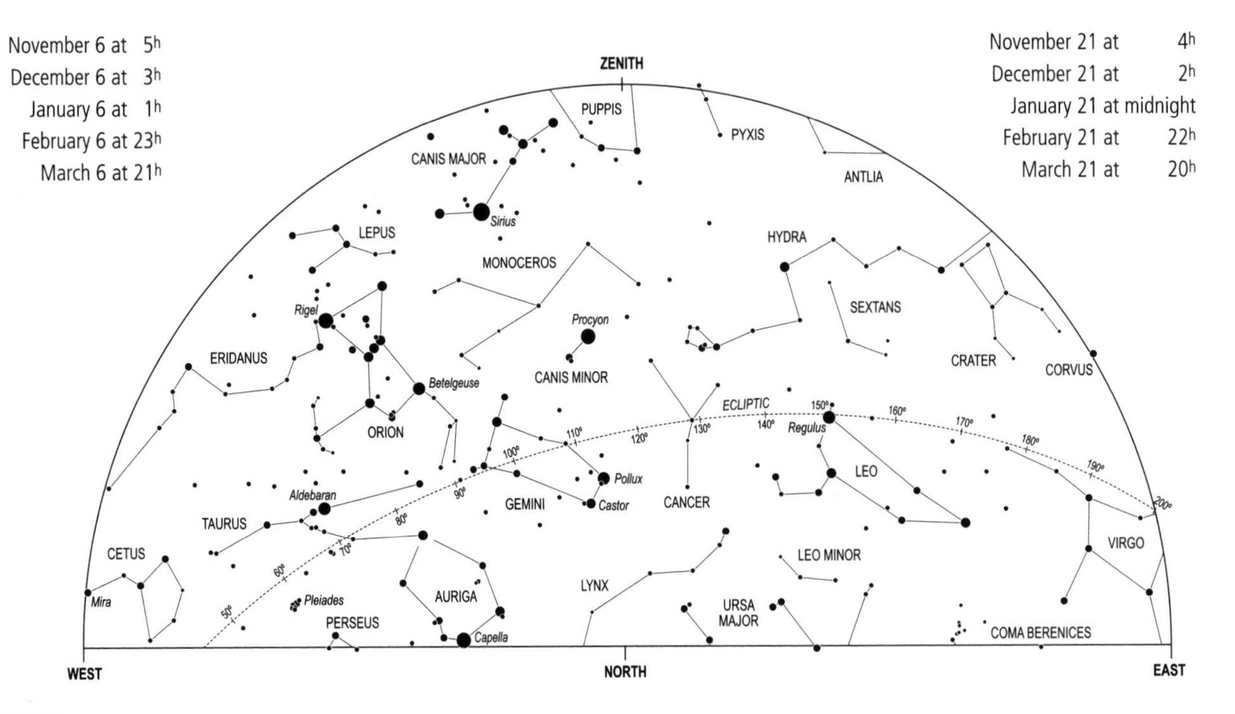
November 6 at 5h
December 6 at 3h
January 6 at 1h
February 6 at 23h
March 6 at 21h
November 21 at 4h
December 21 at 2h
January 21 at midnight
February 21 at 22h
March 21 at 20h
ZENITH
WEST
NORTH
EAST
ECLIPTIC
PUPPIS
PYXIS
ANTLIA
CANIS MAJOR
Sirius
LEPUS
MONOCEROS
HYDRA
SEXTANS
Rigel
Procyon
ERIDANUS
CRATER
CORVUS
CANIS MINOR
Betelgeuse
ORION
Regulus
LEO
Pollux
Castor
GEMINI
CANCER
Aldebaran
TAURUS
CETUS
VIRGO
LEO MINOR
LYNX
AURIGA
Pleiades
Mira
URSA MAJOR
PERSEUS
Capella
COMA BERENICES
50°
60°
70°
80°
90°
100°
110°
120°
130°
140°
150°
160°
170°
180°
190°
200°

2S

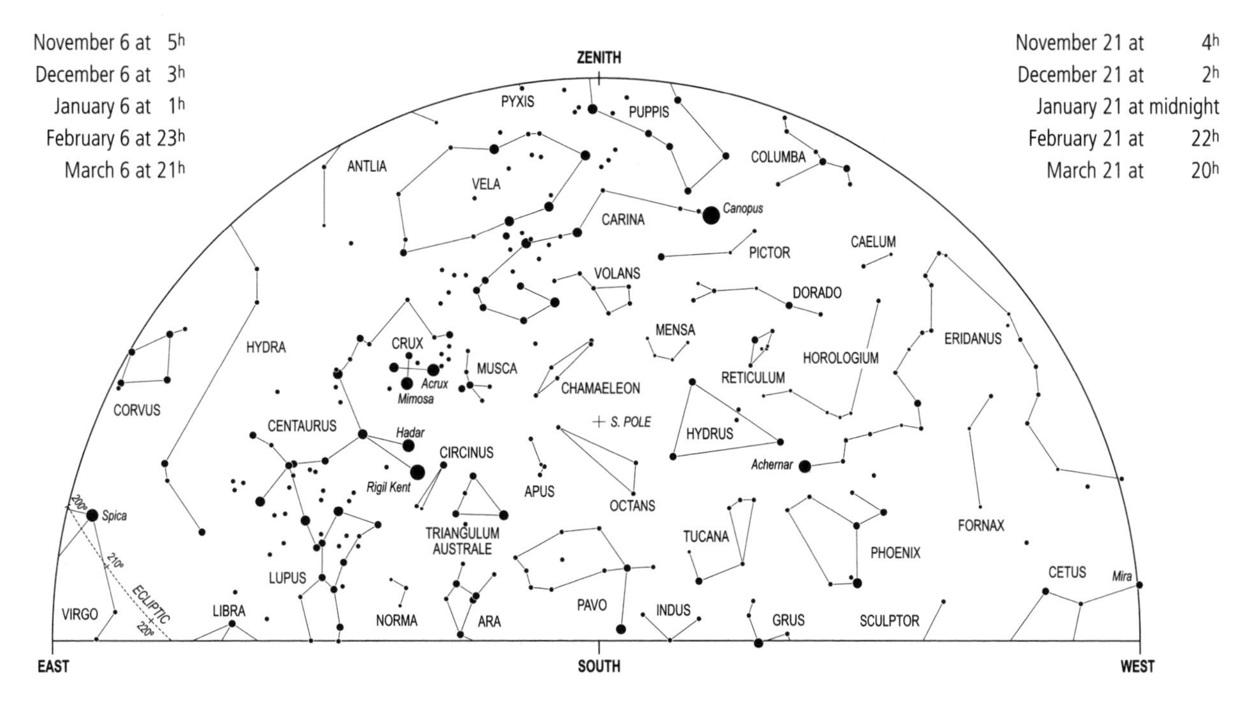
November 6 at 5h
December 6 at 3h
January 6 at 1h
February 6 at 23h
March 6 at 21h
November 21 at 4h
December 21 at 2h
January 21 at midnight
February 21 at 22h
March 21 at 20h
ZENITH
PYXIS
PUPPIS
COLUMBA
ANTLIA
VELA
Canopus
CARINA
CAELUM
PICTOR
VOLANS
DORADO
HYDRA
CRUX
MUSCA
MENSA
ERIDANUS
HOROLOGIUM
RETICULUM
Acrux
Mimosa
CHAMAELEON
CORVUS
S. POLE
CENTAURUS
HYDRUS
Hadar
CIRCINUS
Achernar
Rigil Kent
APUS
OCTANS
Spica
200°
TRIANGULUM AUSTRALE
TUCANA
FORNAX
PHOENIX
210°
LUPUS
CETUS
Mira
ECLIPTIC
220°
VIRGO
LIBRA
NORMA
ARA
PAVO
INDUS
GRUS
SCULPTOR
EAST
SOUTH
WEST

3N

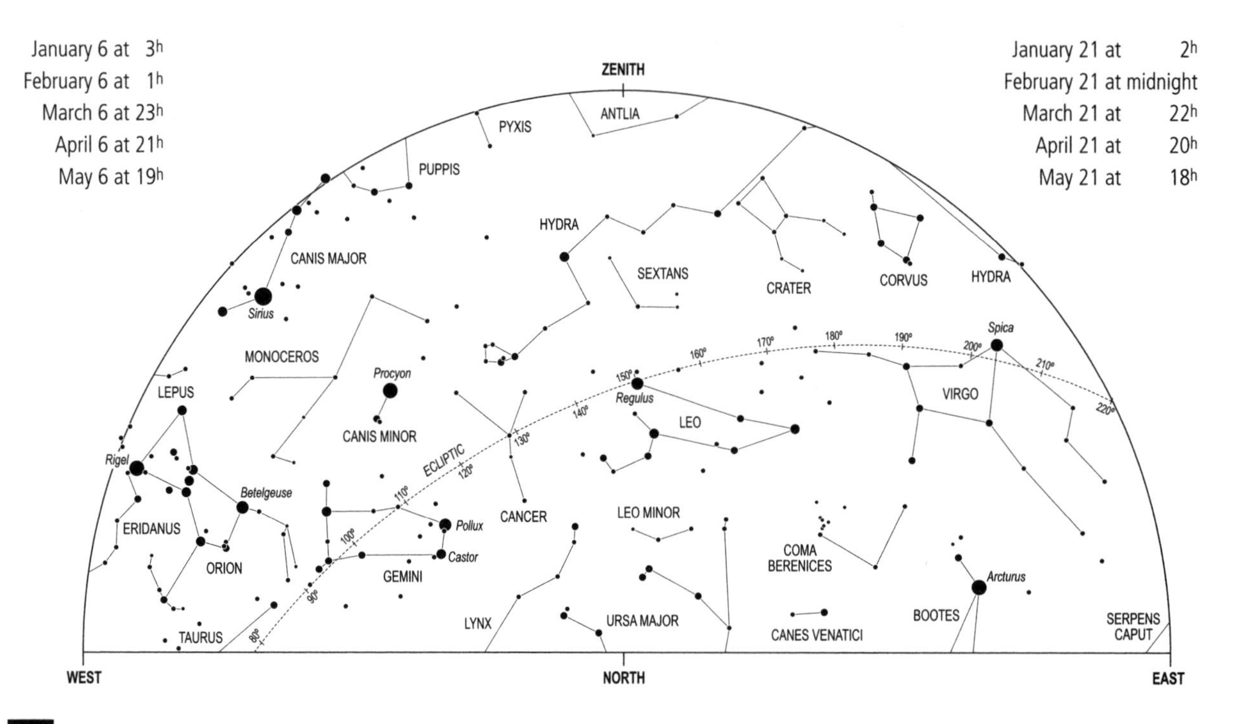
January 6 at 3h
February 6 at 1h
March 6 at 23h
April 6 at 21h
May 6 at 19h
January 21 at 2h
February 21 at midnight
March 21 at 22h
April 21 at 20h
May 21 at 18h
ZENITH
ANTLIA
PYXIS
PUPPIS
HYDRA
CANIS MAJOR
Sirius
SEXTANS
CRATER
CORVUS
HYDRA
MONOCEROS
Procyon
Spica
LEPUS
CANIS MINOR
Regulus
VIRGO
LEO
ECLIPTIC
Rigel
Betelgeuse
ERIDANUS
Pollux
CANCER
LEO MINOR
ORION
Castor
GEMINI
COMA BERENICES
Arcturus
TAURUS
LYNX
URSA MAJOR
CANES VENATICI
BOOTES
SERPENS CAPUT
WEST
NORTH
EAST
80°
90°
100°
110°
120°
130°
140°
150°
160°
170°
180°
190°
200°
210°
220°

3S

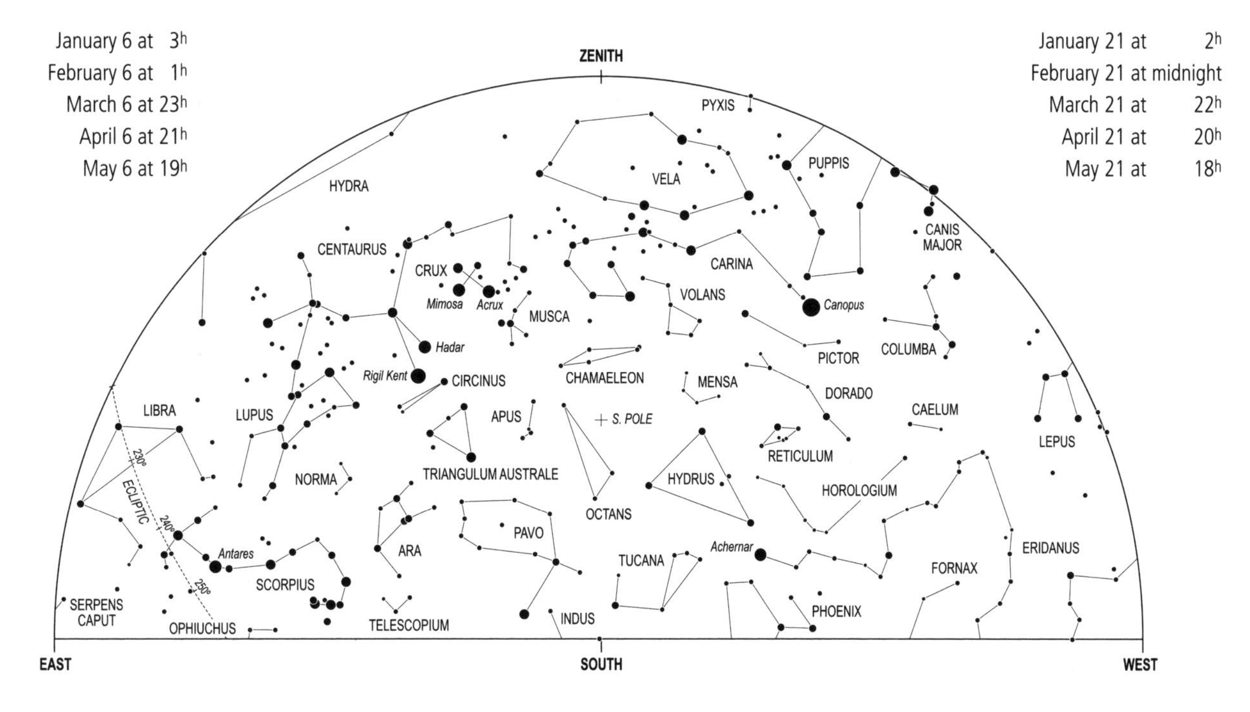
January 6 at 3h
February 6 at 1h
March 6 at 23h
April 6 at 21h
May 6 at 19h
January 21 at 2h
February 21 at midnight
March 21 at 22h
April 21 at 20h
May 21 at 18h
ZENITH
EAST
SOUTH
WEST
PYXIS
HYDRA
VELA
PUPPIS
CANIS MAJOR
CENTAURUS
CRUX
Mimosa
Acrux
CARINA
VOLANS
Canopus
MUSCA
Hadar
PICTOR
COLUMBA
Rigil Kent
CIRCINUS
CHAMAELEON
MENSA
DORADO
CAELUM
LIBRA
LUPUS
APUS
S. POLE
LEPUS
RETICULUM
230°
TRIANGULUM AUSTRALE
NORMA
HYDRUS
HOROLOGIUM
ECLIPTIC
OCTANS
240°
PAVO
Antares
ARA
TUCANA
Achernar
ERIDANUS
FORNAX
250°
SCORPIUS
PHOENIX
SERPENS CAPUT
OPHIUCHUS
TELESCOPIUM
INDUS

4N

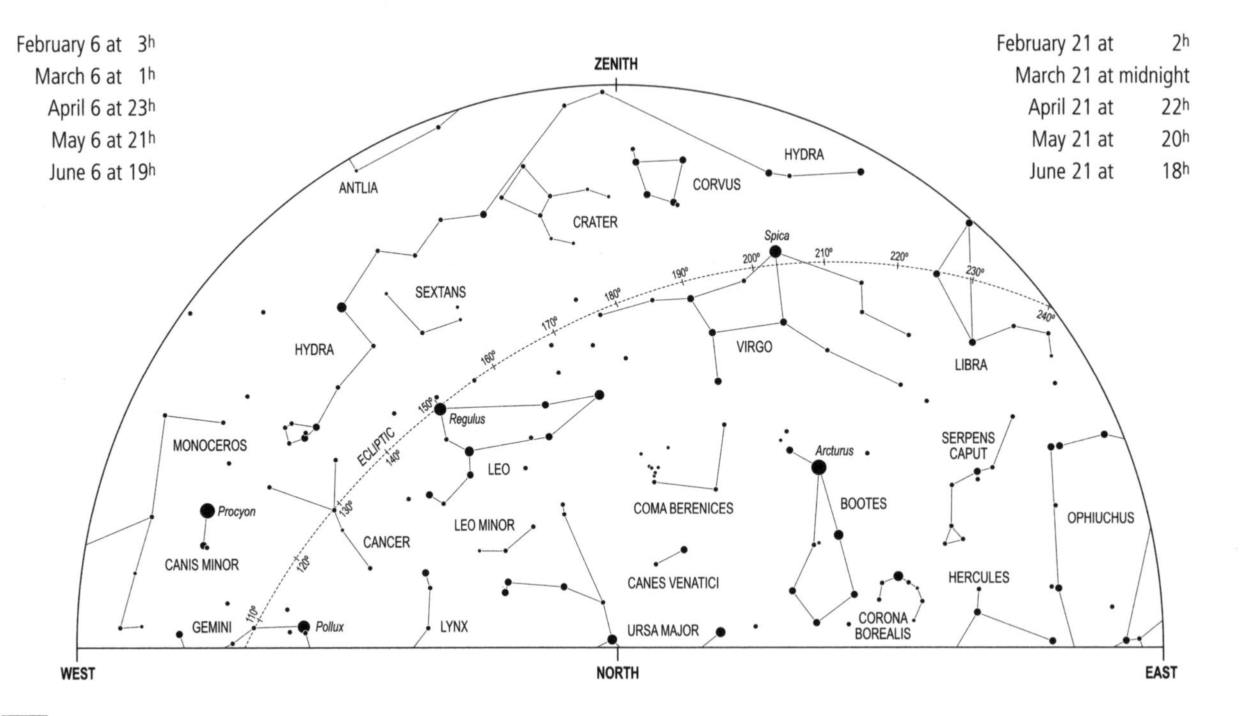

February 6 at 3h
March 6 at 1h
April 6 at 23h
May 6 at 21h
June 6 at 19h
February 21 at 2h
March 21 at midnight
April 21 at 22h
May 21 at 20h
June 21 at 18h
ZENITH
WEST
NORTH
EAST
ECLIPTIC
110°
120°
130°
140°
150°
160°
170°
180°
190°
200°
210°
220°
230°
240°
ANTLIA
CRATER
CORVUS
HYDRA
SEXTANS
HYDRA
VIRGO
Spica
LIBRA
MONOCEROS
Regulus
LEO
Arcturus
SERPENS CAPUT
Procyon
CANIS MINOR
CANCER
LEO MINOR
COMA BERENICES
BOOTES
OPHIUCHUS
CANES VENATICI
HERCULES
GEMINI
Pollux
LYNX
URSA MAJOR
CORONA BOREALIS

4S

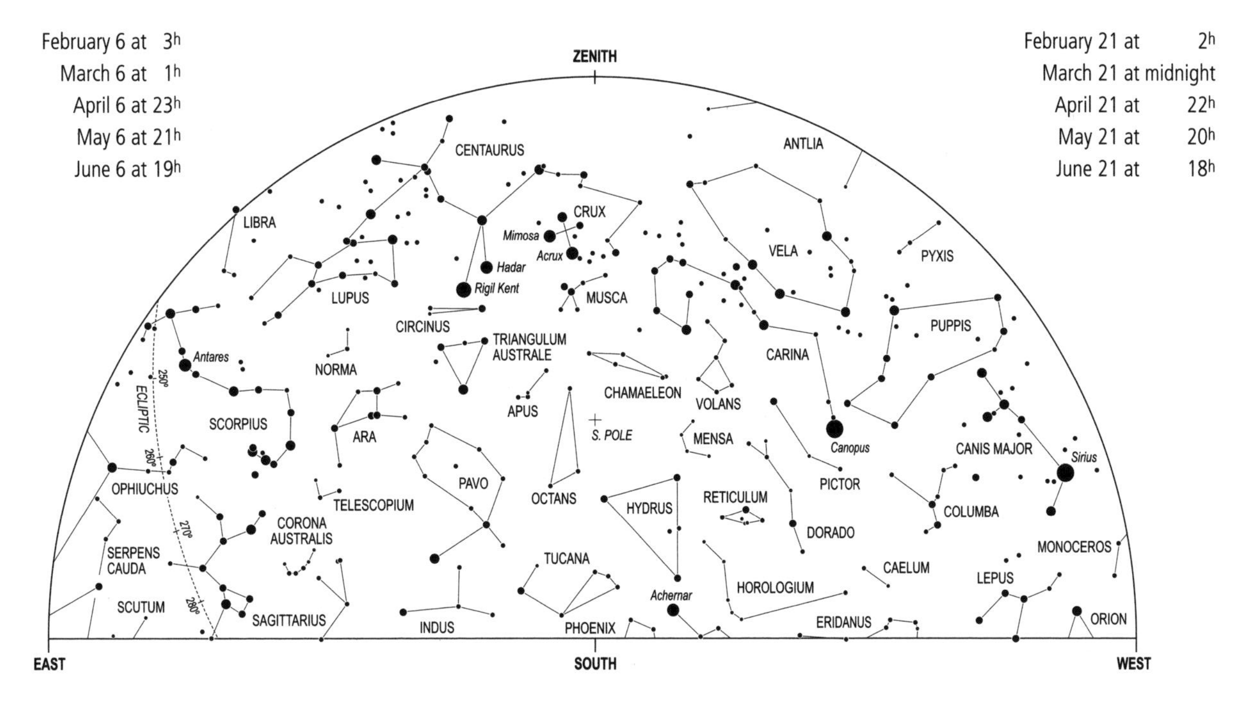
February 6 at 3h
March 6 at 1h
April 6 at 23h
May 6 at 21h
June 6 at 19h
February 21 at 2h
March 21 at midnight
April 21 at 22h
May 21 at 20h
June 21 at 18h
ZENITH
EAST
SOUTH
WEST
S. POLE
ECLIPTIC
250°
260°
270°
280°
CENTAURUS
ANTLIA
LIBRA
CRUX
Mimosa
Acrux
Hadar
Rigil Kent
VELA
PYXIS
LUPUS
MUSCA
PUPPIS
CIRCINUS
TRIANGULUM AUSTRALE
CARINA
Antares
NORMA
CHAMAELEON
VOLANS
APUS
SCORPIUS
ARA
MENSA
Canopus
CANIS MAJOR
Sirius
PAVO
OCTANS
PICTOR
RETICULUM
OPHIUCHUS
HYDRUS
TELESCOPIUM
COLUMBA
CORONA AUSTRALIS
DORADO
MONOCEROS
SERPENS CAUDA
TUCANA
CAELUM
LEPUS
HOROLOGIUM
SCUTUM
Achernar
SAGITTARIUS
INDUS
PHOENIX
ERIDANUS
ORION

5N

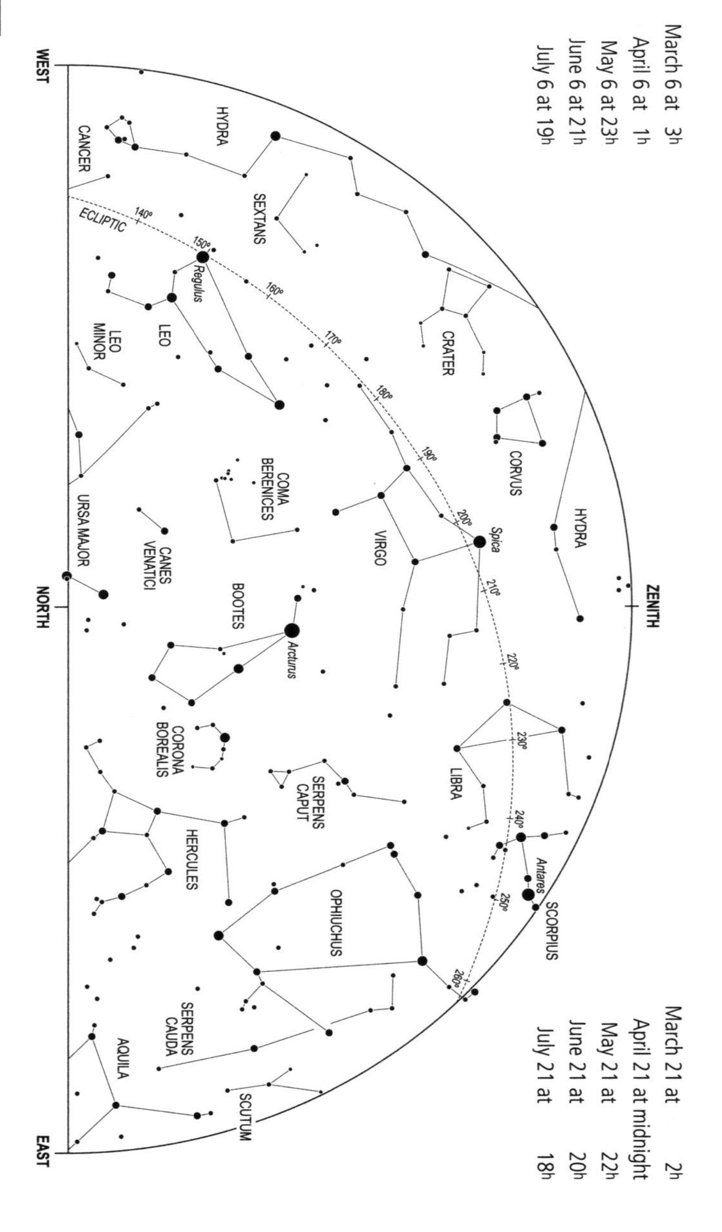
March 6 at 3h
April 6 at 1h
May 6 at 23h
June 6 at 21h
July 6 at 19h
March 21 at 2h
April 21 at midnight
May 21 at 22h
June 21 at 20h
July 21 at 18h
WEST
NORTH
EAST
ZENITH
ECLIPTIC
140°
150°
160°
170°
180°
190°
200°
210°
220°
230°
240°
250°
260°
HYDRA
CANCER
SEXTANS
Regulus
LEO
LEO MINOR
CRATER
CORVUS
COMA BERENICES
URSA MAJOR
CANES VENATICI
VIRGO
Spica
HYDRA
BOOTES
Arcturus
CORONA BOREALIS
SERPENS CAPUT
LIBRA
HERCULES
OPHIUCHUS
Antares
SCORPIUS
SERPENS CAUDA
AQUILA
SCUTUM

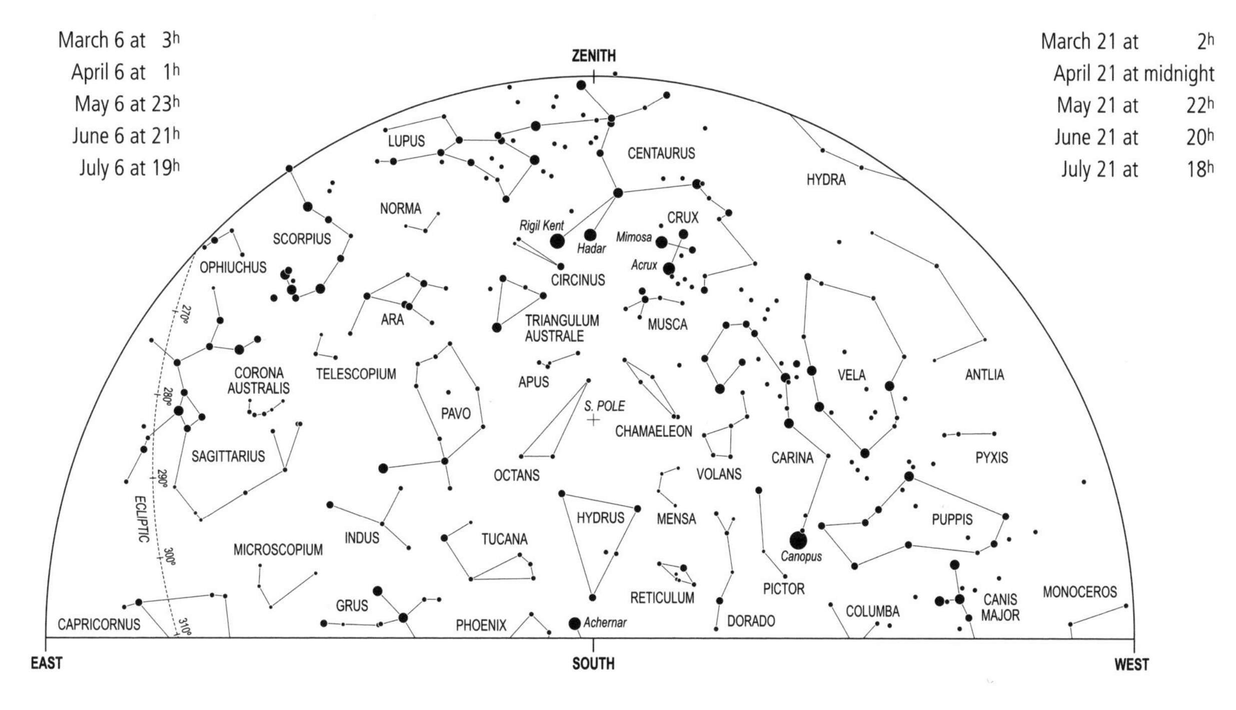

March 6 at 3h
April 6 at 1h
May 6 at 23h
June 6 at 21h
July 6 at 19h
March 21 at 2h
April 21 at midnight
May 21 at 22h
June 21 at 20h
July 21 at 18h
ZENITH
EAST
SOUTH
WEST
LUPUS
CENTAURUS
HYDRA
NORMA
SCORPIUS
OPHIUCHUS
Rigil Kent
Hadar
Mimosa
Acrux
CRUX
CIRCINUS
ARA
TRIANGULUM AUSTRALE
MUSCA
CORONA AUSTRALIS
TELESCOPIUM
APUS
VELA
ANTLIA
S. POLE
CHAMAELEON
PAVO
SAGITTARIUS
OCTANS
VOLANS
CARINA
PYXIS
270°
280°
290°
300°
310°
ECLIPTIC
HYDRUS
MENSA
PUPPIS
INDUS
TUCANA
MICROSCOPIUM
Canopus
PICTOR
RETICULUM
GRUS
PHOENIX
Achernar
DORADO
COLUMBA
CANIS MAJOR
MONOCEROS
CAPRICORNUS

5S

6N

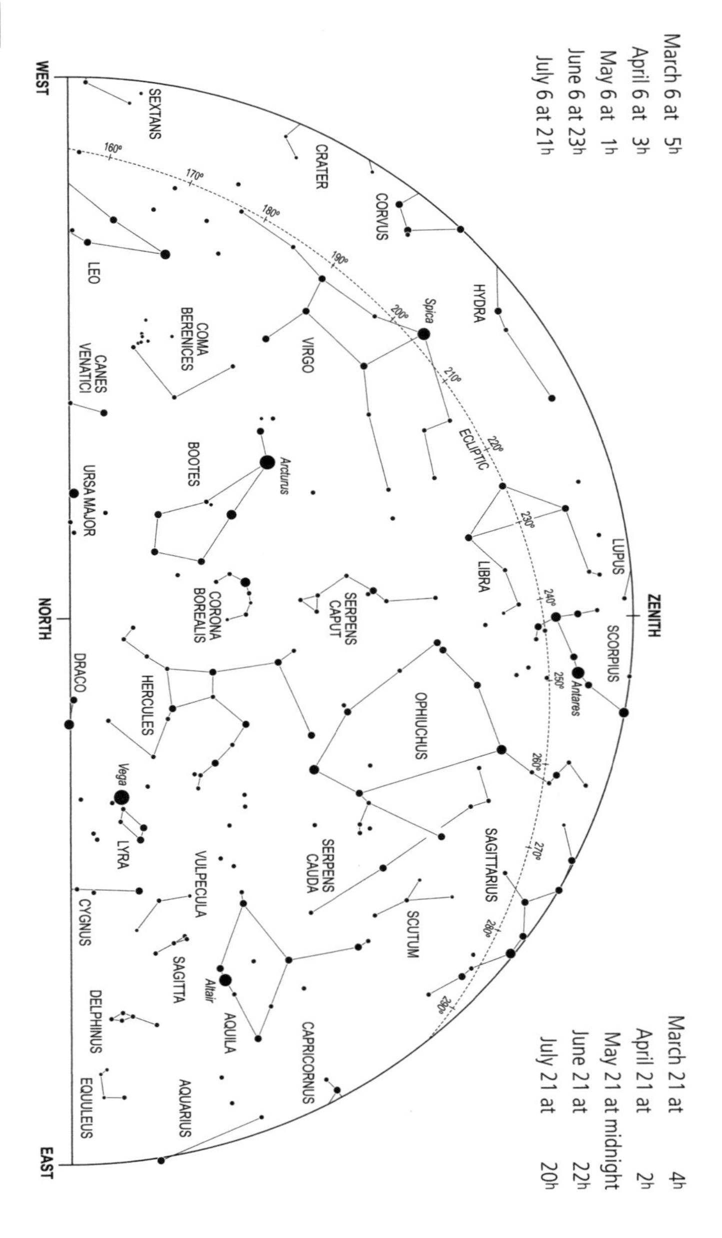
March 6 at 5h
April 6 at 3h
May 6 at 1h
June 6 at 23h
July 6 at 21h
March 21 at 4h
April 21 at 2h
May 21 at midnight
June 21 at 22h
July 21 at 20h
WEST
NORTH
EAST
ZENITH
ECLIPTIC
SEXTANS
CRATER
CORVUS
HYDRA
LEO
COMA BERENICES
VIRGO
Spica
CANES VENATICI
BOOTES
Arcturus
URSA MAJOR
LIBRA
LUPUS
CORONA BOREALIS
SERPENS CAPUT
SCORPIUS
Antares
DRACO
HERCULES
OPHIUCHUS
Vega
LYRA
SERPENS CAUDA
SAGITTARIUS
VULPECULA
CYGNUS
SCUTUM
SAGITTA
Altair
AQUILA
DELPHINUS
CAPRICORNUS
EQUULEUS
AQUARIUS
160°
170°
180°
190°
200°
210°
220°
230°
240°
250°
260°
270°
280°
290°

6S

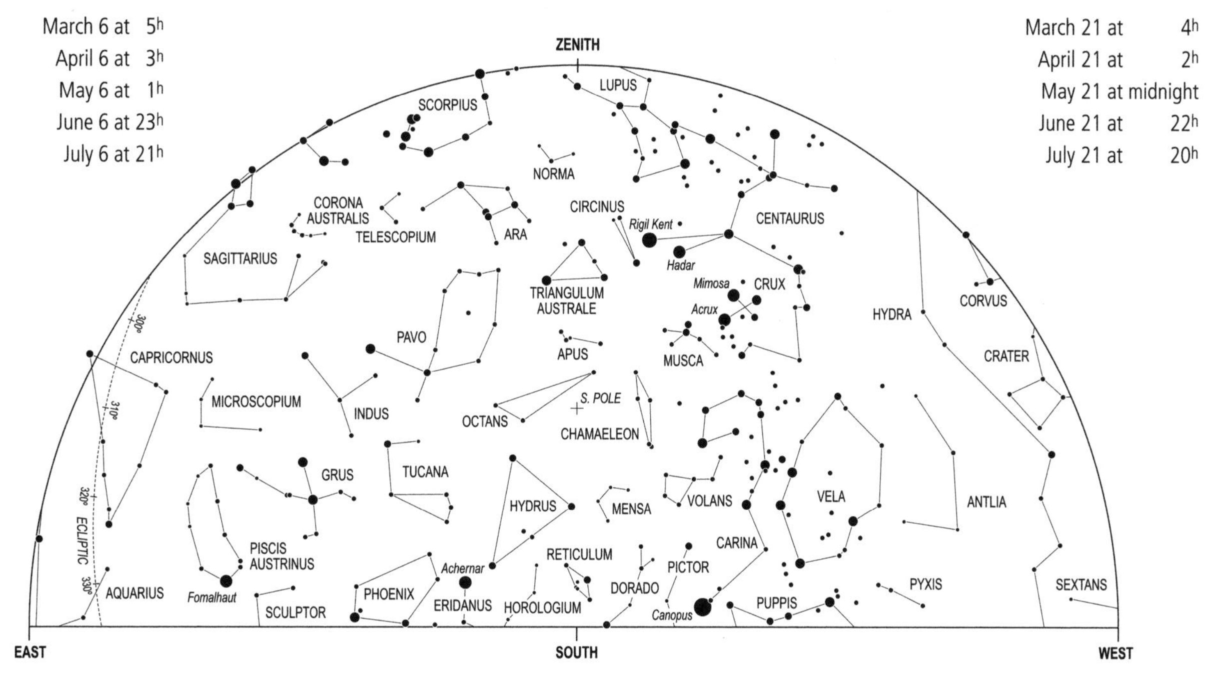
March 6 at 5h
April 6 at 3h
May 6 at 1h
June 6 at 23h
July 6 at 21h
March 21 at 4h
April 21 at 2h
May 21 at midnight
June 21 at 22h
July 21 at 20h
ZENITH
EAST
SOUTH
WEST
SCORPIUS
LUPUS
NORMA
CORONA AUSTRALIS
TELESCOPIUM
ARA
CIRCINUS
Rigil Kent
CENTAURUS
Hadar
SAGITTARIUS
TRIANGULUM AUSTRALE
Mimosa
CRUX
Acrux
HYDRA
CORVUS
PAVO
APUS
MUSCA
CAPRICORNUS
CRATER
MICROSCOPIUM
INDUS
OCTANS
S. POLE
CHAMAELEON
GRUS
TUCANA
HYDRUS
MENSA
VOLANS
VELA
ANTLIA
CARINA
PISCIS AUSTRINUS
RETICULUM
PICTOR
Achernar
DORADO
PYXIS
SEXTANS
AQUARIUS
Fomalhaut
PHOENIX
ERIDANUS
HOROLOGIUM
PUPPIS
SCULPTOR
Canopus
300°
310°
320°
330°
ECLIPTIC

7N

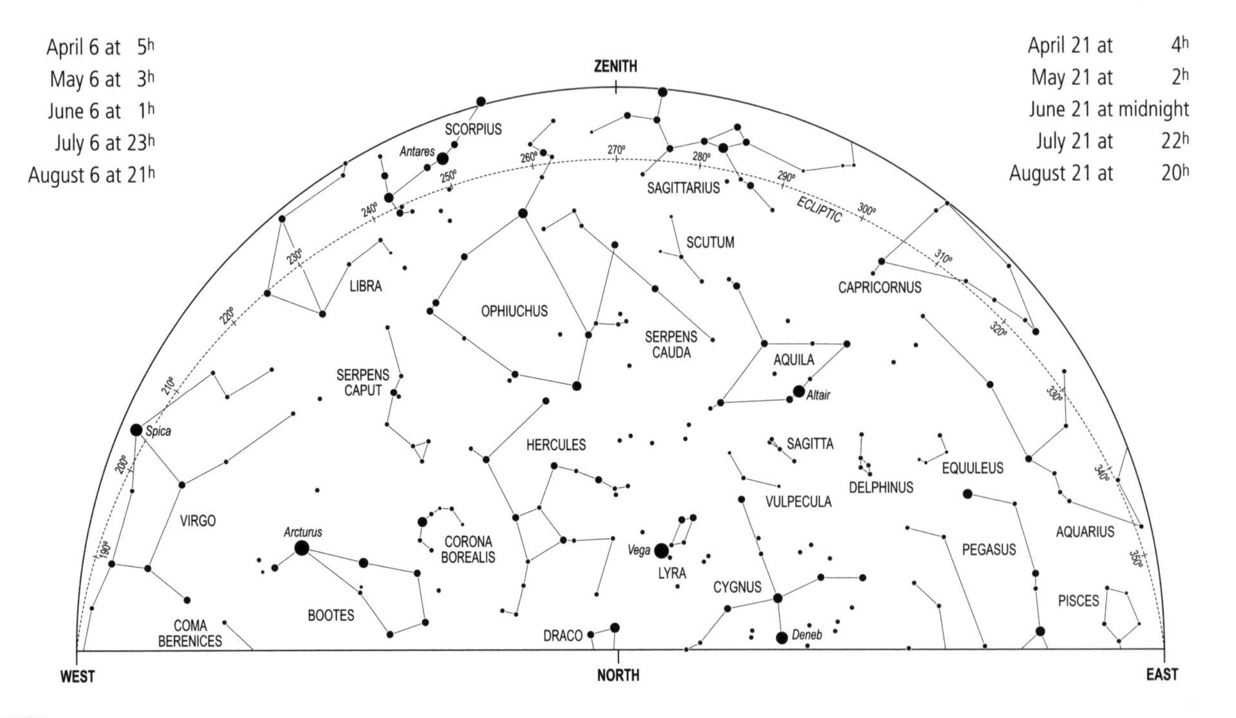
April 6 at 5h
May 6 at 3h
June 6 at 1h
July 6 at 23h
August 6 at 21h
April 21 at 4h
May 21 at 2h
June 21 at midnight
July 21 at 22h
August 21 at 20h
ZENITH
WEST
NORTH
EAST
ECLIPTIC
190°
200°
210°
220°
230°
240°
250°
260°
270°
280°
290°
300°
310°
320°
330°
340°
350°
Spica
Antares
Arcturus
Vega
Altair
Deneb
SCORPIUS
SAGITTARIUS
SCUTUM
LIBRA
OPHIUCHUS
SERPENS CAUDA
CAPRICORNUS
AQUILA
SERPENS CAPUT
HERCULES
SAGITTA
EQUULEUS
DELPHINUS
VULPECULA
VIRGO
CORONA BOREALIS
AQUARIUS
PEGASUS
LYRA
BOOTES
CYGNUS
PISCES
COMA BERENICES
DRACO

7S

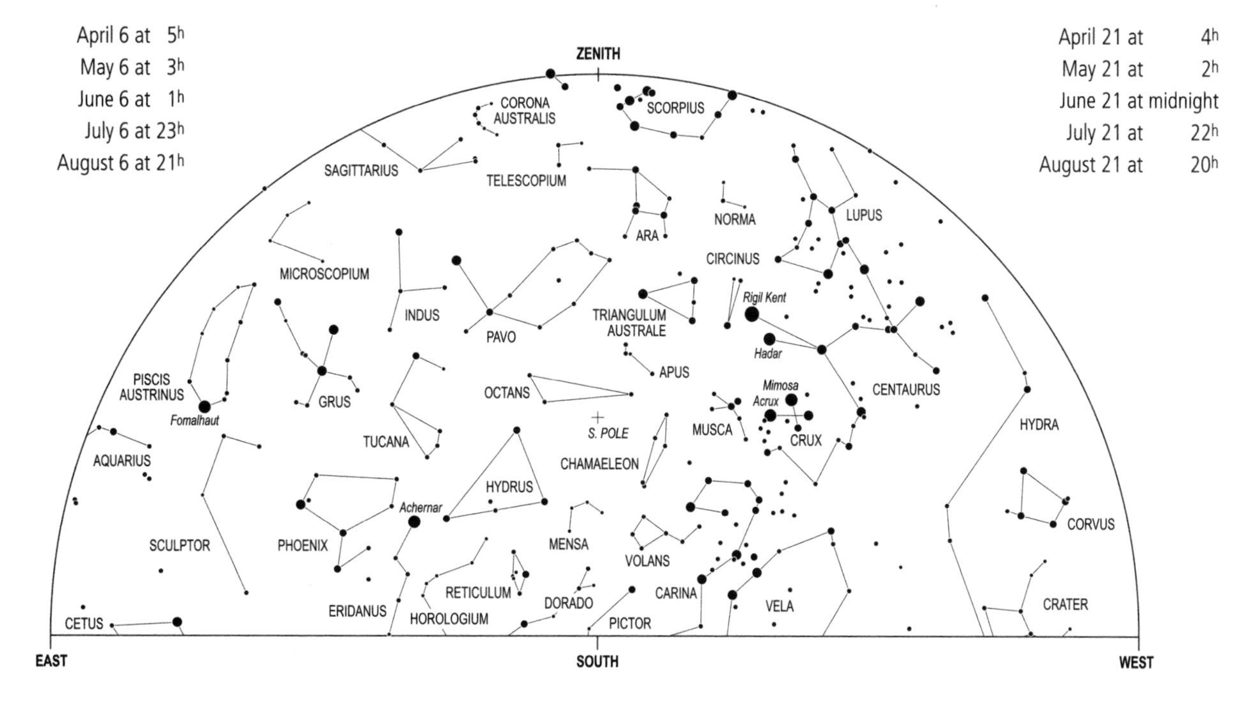

April 6 at 5h
May 6 at 3h
June 6 at 1h
July 6 at 23h
August 6 at 21h
April 21 at 4h
May 21 at 2h
June 21 at midnight
July 21 at 22h
August 21 at 20h
ZENITH
CORONA AUSTRALIS
SCORPIUS
SAGITTARIUS
TELESCOPIUM
NORMA
LUPUS
ARA
CIRCINUS
MICROSCOPIUM
INDUS
PAVO
TRIANGULUM AUSTRALE
Rigil Kent
Hadar
APUS
PISCIS AUSTRINUS
Fomalhaut
GRUS
OCTANS
S. POLE
MUSCA
Mimosa
Acrux
CRUX
CENTAURUS
HYDRA
TUCANA
AQUARIUS
CHAMAELEON
HYDRUS
Achernar
SCULPTOR
PHOENIX
MENSA
VOLANS
CORVUS
RETICULUM
DORADO
CARINA
VELA
CRATER
CETUS
ERIDANUS
HOROLOGIUM
PICTOR
EAST
SOUTH
WEST

8N

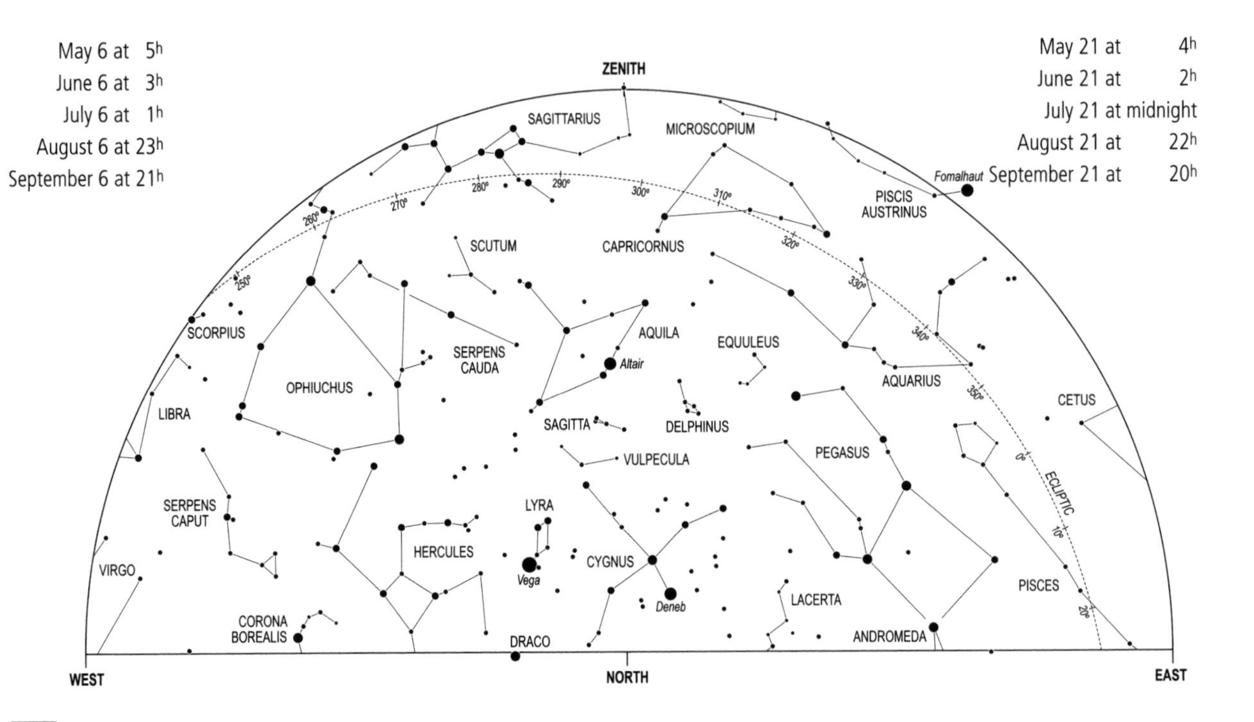

May 6 at 5h
June 6 at 3h
July 6 at 1h
August 6 at 23h
September 6 at 21h
May 21 at 4h
June 21 at 2h
July 21 at midnight
August 21 at 22h
September 21 at 20h
ZENITH
WEST
NORTH
EAST
SAGITTARIUS
MICROSCOPIUM
Fomalhaut
PISCIS AUSTRINUS
SCUTUM
CAPRICORNUS
SCORPIUS
SERPENS CAUDA
AQUILA
Altair
EQUULEUS
AQUARIUS
OPHIUCHUS
LIBRA
SAGITTA
DELPHINUS
CETUS
VULPECULA
PEGASUS
ECLIPTIC
SERPENS CAPUT
LYRA
HERCULES
CYGNUS
Vega
Deneb
VIRGO
LACERTA
PISCES
CORONA BOREALIS
DRACO
ANDROMEDA
250°
260°
270°
280°
290°
300°
310°
320°
330°
340°
350°
0°
10°
20°

May 6 at 5h
June 6 at 3h
July 6 at 1h
August 6 at 23h
September 6 at 21h

May 21 at 4h
June 21 at 2h
July 21 at midnight
August 21 at 22h
September 21 at 20h

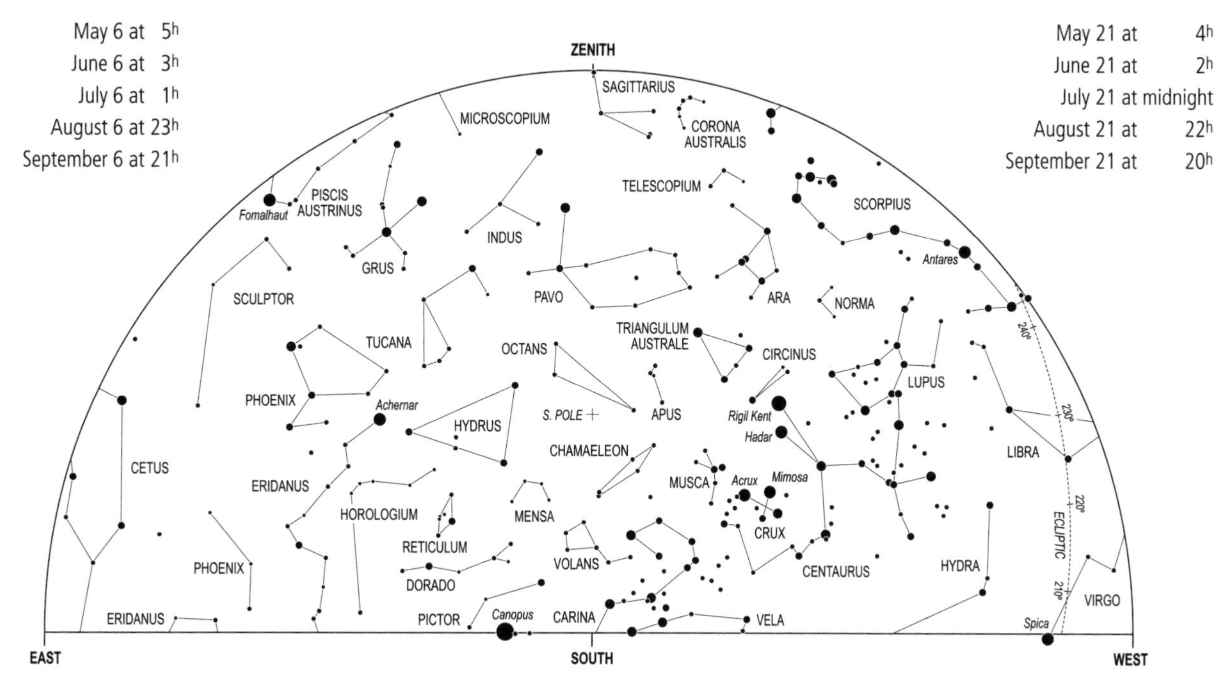

8S

9N

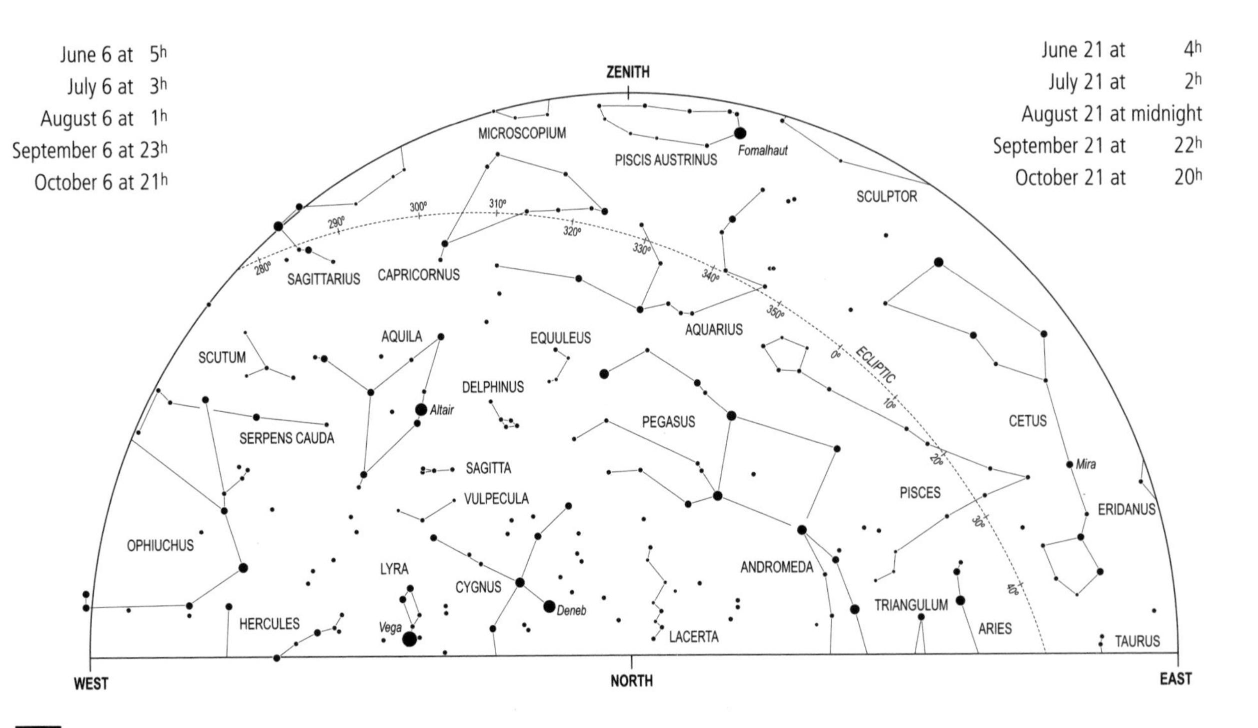
June 6 at 5h
July 6 at 3h
August 6 at 1h
September 6 at 23h
October 6 at 21h
June 21 at 4h
July 21 at 2h
August 21 at midnight
September 21 at 22h
October 21 at 20h
ZENITH
MICROSCOPIUM
PISCIS AUSTRINUS
Fomalhaut
SCULPTOR
SAGITTARIUS
CAPRICORNUS
AQUARIUS
ECLIPTIC
AQUILA
EQUULEUS
SCUTUM
DELPHINUS
Altair
SERPENS CAUDA
PEGASUS
CETUS
SAGITTA
Mira
VULPECULA
PISCES
ERIDANUS
OPHIUCHUS
ANDROMEDA
LYRA
CYGNUS
TRIANGULUM
HERCULES
Deneb
Vega
ARIES
LACERTA
TAURUS
WEST
NORTH
EAST
280°
290°
300°
310°
320°
330°
340°
350°
0°
10°
20°
30°
40°

9S

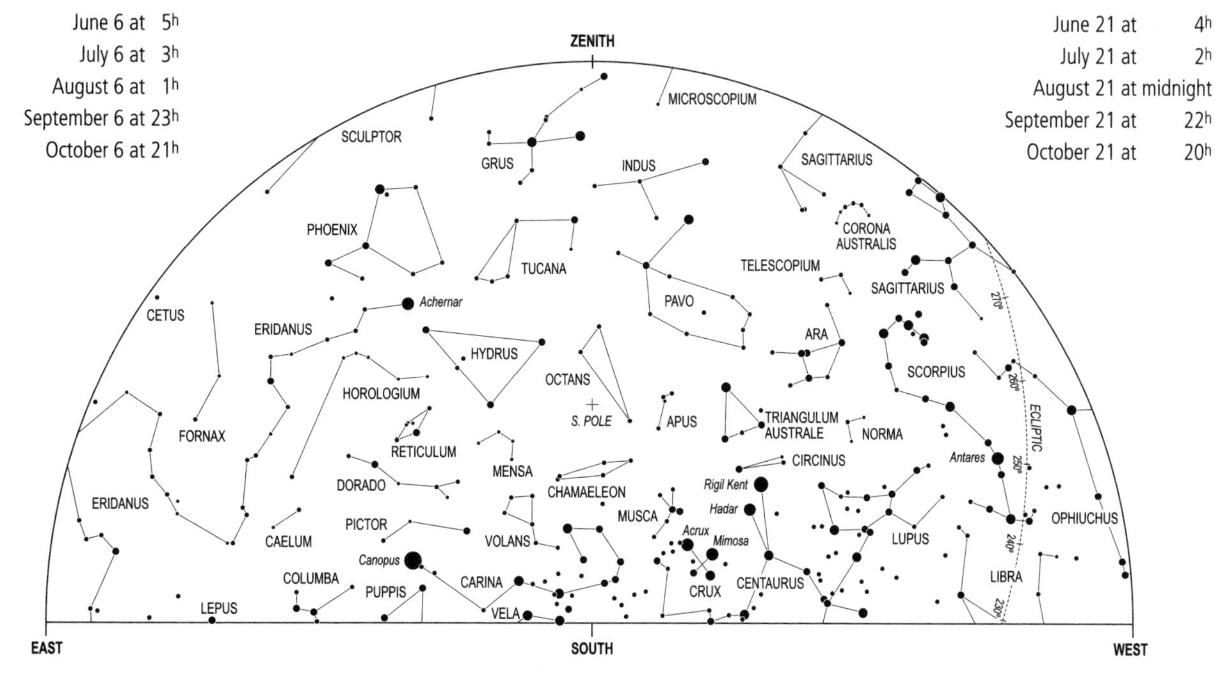
June 6 at 5h
July 6 at 3h
August 6 at 1h
September 6 at 23h
October 6 at 21h
June 21 at 4h
July 21 at 2h
August 21 at midnight
September 21 at 22h
October 21 at 20h
ZENITH
EAST
SOUTH
WEST
MICROSCOPIUM
SCULPTOR
GRUS
INDUS
SAGITTARIUS
PHOENIX
TUCANA
CORONA AUSTRALIS
TELESCOPIUM
SAGITTARIUS
CETUS
Achernar
ERIDANUS
PAVO
270°
ARA
HYDRUS
OCTANS
SCORPIUS
260°
HOROLOGIUM
S. POLE
APUS
TRIANGULUM AUSTRALE
NORMA
ECLIPTIC
FORNAX
RETICULUM
CIRCINUS
Antares
250°
MENSA
DORADO
CHAMAELEON
Rigil Kent
ERIDANUS
Hadar
MUSCA
OPHIUCHUS
PICTOR
LUPUS
CAELUM
VOLANS
Acrux
Mimosa
240°
Canopus
LIBRA
COLUMBA
CARINA
CENTAURUS
PUPPIS
CRUX
LEPUS
VELA
230°

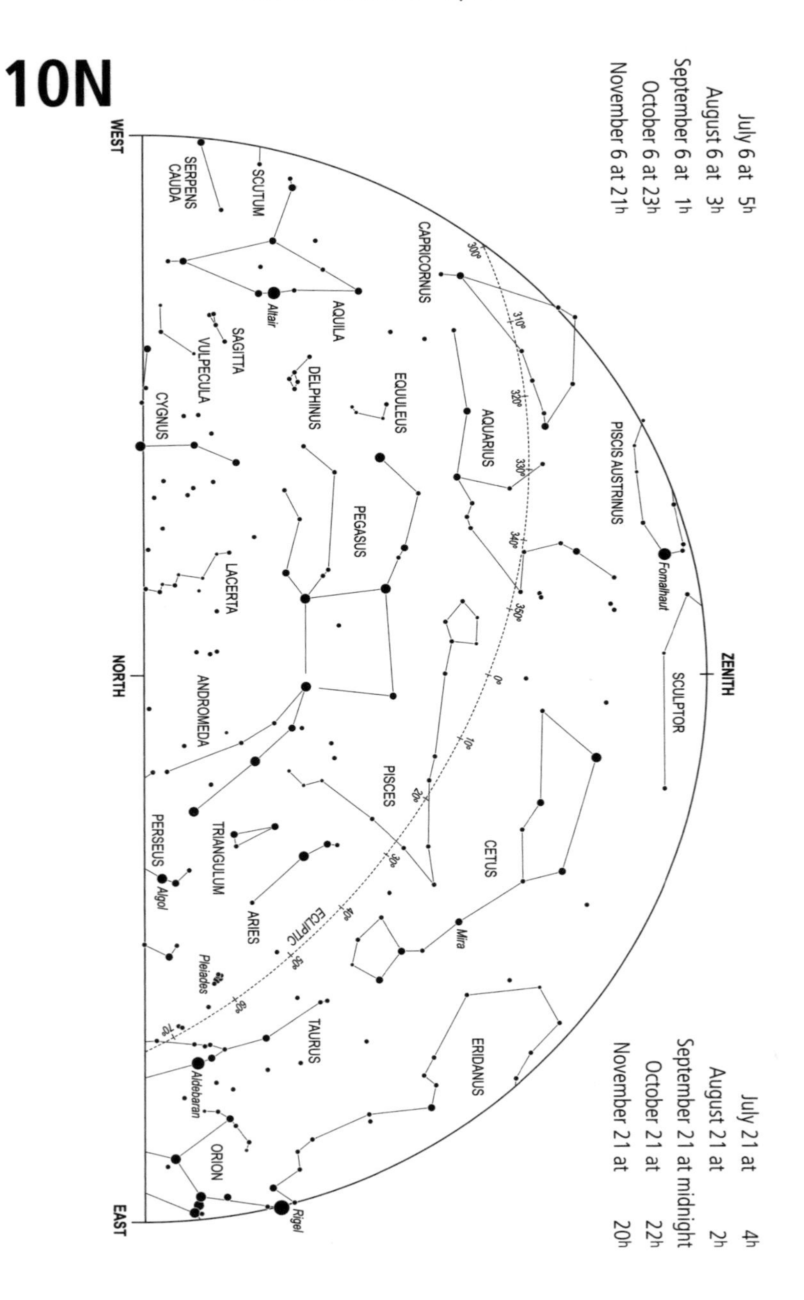
10N
July 6 at 5h
August 6 at 3h
September 6 at 1h
October 6 at 23h
November 6 at 21h
July 21 at 4h
August 21 at 2h
September 21 at midnight
October 21 at 22h
November 21 at 20h
WEST
NORTH
EAST
ZENITH
SERPENS CAUDA
SCUTUM
CAPRICORNUS
AQUILA
Altair
SAGITTA
VULPECULA
CYGNUS
DELPHINUS
EQUULEUS
AQUARIUS
PISCIS AUSTRINUS
Fomalhaut
PEGASUS
LACERTA
ANDROMEDA
SCULPTOR
PISCES
CETUS
Mira
TRIANGULUM
PERSEUS
Algol
ARIES
ECLIPTIC
Pleiades
TAURUS
Aldebaran
ERIDANUS
ORION
Rigel
300°
310°
320°
330°
340°
350°
0°
10°
20°
30°
40°
50°
60°
70°

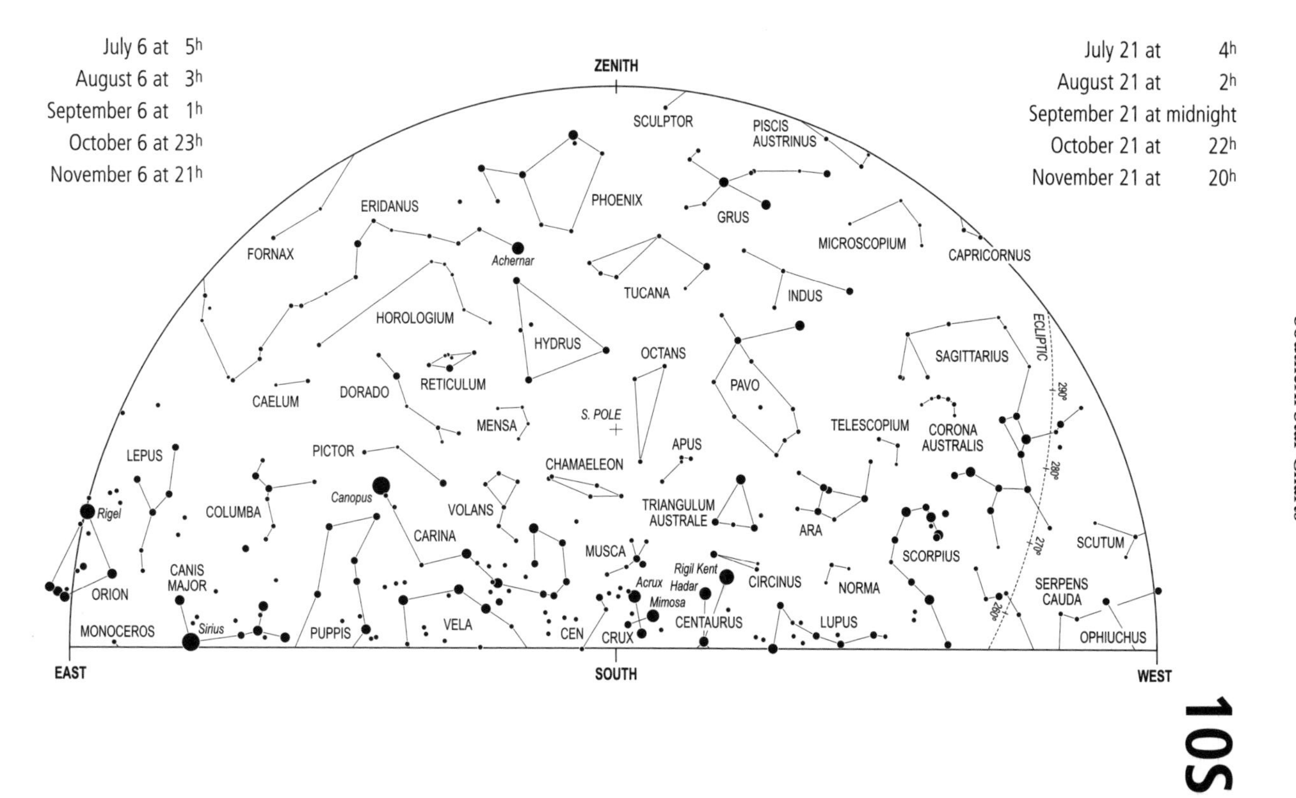
July 6 at 5h
August 6 at 3h
September 6 at 1h
October 6 at 23h
November 6 at 21h
July 21 at 4h
August 21 at 2h
September 21 at midnight
October 21 at 22h
November 21 at 20h
ZENITH
SCULPTOR
PISCIS AUSTRINUS
PHOENIX
GRUS
ERIDANUS
MICROSCOPIUM
FORNAX
Achernar
CAPRICORNUS
TUCANA
INDUS
HOROLOGIUM
HYDRUS
OCTANS
ECLIPTIC
SAGITTARIUS
RETICULUM
PAVO
290°
CAELUM
DORADO
MENSA
S. POLE
TELESCOPIUM
CORONA AUSTRALIS
LEPUS
PICTOR
APUS
CHAMAELEON
280°
Rigel
Canopus
COLUMBA
VOLANS
TRIANGULUM AUSTRALE
ARA
CARINA
MUSCA
SCORPIUS
270°
SCUTUM
Rigil Kent
CANIS MAJOR
Acrux
Hadar
CIRCINUS
NORMA
SERPENS CAUDA
ORION
Mimosa
260°
VELA
CENTAURUS
LUPUS
MONOCEROS
Sirius
PUPPIS
CEN
CRUX
OPHIUCHUS
EAST
SOUTH
WEST
10S

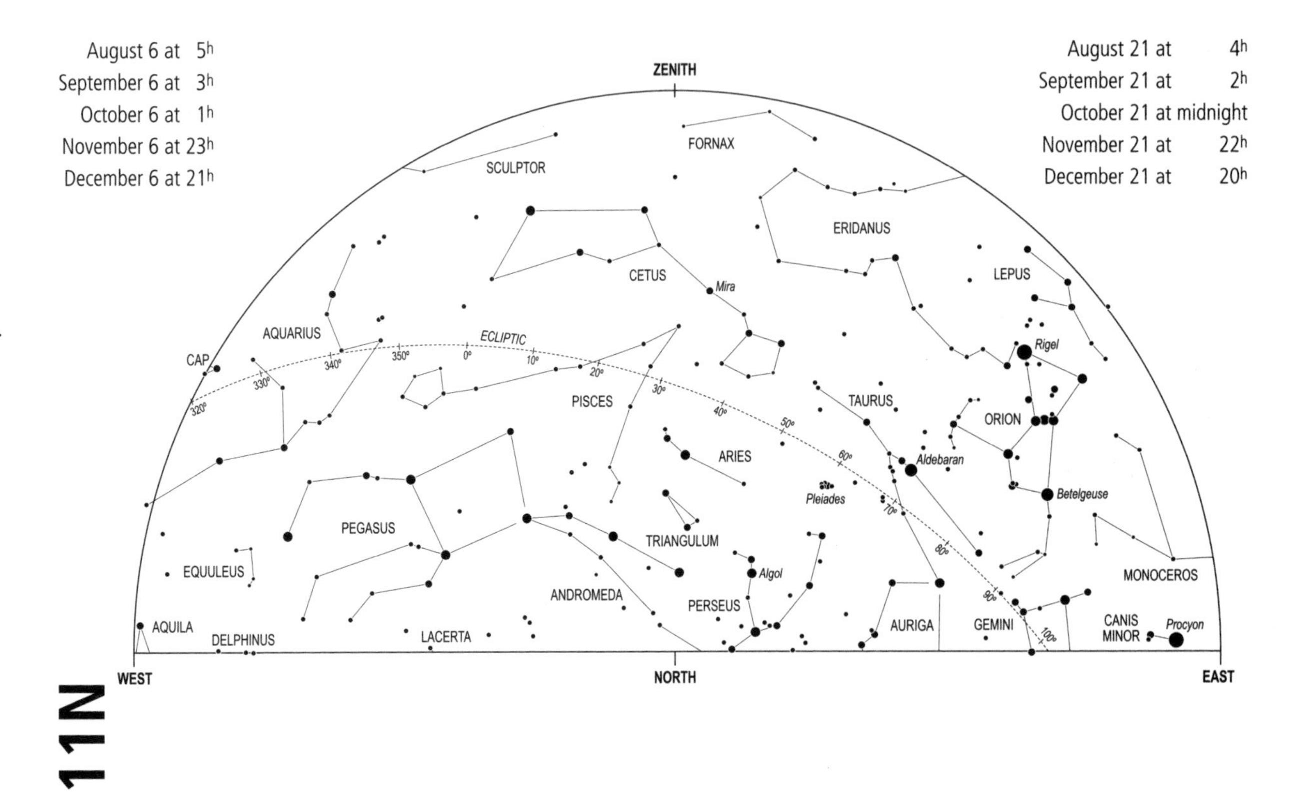
11N
August 6 at 5h
September 6 at 3h
October 6 at 1h
November 6 at 23h
December 6 at 21h
August 21 at 4h
September 21 at 2h
October 21 at midnight
November 21 at 22h
December 21 at 20h
ZENITH
WEST
NORTH
EAST
SCULPTOR
FORNAX
ERIDANUS
CETUS
Mira
LEPUS
AQUARIUS
CAP
ECLIPTIC
320°
330°
340°
350°
0°
10°
20°
30°
40°
50°
60°
70°
80°
90°
100°
PISCES
TAURUS
Rigel
ORION
ARIES
Aldebaran
Pleiades
Betelgeuse
PEGASUS
TRIANGULUM
Algol
EQUULEUS
ANDROMEDA
PERSEUS
MONOCEROS
AQUILA
DELPHINUS
LACERTA
AURIGA
GEMINI
CANIS MINOR
Procyon

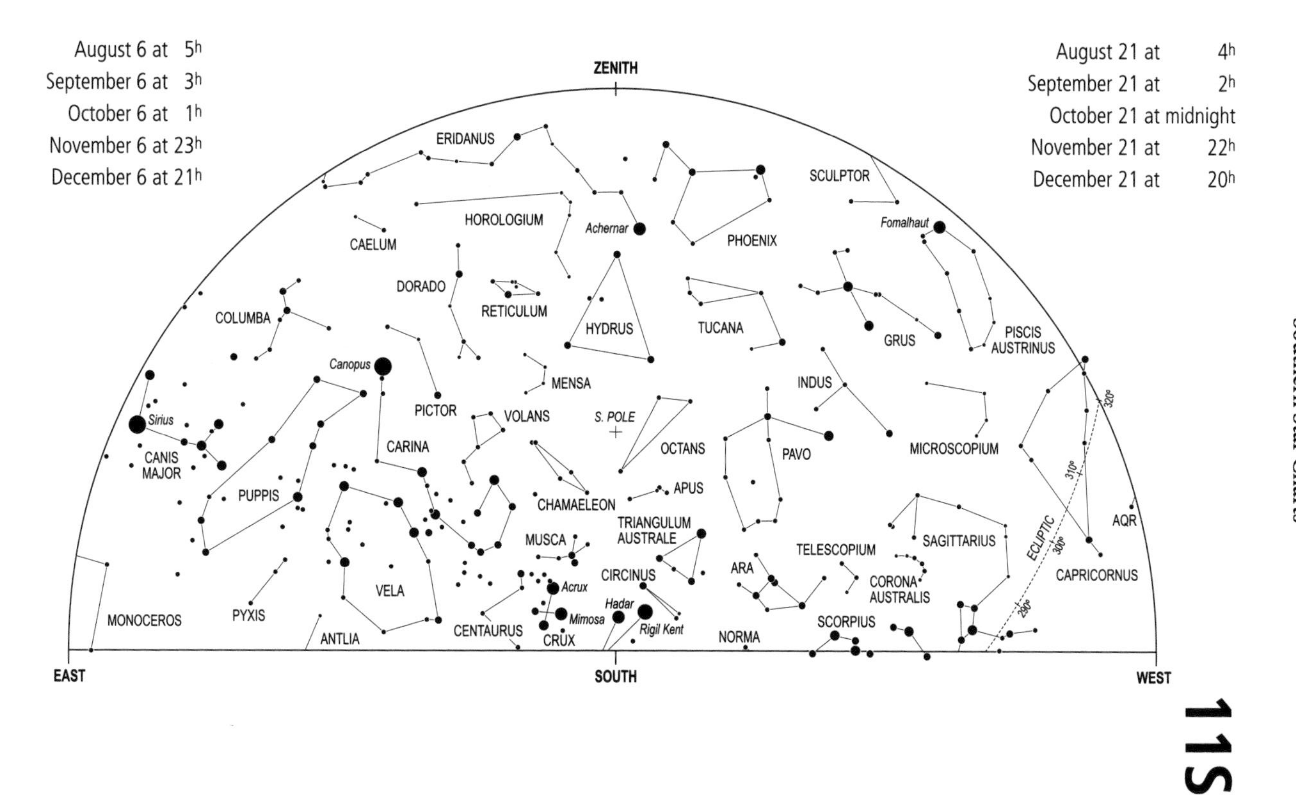
August 6 at 5h
September 6 at 3h
October 6 at 1h
November 6 at 23h
December 6 at 21h
August 21 at 4h
September 21 at 2h
October 21 at midnight
November 21 at 22h
December 21 at 20h
ZENITH
ERIDANUS
SCULPTOR
HOROLOGIUM
Achernar
Fomalhaut
CAELUM
PHOENIX
DORADO
RETICULUM
HYDRUS
TUCANA
GRUS
PISCIS AUSTRINUS
COLUMBA
Canopus
MENSA
INDUS
PICTOR
VOLANS
S. POLE
Sirius
OCTANS
PAVO
MICROSCOPIUM
CANIS MAJOR
CARINA
320°
310°
APUS
PUPPIS
CHAMAELEON
AQR
TRIANGULUM AUSTRALE
MUSCA
SAGITTARIUS
TELESCOPIUM
ECLIPTIC
300°
CIRCINUS
ARA
CAPRICORNUS
Acrux
CORONA AUSTRALIS
VELA
290°
Hadar
MONOCEROS
PYXIS
Mimosa
SCORPIUS
Rigil Kent
ANTLIA
CENTAURUS
NORMA
CRUX
EAST
SOUTH
WEST
11S

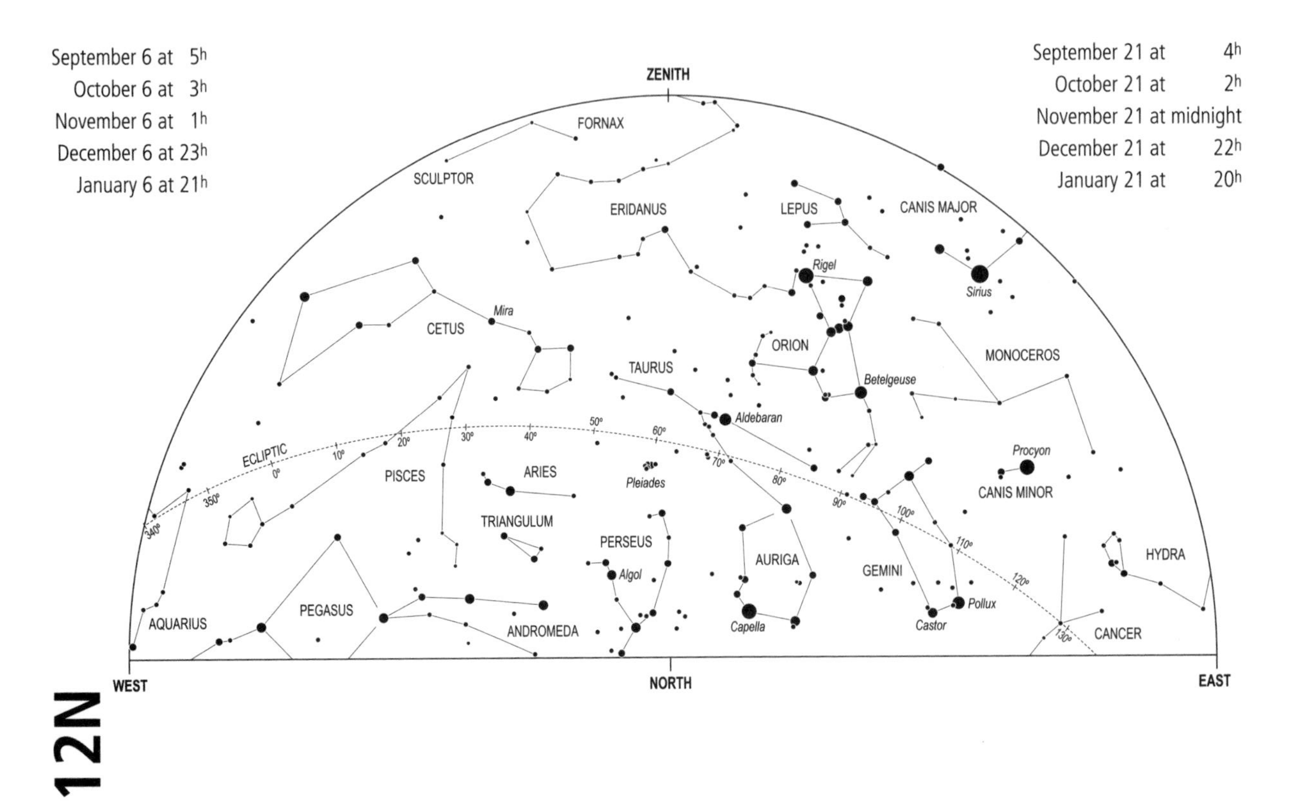
12N
September 6 at 5h
October 6 at 3h
November 6 at 1h
December 6 at 23h
January 6 at 21h
September 21 at 4h
October 21 at 2h
November 21 at midnight
December 21 at 22h
January 21 at 20h
ZENITH
WEST
NORTH
EAST
ECLIPTIC
340°
350°
0°
10°
20°
30°
40°
50°
60°
70°
80°
90°
100°
110°
120°
130°
SCULPTOR
FORNAX
ERIDANUS
LEPUS
CANIS MAJOR
Rigel
Sirius
CETUS
Mira
ORION
MONOCEROS
TAURUS
Betelgeuse
Aldebaran
Procyon
PISCES
ARIES
Pleiades
CANIS MINOR
TRIANGULUM
PERSEUS
AURIGA
GEMINI
HYDRA
Algol
PEGASUS
AQUARIUS
ANDROMEDA
Capella
Castor
Pollux
CANCER

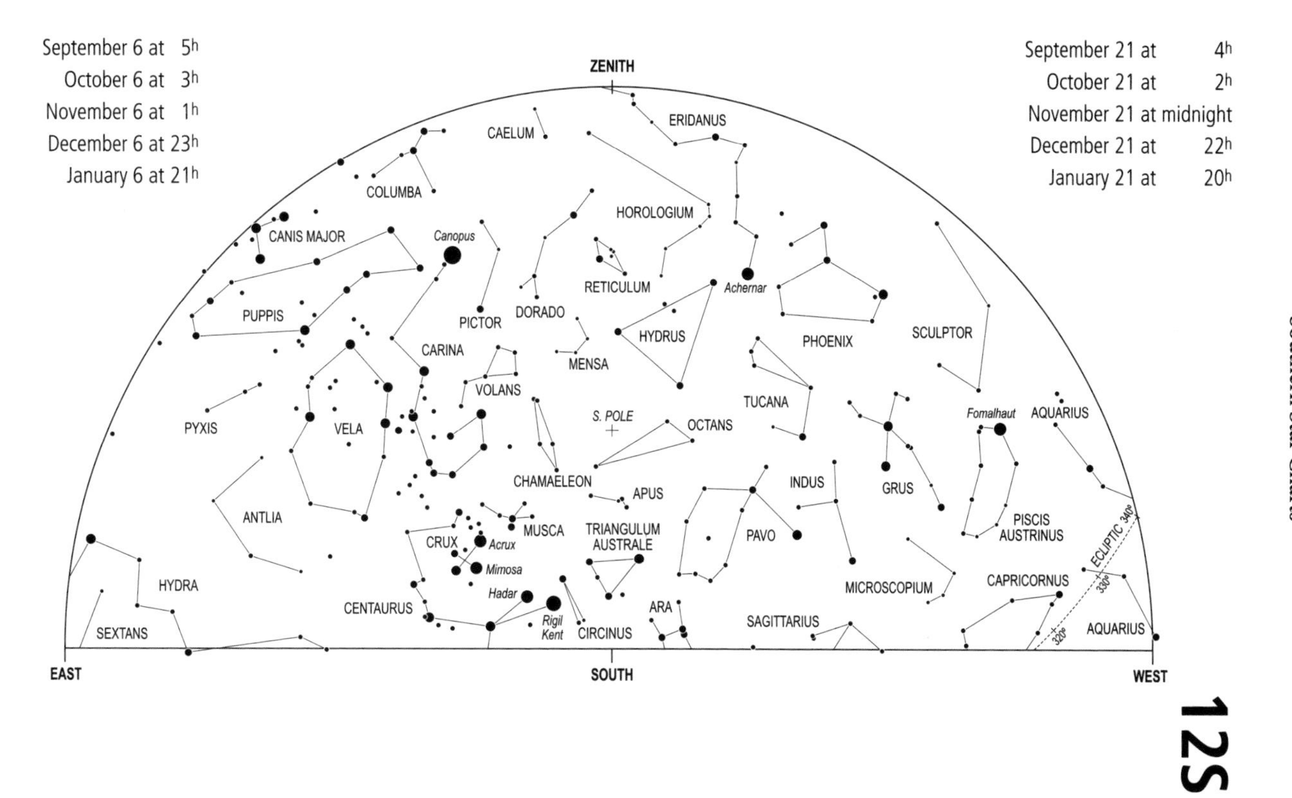
September 6 at 5h
October 6 at 3h
November 6 at 1h
December 6 at 23h
January 6 at 21h
September 21 at 4h
October 21 at 2h
November 21 at midnight
December 21 at 22h
January 21 at 20h
ZENITH
CAELUM
ERIDANUS
COLUMBA
HOROLOGIUM
CANIS MAJOR
Canopus
RETICULUM
Achernar
PUPPIS
PICTOR
DORADO
HYDRUS
PHOENIX
SCULPTOR
CARINA
MENSA
VOLANS
TUCANA
PYXIS
VELA
S. POLE
OCTANS
Fomalhaut
AQUARIUS
CHAMAELEON
APUS
INDUS
GRUS
ANTLIA
MUSCA
TRIANGULUM AUSTRALE
PAVO
PISCIS AUSTRINUS
CRUX
Acrux
Mimosa
ECLIPTIC 340°
HYDRA
Hadar
MICROSCOPIUM
CAPRICORNUS
330°
CENTAURUS
ARA
Rigil Kent
CIRCINUS
SAGITTARIUS
SEXTANS
320°
AQUARIUS
EAST
SOUTH
WEST
12S

The Planets and the Ecliptic

The paths of the planets about the Sun all lie close to the plane of the ecliptic, which is marked for us in the sky by the apparent path of the Sun among the stars, and is shown on the star charts by a broken line. The Moon and naked-eye planets will always be found close to this line, never departing from it by more than about 7°. Thus the planets are most favourably placed for observation when the ecliptic is well displayed, and this means that it should be as high in the sky as possible. This avoids the difficulty of finding a clear horizon, and also overcomes the problem of atmospheric absorption, which greatly reduces the light of the stars. Thus a star at an altitude of 10° suffers a loss of 60 per cent of its light, which corresponds to a whole magnitude; at an altitude of only 4°, the loss may amount to two magnitudes.

The position of the ecliptic in the sky is therefore of great importance, and since it is tilted at about 23.5° to the Equator, it is only at certain times of the day or year that it is displayed to the best advantage. It will be realized that the Sun (and therefore the ecliptic) is at its highest in the sky at noon in midsummer, and at its lowest at noon in midwinter. Allowing for the daily motion of the sky, it follows that the ecliptic is highest at midnight in the winter, at sunset in the spring, at noon in the summer and at sunrise in the autumn. Hence these are the best times to see the planets. Thus, if Venus is an evening object in the western sky after sunset, it will be seen to best advantage if this occurs in the spring, when the ecliptic is high in the sky and slopes down steeply to the horizon. This means that the planet is not only higher in the sky, but also will remain for a much longer period above the horizon. For similar reasons, a morning object will be seen at its best on autumn mornings before sunrise, when the ecliptic is high in the east. The outer planets, which can come to opposition (i.e. opposite the Sun), are best seen when opposition occurs in the winter months, when the ecliptic is high in the sky at midnight.

The seasons are reversed in the Southern Hemisphere, spring beginning at the September equinox, when the Sun crosses the Equator on its way south, summer beginning at the December solstice, when the

Sun is highest in the southern sky, and so on. Thus, the times when the ecliptic is highest in the sky, and therefore best placed for observing the planets, may be summarized as follows:

	Midnight	**Sunrise**	**Noon**	**Sunset**
Northern latitudes	December	September	June	March
Southern latitudes	June	March	December	September

In addition to the daily rotation of the celestial sphere from east to west, the planets have a motion of their own among the stars. The apparent movement is generally *direct*, i.e. to the east, in the direction of increasing longitude, but for a certain period (which depends on the distance of the planet) this apparent motion is reversed. With the outer planets this *retrograde* motion occurs about the time of opposition. Owing to the different inclination of the orbits of these planets, the actual effect is to cause the apparent path to form a loop, or sometimes an S-shaped curve. The same effect is present in the motion of the inferior planets, Mercury and Venus, but it is not so obvious, since it always occurs at the time of inferior conjunction.

The *inferior planets*, Mercury and Venus, move in smaller orbits than that of the Earth, and so are always seen near the Sun. They are most obvious at the times of greatest angular distance from the Sun (greatest elongation), which may reach 28° for Mercury, and 47° for Venus. They are seen as evening objects in the western sky after sunset (at eastern elongations) or as morning objects in the eastern sky before sunrise (at western elongations). The succession of phenomena, conjunctions and elongations, always follows the same order, but the intervals between them are not equal. Thus, if either planet is moving round the far side of its orbit, its motion will be to the east, in the same direction in which the Sun appears to be moving. It therefore takes much longer for the planet to overtake the Sun – that is, to come to superior conjunction – than it does when moving round to inferior conjunction, between Sun and Earth. The intervals given in the table at the top of p. 60 are average values; they remain fairly constant in the case of Venus, which travels in an almost circular orbit. In the case of Mercury, however, conditions vary widely because of the great eccentricity and inclination of the planet's orbit.

		Mercury	Venus
Inferior Conjunction	to Elongation West	22 days	72 days
Elongation West	to Superior Conjunction	36 days	220 days
Superior Conjunction	to Elongation East	35 days	220 days
Elongation East	to Inferior Conjunction	22 days	72 days

The greatest brilliancy of Venus always occurs about thirty-six days before or after inferior conjunction. This will be about a month after greatest eastern elongation (as an evening object), or a month before greatest western elongation (as a morning object). No such rule can be given for Mercury, because its distances from the Earth and the Sun can vary over a wide range.

Mercury is not likely to be seen unless a clear horizon is available. It is seldom as much as 10° above the horizon in the twilight sky in northern temperate latitudes, but this figure is often exceeded in the Southern Hemisphere. This favourable condition arises because the maximum elongation of 28° can occur only when the planet is at aphelion (furthest from the Sun), and it then lies well south of the Equator. Northern observers must be content with smaller elongations, which may be as little as 18° at perihelion. In general, it may be said that the most favourable times for seeing Mercury as an evening object will be in spring, some days before greatest eastern elongation; in autumn, it may be seen as a morning object some days after greatest western elongation.

Venus is the brightest of the planets and may be seen on occasions in broad daylight. Like Mercury, it is alternately a morning and an evening object, and it will be highest in the sky when it is a morning object in autumn, or an evening object in spring. Venus is to be seen at its best as an evening object in northern latitudes when eastern elongation occurs in June. The planet is then well north of the Sun in the preceding spring months, and is a brilliant object in the evening sky over a long period. In the Southern Hemisphere a November elongation is best. For similar reasons, Venus gives a prolonged display as a morning object in the months following western elongation in October (in northern latitudes) or in June (in the Southern Hemisphere).

The *superior planets*, Mars, Jupiter, Saturn, Uranus and Neptune, which travel in orbits larger than that of the Earth, differ from Mercury and Venus in that they can be seen opposite the Sun in the sky. The superior planets are morning objects after conjunction with the Sun, rising earlier each day until they come to opposition. They will then be nearest to the Earth (and therefore at their brightest), and will be on the meridian at midnight, due south in northern latitudes, but due north in the Southern Hemisphere. After opposition they are evening objects, setting earlier each evening until they set in the west with the Sun at the next conjunction. The difference in brightness from one opposition to another is most noticeable in the case of Mars, whose distance from Earth can vary considerably and rapidly. The other superior planets are at such great distances that there is very little change in brightness from one opposition to the next. The effect of altitude is, however, of some importance, for at a December opposition in northern latitudes the planets will be among the stars of Taurus or Gemini, and can then be at an altitude of more than 60° in southern England. At a summer opposition, when a planet is in Sagittarius, it may only rise to about 15° above the southern horizon, and so makes a less impressive appearance. In the Southern Hemisphere the reverse conditions apply, a June opposition being the best, with the planet in Sagittarius at an altitude which can reach 80° above the northern horizon for observers in South Africa.

Mars, whose orbit is appreciably eccentric, comes nearest to the Earth at oppositions at the end of August. It may then be brighter even than Jupiter, but rather low in the sky in Aquarius for northern observers, though very well placed for those in southern latitudes. These favourable oppositions occur every fifteen or seventeen years (e.g. in 1988, 2003 and 2018). In the Northern Hemisphere the planet is probably better seen at oppositions in the autumn or winter months, when it is higher in the sky – such as in 2005 when opposition was in early November. Oppositions of Mars occur at an average interval of 780 days, and during this time the planet makes a complete circuit of the sky.

Jupiter is always a bright planet, and comes to opposition a month later each year, having moved, roughly speaking, from one Zodiacal constellation to the next.

Saturn moves much more slowly than Jupiter, and may remain in the same constellation for several years. The brightness of Saturn depends on the aspects of its rings, as well as on the distance from Earth and Sun. The Earth passed through the plane of Saturn's rings in 1995 and 1996, when they appeared edge-on; we saw them at maximum opening, and Saturn at its brightest, in 2002. The rings last appeared edge-on in 2009, and they are now opening nicely once again.

Uranus and *Neptune* are both visible with binoculars or a small telescope, but you will need a finder chart to help you to locate them (such as those reproduced in this *Yearbook* on pages 135 and 128). *Pluto* (now officially classified as a 'dwarf planet') is hardly likely to attract the attention of observers without adequate telescopes.

Phases of the Moon in 2015

NICK JAMES

New Moon	d	h	m	First Quarter	d	h	m	Full Moon	d	h	m	Last Quarter	d	h	m
								Jan	5	04	53	Jan	13	09	47
Jan	20	13	14	Jan	27	04	48	Feb	3	23	09	Feb	12	03	50
Feb	18	23	47	Feb	25	17	14	Mar	5	18	05	Mar	13	17	48
Mar	20	09	36	Mar	27	07	43	Apr	4	12	06	Apr	12	03	45
Apr	18	18	57	Apr	25	23	55	May	4	03	42	May	11	10	36
May	18	04	13	May	25	17	19	June	2	16	19	June	9	15	42
June	16	14	05	June	24	11	03	July	2	02	20	July	8	20	24
July	16	01	24	July	24	04	04	July	31	10	43	Aug	7	02	03
Aug	14	14	53	Aug	22	19	31	Aug	29	18	35	Sept	5	09	54
Sept	13	06	41	Sept	21	08	59	Sept	28	02	51	Oct	4	21	06
Oct	13	00	06	Oct	20	20	31	Oct	27	12	05	Nov	3	12	24
Nov	11	17	47	Nov	19	06	27	Nov	25	22	44	Dec	3	07	40
Dec	11	10	29	Dec	18	15	14	Dec	25	11	12	Jan	2	05	31

All times are UTC (GMT)

Longitudes of the Sun, Moon and Planets in 2015

NICK JAMES

Date		Sun	Moon	Venus	Mars	Jupiter	Saturn	Uranus	Neptune
		°	°	°	°	°	°	°	°
Jan	6	285	114	303	325	141	241	13	336
	21	301	307	322	337	140	243	13	336
Feb	6	317	159	342	349	138	244	13	337
	21	332	0	0	1	136	245	14	337
Mar	6	345	168	16	11	134	245	15	338
	21	0	9	34	22	133	245	16	338
Apr	6	16	212	53	34	133	245	16	339
	21	31	60	71	45	133	244	17	339
May	6	45	247	88	56	134	243	18	339
	21	60	96	104	66	135	242	19	340
June	6	75	297	120	77	137	241	20	340
	21	89	141	134	88	140	240	20	340
July	6	104	336	145	98	143	239	20	340
	21	118	173	150	108	145	238	20	339
Aug	6	133	29	148	118	149	238	20	339
	21	148	217	140	128	152	239	20	339
Sept	6	163	80	134	138	156	239	20	338
	21	178	263	138	147	159	240	19	338
Oct	6	192	116	148	157	162	241	19	338
	21	207	299	161	166	165	243	18	337
Nov	6	223	161	177	176	168	245	18	337
	21	238	351	194	185	170	246	17	337
Dec	6	254	192	211	194	172	248	17	337
	21	269	30	229	202	173	250	17	337

Moon: Longitude of the ascending node: Jan 1: 195° Dec 31: 176°

Mercury moves so quickly among the stars that it is not possible to indicate its position on the star charts at convenient intervals. The monthly notes should be consulted for the best times at which the planet may be seen.

The positions of the Sun, Moon and planets other than Mercury are given in the table on p. 64. These objects move along paths which remain close to the ecliptic and this list shows the apparent ecliptic longitude for each object on dates which correspond to those of the star charts. This information can be used to plot the position of the desired object on the selected chart.

EXAMPLES

Three planets are visible in the eastern morning sky in late October. What are they?

> The northern star chart 1S shows the eastern sky on 21 October at 04h. The ecliptic longitude visible to the east ranges from 150° to 170°. With reference to the table on p. 64 it can be seen that three planets are in this range on 21 October. Venus is at longitude 161°, Jupiter is at 165° and Mars is at 166°. The planets are therefore Jupiter, Venus and Mars in Leo. In fact, Jupiter and Mars come to within 1° of each other on the morning of 18 October and Venus and Jupiter pass with a similar separation on the morning of 26 October.

The positions of the Sun and Moon can be plotted on the star maps in the same way as the planets. This is straightforward for the Sun since it always lies on the ecliptic and it moves on average at only 1° per day. The Moon is more difficult since it moves rapidly at an average of 13° per day and it moves up to 5° north or south of the ecliptic during the month. A rough indication of the Moon's position relative to the ecliptic may be obtained by considering its longitude relative to that of the ascending node. The longitude of the ascending node decreases by around 1.6° per month as will be seen from the values for the first and last day of the year given on p. 64.

If d is the difference in longitude between the Moon and its ascending node, then the Moon is on the ecliptic when d = 0°, 180° or 360°.

The Moon is 5° north of the ecliptic if d = 90°, and the Moon is 5° south of the ecliptic if d = 270°.

As an example, from the table on p. 63, it can be seen that the Moon is full on the morning of 4 May. The table on p. 64 shows that the Moon's longitude is 247° at 0h on 6 May. Moving backwards two days at 13° per day we estimate the Moon's longitude at 0h on 4 May to be around 221°. At this time the longitude of the ascending node is found by interpolation to be around 187°. Thus d = 34° and the Moon is north of the ecliptic moving north. Its position may be plotted on northern star chart 6S where it is found low in the south on the border of Virgo and Libra.

Some Events in 2015

Jan	4	*Earth* at Perihelion (closest to the Sun)
	5	Full Moon
	9	Moon at Apogee (furthest from the Earth) (405,410 km)
	14	*Mercury* at Greatest Eastern Elongation (19°)
	20	New Moon
	21	Moon at Perigee (closest to the Earth) (359,640 km)
	30	*Mercury* in Inferior Conjunction
Feb	3	Full Moon
	6	Moon at Apogee (406,155 km)
	6	*Jupiter* at Opposition in Cancer
	18	New Moon
	19	Moon at Perigee (356,990 km)
	24	*Mercury* at Greatest Western Elongation (27°)
	26	*Neptune* in Conjunction with Sun
Mar	5	Full Moon
	5	Moon at Apogee (406,385 km)
	19	Moon at Perigee (357,585 km)
	20	Equinox (Spring Equinox in Northern Hemisphere)
	20	New Moon
	20	Total Eclipse of the Sun
	29	Summer Time Begins in the UK
Apr	1	Moon at Apogee (406,010 km)
	4	Full Moon
	4	Total Eclipse of the Moon
	6	*Uranus* in Conjunction with Sun
	10	*Mercury* in Superior Conjunction
	17	Moon at Perigee (361,025 km)
	18	New Moon
	29	Moon at Apogee (405,085 km)

May	4	Full Moon
	7	*Mercury* at Greatest Eastern Elongation (21°)
	15	Moon at Perigee (366,025 km)
	18	New Moon
	23	*Saturn* at Opposition in Libra
	26	Moon at Apogee (404,245 km)
	30	*Mercury* in Inferior Conjunction
June	2	Full Moon
	6	*Venus* at Greatest Eastern Elongation (45°)
	10	Moon at Perigee (369,710 km)
	14	*Mars* in Conjunction with Sun
	16	New Moon
	21	Solstice (Summer Solstice in Northern Hemisphere)
	23	Moon at Apogee (404,130 km)
	24	*Mercury* at Greatest Western Elongation (22°)
July	2	Full Moon
	5	Moon at Perigee (367,095 km)
	6	*Earth* at Aphelion (furthest from the Sun)
	6	*Pluto* at Opposition in Sagittarius
	12	*Venus* attains greatest brilliancy (mag. −4.5)
	16	New Moon
	21	Moon at Apogee (404,835 km)
	23	*Mercury* in Superior Conjunction
	31	Full Moon
Aug	2	Moon at Perigee (362,135 km)
	14	New Moon
	15	*Venus* in Inferior Conjunction
	18	Moon at Apogee (405,850 km)
	26	*Jupiter* in Conjunction with Sun
	29	Full Moon
	30	Moon at Perigee (358,290 km)
Sept	1	*Neptune* at Opposition in Aquarius
	4	*Mercury* at Greatest Eastern Elongation (27°)
	13	New Moon
	13	Partial Eclipse of Sun

	14	Moon at Apogee (406,465 km)
	20	*Venus* attains greatest brilliancy (mag. −4.5)
	23	Equinox (Autumnal Equinox in Northern Hemisphere)
	28	Full Moon
	28	Moon at Perigee (356,875 km)
	28	Total Eclipse of Moon
	30	*Mercury* in Inferior Conjunction
Oct	11	Moon at Apogee (406,390 km)
	12	*Uranus* at Opposition in Pisces
	13	New Moon
	16	*Mercury* at Greatest Western Elongation (18°)
	26	Moon at Perigee (358,465 km)
	26	*Venus* at Greatest Western Elongation (46°)
	25	Summer Time Ends in the UK
	27	Full Moon
Nov	7	Moon at Apogee (405,720 km)
	11	New Moon
	17	*Mercury* in Superior Conjunction
	20	*Mars* at aphelion
	23	Moon at Perigee (362,815 km)
	25	Full Moon
	30	*Saturn* in Conjunction with Sun
Dec	5	Moon at Apogee (404,800 km)
	11	New Moon
	21	Moon at Perigee (368,415 km)
	22	Solstice (Winter Solstice in Northern Hemisphere)
	25	Full Moon
	29	*Mercury* at Greatest Eastern Elongation (20°)

Monthly Notes 2015

January

Full Moon: 5 January *New Moon:* 20 January

EARTH is at perihelion (nearest to the Sun) on 4 January at a distance of 147 million kilometres (91.4 million miles).

MERCURY attains its greatest eastern elongation (19°) from the Sun on 14 January. For observers in northern and equatorial latitudes, Mercury may be glimpsed low above the south-western horizon for the first three weeks of the month, though for observers in southern latitudes, the period of visibility is reduced to the first half of the month. Figure 1 shows, for observers in latitude 52°N, the changes in azimuth (true bearing from the north through east, south and west) and altitude of Mercury on successive evenings when the Sun is 6° below the horizon. This condition is known as the end-of-evening civil twilight and in this latitude and at this time of year occurs about thirty-five minutes after sunset. The changes in the brightness of the planet are indicated by the relative sizes of the circles marking Mercury's position at five-day intervals.

The much more brilliant planet Venus will be a useful guide to locating the elusive Mercury from about 9–12 January, as they are close together in the same part of the sky; Mercury magnitude −0.6, Venus −3.9. The position of Venus for observers in latitude 52°N is also shown on Figure 1. Mercury may be located slightly further west and slightly below Venus from the 9th to the 11th and at about the same altitude on the 12th. At the beginning of January, Mercury has a magnitude of −0.8 but this has faded to +0.6 three weeks later; thus Mercury is at its brightest before it reaches greatest eastern elongation. Mercury is unobservable towards the end of the month as it passes through inferior conjunction on 30 January.

VENUS is an evening object, but only visible for a short while after sunset at the beginning of January, low above the south-western horizon for observers in the latitudes of the British Isles, but more towards

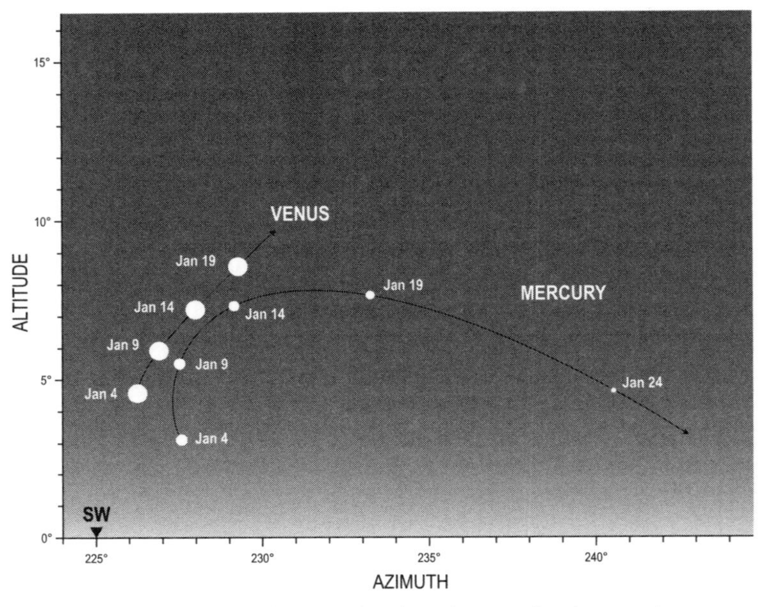

Figure 1. Evening apparition of Mercury from latitude 52°N. The planet reaches greatest eastern elongation on 14 January 2015. It is at its brightest at the beginning of January, before elongation. The chart also shows the positions of the brilliant Venus, in relation to Mercury, between 4 and 19 January. Venus is a useful guide to locating the much fainter Mercury, particularly between 9 and 12 January. The angular diameters of Mercury and Venus are not drawn to scale.

the west for observers south of the Equator. Its magnitude is –3.9. Figure 1 shows, for observers in latitude 52°N, the changes in azimuth and altitude of Venus on successive evenings at the end of evening civil twilight. The planet's proximity to the much fainter Mercury in the sky in the second week of January has already been mentioned above. The elongation of Venus from the Sun is slowly increasing, and by the end of the month the planet is setting over two hours after the Sun from northern temperate latitudes, although considerably less from locations in the Southern Hemisphere.

MARS, magnitude +1.1, is an early evening object, setting in the south-west in the mid-evening. The planet moves rapidly from Capricornus

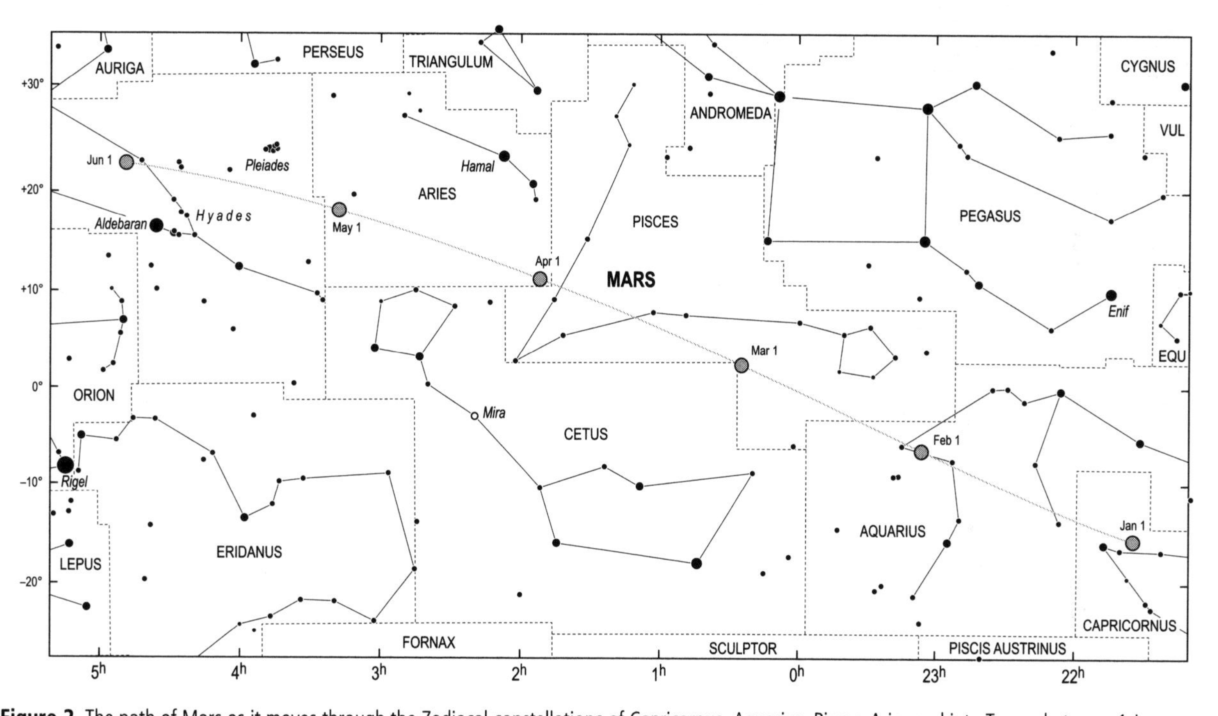

Figure 2. The path of Mars as it moves through the Zodiacal constellations of Capricornus, Aquarius, Pisces, Aries and into Taurus between 1 January and 1 June 2015. The planet is in conjunction with the Sun on 14 June 2015.

into Aquarius during the month. There is no opposition of Mars in 2015 and the planet will not be a prominent object during the year. The path of Mars from January to the end of May 2015 is shown in Figure 2.

JUPITER, magnitude –2.5, is a brilliant object in the night sky, and visible for the greater part of the hours of darkness since it comes to opposition early in February. For observers in northern temperate latitudes the planet is very well placed due to its high northerly declination (angle north of the celestial equator) in the constellation of Leo, near the border with Cancer. For those living in the Southern Hemisphere, the planet is not so well situated. The path of Jupiter among the stars is shown in Figure 4, given with the notes for February.

SATURN is an early morning object, moving direct (i.e. towards the east) among the stars of Libra and into Scorpius during the month. By the end of January it is rising at about 03h 30m for observers in northern temperate latitudes; about three hours earlier for those in the Southern Hemisphere. Its magnitude is steady at +0.6.

Earth's Orbit. On 4 January, the Earth is at perihelion, its closest distance to the Sun, at 147 million kilometres. At aphelion, in early July, the distance will be 152 million kilometres. The difference, therefore, amounts to approximately 5 million kilometres, which is not very much in comparison with the size of the orbit because Earth's orbit is not far from circular. It is always said that the orbit of the Earth round the Sun is not a circle, but an ellipse. Of course, this is quite true – as was demonstrated originally by Johannes Kepler; before his time, it had been tacitly assumed that all celestial orbits must be circular, since the circle is the 'perfect' form, and nothing short of perfection can be allowed in the heavens!

The eccentricity of an orbit is a measurement of how much the orbit deviates from a perfect circle, but it should be remembered that a circle is simply an ellipse with eccentricity zero. In the case of the Earth, the departure from circularity is not great. The eccentricity is only 0.0167, so that if the orbit were drawn on a scale to fit this page, it would be hard to detect any ellipticity. Of the principal planets, only Venus and Neptune have orbits which are less elliptical than that of the Earth. The values for the eight planets are:

Planet	Orbital eccentricity	Perihelion distance (km)	Aphelion distance (km)
Mercury	0.2056	46 million	70 million
Venus	0.0068	107 million	109 million
Earth	0.0167	147 million	152 million
Mars	0.0934	207 million	249 million
Jupiter	0.0483	741 million	817 million
Saturn	0.0560	1,350 million	1,510 million
Uranus	0.0461	2,750 million	3,000 million
Neptune	0.0097	4,450 million	4,550 million

Pluto's orbit is of eccentricity 0.248, so at perihelion its distance from the Sun is 4,440 million kilometres (actually closer than Neptune, though there is no chance of a collision), but at aphelion it is 7,380 million kilometres. When Pluto was reclassified as a dwarf planet, Mercury became the planet with the most eccentric orbit.

The Earth's changing distance from the Sun makes very little difference to the seasons, particularly since the greater amount of sea in the Southern Hemisphere tends to stabilize the temperature there. However, the difference is much more important for Mars, which has an eccentricity of 0.0934, making it the second most eccentric orbit of the major planets. Its distance from the Sun ranges between 249 million km and only 207 million km, and since Mars, like Earth, has its southern summer when at perihelion (and there are no oceans on Mars), the Southern Hemisphere temperatures on Mars have a much greater range than those in the north.

The fact that Mars has an orbit more eccentric than ours also means that its magnitude at opposition can vary considerably. At its best Mars may outshine every planet apart from Venus, as it will next do in July 2018 when the magnitude will be –2.8; at the least favourable oppositions, such as in March 2012, the magnitude is only –1.2, so that it cannot even match Sirius. When furthest from Earth, Mars is not much brighter than the Pole Star.

The First-Magnitude Stars. The January evening sky boasts many bright stars and easily recognizable constellations, such as Taurus, Auriga, Orion, Gemini, Canis Major and Canis Minor. Conventionally, the twenty-one brightest stars in the sky are classed as being of the first magnitude; they range from Sirius in Canis Major (magnitude –1.44) down to Regulus in Leo (+1.36). Next in order comes Adhara (Epsilon Canis Majoris), which is of magnitude +1.50 and is not reckoned among the elite; neither is Castor, the senior but fainter of the Twins, at magnitude +1.58. Here is the list of first-magnitude stars according to data from the European Space Agency's Hipparcos satellite.

Rank	Name	Designation	Visual Mag.
1	Sirius	Alpha Canis Majoris	–1.44
2	Canopus	Alpha Carinae	–0.62
3	Rigil Kent	Alpha Centauri	–0.28c
4	Arcturus	Alpha Boötis	–0.05v
5	Vega	Alpha Lyrae	0.03v
6	Capella	Alpha Aurigae	0.08v
7	Rigel	Beta Orionis	0.18v
8	Procyon	Alpha Canis Minoris	0.40
9	Achernar	Alpha Eridani	0.45v
10	Betelgeuse	Alpha Orionis	0.45v
11		Beta Centauri	0.61v
12	Altair	Alpha Aquilae	0.76v
13	Acrux	Alpha Crucis	0.77c
14	Aldebaran	Alpha Tauri	0.87
15	Spica	Alpha Virginis	0.98v
16	Antares	Alpha Scorpii	1.06v
17	Pollux	Beta Geminorum	1.16
18	Fomalhaut	Alpha Piscis Austrinus	1.17
19		Beta Crucis	1.25v
20	Deneb	Alpha Cygni	1.25v
21	Regulus	Alpha Leonis	1.36

c = combined magnitude
v= variable

Many of these first-magnitude stars are sufficiently close to the Earth's equator to be seen from every inhabited country, but there are some which are not, and it may be interesting to list those which are at declinations of ±40° or higher. (Remember that 'declination' is the angular distance of a star north or south of the celestial equator and is akin to latitude on the surface of the Earth.)

There are only two such stars in the far north – Vega, at declination +38° 47', just missed out – but no less than six in the far south:

(North)
Capella, +46° 00'
Deneb, +45° 17'

(South)
Acrux, –63° 06'
Alpha Centauri, –60° 50'
Agena, –60° 22'
Beta Crucis, –59° 41'
Achernar, –57° 14'
Canopus, –52° 42'

All these far-southern stars are invisible from Britain; go to the southern part of New Zealand, and you will lose Deneb and Capella (as well as Vega).

To find your limiting declination, simply take your latitude on the Earth's surface and subtract it from 90 degrees. This gives your co-latitude. The latitude of Athens, for example, is 38°N. So 90 – 38 = 52; therefore any star south of declination –52° will never rise, and any star north of declination +52° will never set. Thus you will never see Achernar, marking the southern end of Eridanus, the River, but Dubhe in the Great Bear, declination +61° 45', will be circumpolar. Falkland Islanders (latitude –51°) will see Canopus all the time, but Capella not at all.

The southernmost of the first-magnitude stars observable from the British Isles is Fomalhaut (declination –29° 37'). During January evenings, Capella is near the zenith or overhead point from Britain, but rather low down in the north from countries such as Australia and South Africa. From London, Capella is just circumpolar, though on summer evenings it skirts the northern horizon, so that any mist or light pollution will usually hide it.

February

Full Moon: 3 February *New Moon:* 18 February

MERCURY, although it reaches greatest western elongation (27°) on 24 February, is not suitably placed for observation from the latitudes of the British Isles. For observers in equatorial latitudes and the Southern Hemisphere this will be the most favourable morning apparition of the year, with the planet being visible in the dawn twilight sky for much of February and March. Figure 3 shows, for observers in latitude 35°S, the changes in azimuth (true bearing from the north through east, south and west) and altitude of Mercury on successive mornings when the Sun is 6° below the horizon. This condition is known as the beginning of morning civil twilight and in this latitude and at this time of year occurs about twenty-five minutes before sunrise. The changes in the brightness of the planet are indicated by the relative sizes of the circles marking Mercury's position at five-day intervals. Mercury is at its brightest after it reaches greatest western elongation, brightening from magnitude +0.9 on 10 February to +0.1 by the end of the month. The planet may first be glimpsed low above the eastern horizon in the second week of February, but will be much better placed in the last week of February and the first week of March.

VENUS continues to be visible as a brilliant object in the west-south-western sky at dusk. Because of its rapid motion northwards in declination, observers in the latitudes of the British Isles will see it for a little longer each evening and by the end of the month it will be setting nearly three hours after the Sun. The planet brightens slightly from magnitude –3.9 to –4.0 and the phase decreases from 92 per cent to 87 per cent during the month. Venus passes less than half a degree south of Mars on the evening of 21 February, with the waxing crescent Moon making a lovely grouping with the two planets on the evening of the 20th.

MARS, magnitude +1.2, continues to set in the west in the mid-evening. Its rapid eastwards motion against the background stars

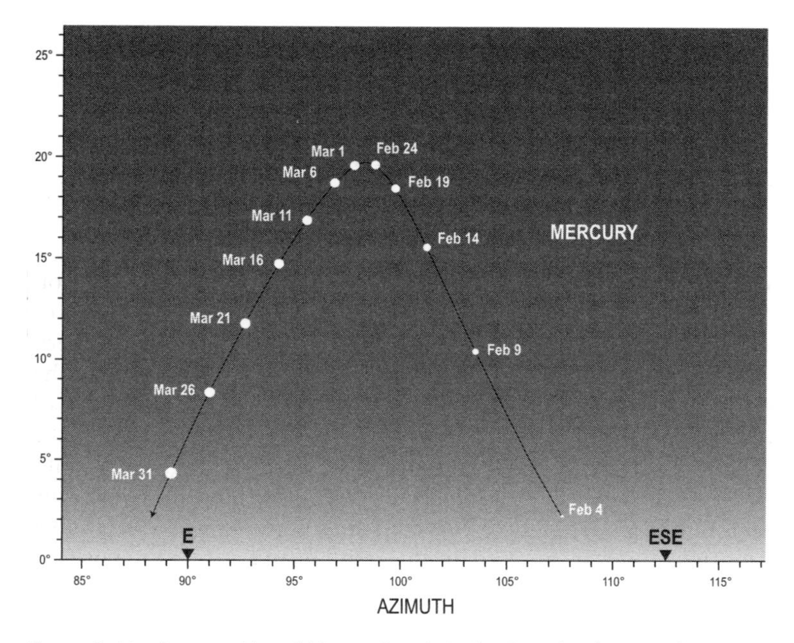

Figure 3. Morning apparition of Mercury from latitude 35°S. The planet reaches greatest western elongation on 24 February 2015. It will be at its brightest in mid-March, after elongation.

carries it from Aquarius into neighbouring Pisces on 11 February. Mars will be very close to the far more brilliant Venus on the evening of 21 February, an aid to locating the fainter planet.

JUPITER is at opposition on 6 February and is visible all night long. The planet is moving retrograde (i.e. towards the west) in Cancer, quite near the border with Leo, and is a brilliant object at magnitude –2.6. Figure 4 shows the path of Jupiter against the background stars during 2015. The planet's high northerly declination (+16° 28'), although not as good as last year, is still very favourable for observers in northern temperate latitudes. At opposition the planet's apparent diameter is 45.5 arc seconds and it is 650 million kilometres (404 million miles) from the Earth. The dark evenings afford plenty of opportunities for observing the movements of the four Galilean satellites, Io, Europa,

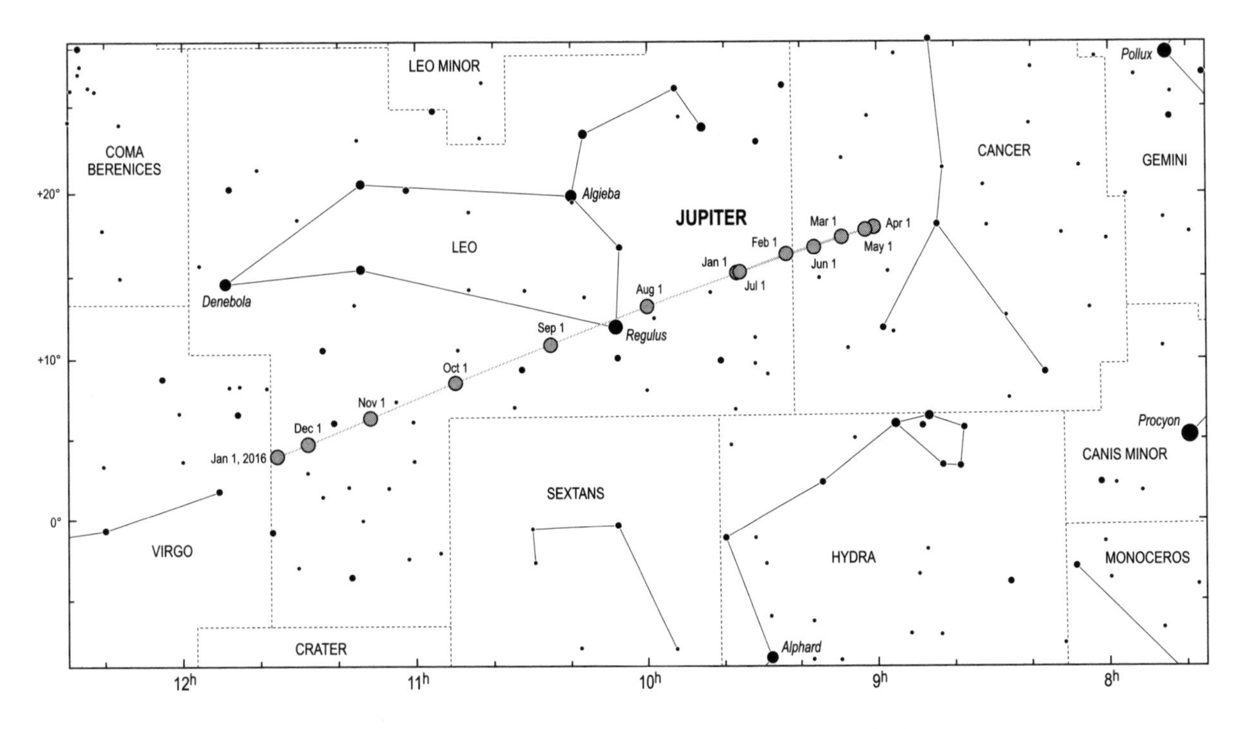

Figure 4. The path of Jupiter as it moves against the background stars of Leo, Cancer and back into Leo during 2015.

Ganymede and Callisto. The full Moon will make a striking configuration with Jupiter on the night of 3/4 February.

SATURN continues to move direct in Scorpius. By the end of the month it still only rises at about 01h 40m for observers in northern temperate latitudes, but about three hours earlier for those in the Southern Hemisphere. Its brightness increases very slightly from magnitude +0.6 to +0.5 during February.

Equatorial Stars. Throughout February, Orion is dominant in the evening sky. This is true for both the Northern and Southern Hemispheres of the globe, because Orion is cut by the celestial equator; thus of its two leaders, Betelgeuse is well north of the Equator and Rigel well south. Orion is therefore visible from every inhabited country – though from an observatory established right at the South Pole, observers there would find that Rigel is above the horizon all the time and Betelgeuse never!

It is interesting to check and see which bright stars are very close to the celestial equator. There are, in fact, fewer of them than might be expected. In fact, there are only twelve stars above the fourth magnitude which are within two degrees of the equator; they are as follows (positions given for epoch J2000.0):

Star	Declination			Magnitude
	°	′	″	
Alpha Aquarii	–00	19	11	2.9
Gamma Aquarii	–01	23	14	3.9
Zeta Aquarii	–00	01	13	3.7
Theta Aquilae	–00	49	17	3.2
Iota Hydrae	–01	08	34	3.9
Lambda Ophiuchi	+01	59	02	3.8
Delta Orionis	–00	17	57	2.2
Epsilon Orionis	–01	12	07	1.7
Zeta Orionis	–01	56	34	1.8
Gamma Virginis	–01	26	58	2.7
Zeta Virginis	–00	35	45	3.4
Eta Virginis	–00	40	01	3.9

Taking the magnitude limit down to 4.75, we can add fifteen more stars: Pi Aquarii (4.7), Eta Aquarii (4.0), Nu Aquilae (4.7), Iota Aquilae (4.4), Delta Ceti (4.1), Tau Hydrae (4.6), Upsilon Leonis (4.3), Delta Monocerotis (4.1), 28 Monocerotis (4.7), 41 Ophiuchi (4.7), 68 Ophiuchi (4.4), Pi^6 Orionis (4.5), Omicron Orionis (4.7), 31 Orionis (4.7) and Alpha Sextantis (4.5).

Equatorial stars may be used to find the diameter of a telescopic field. Observe the time taken for the star to pass centrally across the field of view of your eyepiece from the edge on one side to the edge on the other. Express this time in minutes and seconds; multiply by 15, and you will have the diameter of the field in minutes and seconds of arc. For example, if your chosen equatorial star takes 1 minute and 2 seconds to drift across the field of view, the diameter of the telescopic field is 15.5 arc minutes.

Columba and Caelum. Two small southern constellations are on view during February evenings in the general area of Orion and Canis Major (Figure 5). It is often said that Columba, the Dove, was introduced to the sky in 1679 by an otherwise obscure French architect and astronomer named Augustine Royer, by splitting off part of the constellation Canis Major. Royer is also credited with being the first to introduce the Southern Cross as a separate constellation (previously the Cross had been included in Centaurus, which practically surrounds it). In fact, it seems that both Columba and Crux were introduced by the Flemish astronomer and cartographer Petrus Plancius, and depicted by him on celestial globes and a large wall map between 1589 and 1592. They were subsequently included by Johann Bayer in his sky atlas, the *Uranometria*, in 1603.

Columba seems to represent the dove which Noah released from the Ark, and was originally Columba Noachi, Noah's Dove. To find Columba, come south from the bottom of Orion (marked by Rigel and Saiph), through the distinctive pattern of Lepus (the Hare) and Columba lies about fifteen degrees below. It is not at all prominent, and even from the latitudes of Southern England it is always very low down; in fact, the star Eta Columbae (magnitude 3.9) never rises at all from anywhere in the British Isles. The brightest stars are Alpha or Phakt (magnitude 2.7) and Beta or Wazn (3.1).

Just north and slightly east of Alpha is Mu Columbae (magnitude 5.2), a hot O9.5-type star and one of the few of its class visible to the

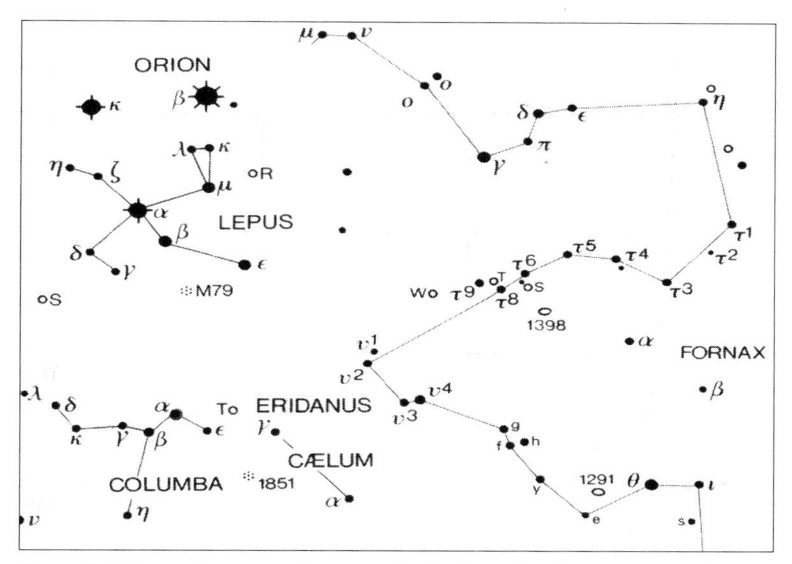

Figure 5. To locate the two small patterns of Columba and Caelum, come south from the bottom of Orion, marked by Kappa (κ) and Beta (β) Orionis, through the distinctive pattern of Lepus (the Hare), and Columba lies about fifteen degrees below, with Caelum (the Graving Tool) adjoining Columba on its western side. Both patterns are inconveniently low from northern temperate latitudes.

unaided eye. Mu Columbae and its 'partner', AE Aurigae, are 'runaway stars' from the Orion complex. Mu is moving at 117 kilometres per second relative to the Sun and directly away from AE Aurigae at 200 kilometres per second. By tracing the paths of the two runaways back in time, it appears that they were both near the Trapezium star cluster (Theta^1 Orionis) in the Orion Nebula about 2.5 million years ago. The third star apparently connected with our two runaways is Iota Orionis, a multiple star containing a very close double with a highly eccentric orbit. It seems that 2.5 million years ago – long before the Trapezium itself had even formed – two double stars collided, swapped two members and ejected the other two at very high speeds – producing our two runaway stars and making a fascinating connection between Orion, Auriga and Columba. There is little else of interest in Columba to the user of a small telescope.

Adjoining Columba, on its western side, is Caelum, the Graving

Tool – originally Caelum Sculptoris. It was created by Abbé Nicolas Louis de Lacaille in 1752, but for no good reason, because it has no star above the fourth magnitude and only two above the fifth. It is devoid of interesting objects, and seems to have no justification for separate identity. It is so far south from Britain that it is to all intents and purposes inaccessible.

Caelum is one of a number of small, dim, 'modern' constellations. One tends to sympathize with the great last-century astronomer Sir John Herschel, who commented that such constellation patterns seemed to have been devised so as to cause as much confusion and inconvenience as possible.

March

Full Moon: 5 March *New Moon:* 20 March

Equinox: 20 March

Summer Time in the United Kingdom commences on 29 March.

MERCURY, for observers in the latitudes of the British Isles, remains unsuitably placed for observation throughout the month. But for observers in equatorial latitudes and the Southern Hemisphere, Mercury continues to be visible in the early morning twilight sky, about twenty-five minutes before sunrise, until the last few days of March, as shown in Figure 3, given with the notes for February. During this period, the planet brightens from magnitude +0.1 to –0.9.

VENUS, magnitude –4.0, continues to be visible as a brilliant object in the western sky in the evenings, and for observers in the latitudes of the British Isles it is visible for more than three hours after sunset by the end of March. Those in the Southern Hemisphere will be restricted to only about half this time. The phase of Venus decreases from 87 per cent to 78 per cent during the month. The waxing crescent Moon will make a lovely pairing with Venus in the evening twilight on 22 March.

MARS continues to set in the west about two hours after the Sun, but at magnitude +1.3, it resembles a first-magnitude star and is not particularly prominent. Its continued rapid eastwards motion carries it from Pisces into Aries during the month. The waxing crescent Moon will lie fairly close to Mars in the western twilight sky on 21 March.

JUPITER continues to be visible as a conspicuous object in the south-eastern sky as soon as darkness falls. As it was at opposition early in February, it is still visible in the southern skies until the early morning hours. Reference to Figure 4, given with the notes for February, shows that Jupiter is moving very slowly retrograde in Cancer. The planet fades slightly from magnitude –2.5 to –2.3 during the month.

SATURN reaches a stationary point this month (on 14 March) and from this date its motion is retrograde in Scorpius. The planet continues to brighten slowly from magnitude +0.5 to +0.3 during the month. By the end of March, Saturn is rising before midnight for observers in northern temperate latitudes, but nearly three hours earlier for those further south.

Mercury – According to Antoniadi. Today, thanks to the American MESSENGER spacecraft, which has been in orbit around Mercury since March 2011, and is now in its second extended mission (which continues science operations through to March 2015), we know that the planet is totally barren, with a maximum surface temperature of about 430 °C, and a virtually non-existent atmosphere which is technically an exosphere; the pressure is probably about 10^{-15} bar (a million million times less than the pressure of a feather sitting on your head). Craters are abundant, ranging from vast ringed basins – such as the Caloris Basin, which is 1,550 kilometres across – to tiny bowl-shaped depressions, and they vary from bright, young ray craters to those which are barely visible remnants (Figure 6). There are also ridges, highlands, mountains, plains, escarpments and valleys. Some regions are heavily cratered while others are almost crater-free. In the distant past the planet was volcanically active and the impact basins filled with magma, producing smooth plains. On the opposite side of Mercury to the Caloris Basin there is a large hilly area called the 'Weird Terrain' which may have been pushed out by the shock waves created by the gigantic impact on the other side of the planet.

Before the Space Age, the best map of Mercury was provided by the Greek astronomer Eugenios Antoniadi, who spent much of his life in France and observed with the great 33-inch Meudon refractor. Antoniadi's book about Mercury was published in 1934, though it was not until 1974 that it appeared in English (the late Patrick Moore undertook the translation). Antoniadi's views about the planet sound strange today. He wrote:

> The dawn and dusk so peculiar to Mercury would seem excessively pale . . . A more-than-glacial breeze would be extremely punishing

Figure 6. Much of Mercury's surface is heavily cratered, so overlapping craters can be seen in many places, including the areas around Camoes crater seen here. This image was acquired by the Narrow Angle Camera (NAC) of the Mercury Dual Imaging System (MDIS) on the MESSENGER spacecraft. Camoes, the largest crater here, is about 105 kilometres across and is superimposed by three different craters that each have central peaks. The largest of these superimposed craters is overlapped by one of the others, creating a small crater partially inside another which, in turn, has another smaller crater partially inside it. (Image courtesy of NASA/Johns Hopkins University Applied Physics Laboratory/Carnegie Institution of Washington.)

during the night, while a wind incomparably more scorching than the desert simoom would probably give rise to a spectacle of fuming dunes, which during the day would raise eddies of greyish dust which would cover the sky, and conceal even the Sun with sinister, all-absorbing clouds . . . The constellations will, of course, present the same aspect from Mercury as from the Earth; but they will show up more brilliantly, and the Milky Way will appear so low in altitude that it would seem to be nearly on top of the observer. Star-twinkling will be unknown . . . The Zodiacal Light, in which Mercury is always plunged, will seem to be more luminous but more diffuse than it

> does to us; and since it will fill most of the sky, it will form a broad band, somewhat ill-defined . . . Seen from Mercury, all the planets are naturally superior, coming to opposition with the Sun. Venus will be the most brilliant star in the sky, much more brilliant than it appears from Earth – a veritable celestial diamond, casting shadows with diffraction fringes. Its disk at opposition will attain a diameter of 70″ when Mercury is near aphelion. Our Earth will come next, a magnificent star of the first magnitude, whose maximum diameter will attain 33″, and which, with its bright surface, should be comparable with our own view of Venus, casting comparable shadows. Keen eyesight would also show our Moon, slowly oscillating from one side of the Earth to the other. Mars, as seen from Mercury, would never fall to the 2nd magnitude; Jupiter and Saturn would be of the first magnitude for most of the time . . . Mercury would be an ideal planet from which to admire and study comets. Indeed, these 'hairy stars' would appear, near their perihelion, with a grandeur which would make our comets of 1811, 1843, 1858 and 1882 seem very feeble . . . Meteor showers ought to be rich from Mercury, but will, in general, describe very rapid trajectories in the rarefied Mercurian atmosphere.

Of course, we now know that Mercury's exosphere is far too thin to produce luminous meteors, and there are no winds. Antoniadi's description does at least show us how little we knew about the innermost planet only eighty years ago!

Star Clusters in Cancer. This month the giant planet Jupiter is a brilliant object moving slowly westwards against the background stars of Cancer (the Crab). When not graced by the presence of a bright planet, Cancer is one of the faintest of the Zodiacal constellations (Figure 7). There are only two stars above the fourth magnitude, and its brightest star, Beta (Altarf), is only of magnitude 3.5. However, the constellation is easy to identify, as it lies within the large triangle formed by Procyon, Pollux and Regulus.

There is a mythological legend attached to it. It represents a crab which the goddess Juno, queen of Olympus, sent to attack Hercules, who was doing battle with the multi-headed Hydra. Not unnaturally, Hercules trod on the crab and squashed it! However, as a reward for the crab's efforts, Juno placed it in the sky.

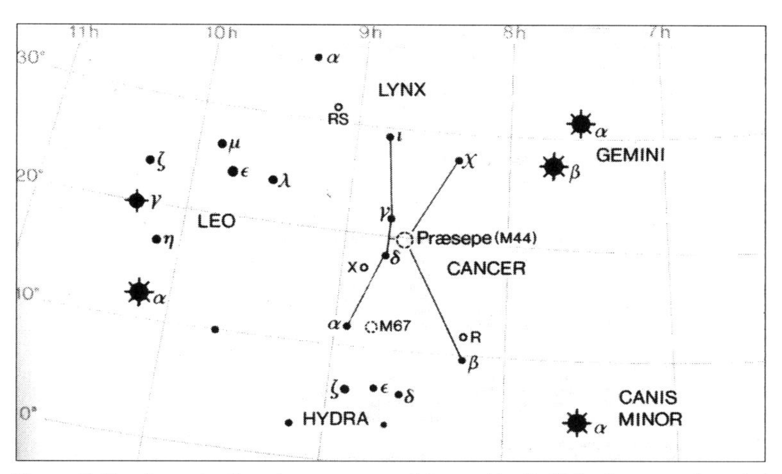

Figure 7. The dim and rather obscure pattern of Cancer (the Crab) lies in an area bounded by Leo to the east, Gemini to the west, the equally obscure Lynx to the north and Hydra and Canis Minor to the south. The locations of the two open star clusters Messier 44 (M44) and Messier 67 (M67) are also shown here.

Much the most interesting features of Cancer are the two open clusters M44 (Praesepe) and M67. Praesepe (Figure 8, overleaf) is nicknamed the Beehive, though it has also been referred to as the Manger, so that the two stars flanking it, Delta (magnitude 3.9) and Gamma (4.7), are known as the Asses – Asellus Australis and Asellus Borealis, respectively. Praesepe was once given a Greek letter, Epsilon, but this has never been used. The cluster is easily visible with the naked eye, but is best viewed with binoculars. It has been known since very early times; around 130 BC, Hipparchus referred to it as 'a cloudy star'. The cluster contains at least 1,000 gravitationally bound stars and it is over 590 light years away. There is no nebulosity and there are not only some red giants and white dwarf stars, but also Main Sequence stars including a large number of red dwarf stars, yellow and orange Sun-like stars of spectral classes F, G and K and a few orange giants, so that Praesepe is clearly older than clusters such as the Pleiades in Taurus. There seems to be general agreement that the age of the Beehive is about 600 million years.

M67, near Acubens or Alpha Cancri (magnitude 4.3), is on the fringe of naked-eye visibility. It appears more compact and richer than

Figure 8. The open cluster M44 in Cancer, also known as Praesepe or the Beehive Cluster. This image was acquired by the Two Micron All Sky Survey (2MASS) as part of the 2MASS Atlas Image Gallery: The Messier Catalogue. (Image courtesy of 2MASS, a joint project of the University of Massachusetts and the Infrared Processing and Analysis Center/California Institute of Technology, funded by the National Aeronautics and Space Administration and the National Science Foundation.)

Praesepe, and again binoculars show it well. There are at least 500 stars above the 16th magnitude. It is about 2,700 light years distant, and seems to be one of the oldest of all galactic clusters with an estimated age of about 4,000 million years. It lies well over 1,000 light years above the main plane of the Galaxy, and so is relatively immune to disruption from field stars.

The variable star X Cancri lies in the same low-power binocular field with Delta. It is a semi-regular star, with a rough period of around 195 days; the spectral type is N so that the red colour is very pronounced. The late Patrick Moore wrote of this star:

> In the same binocular field with Delta [Cancri] you will find one of the reddest stars in the sky: X Cancri. It is a semi-regular variable; at maximum it rises to magnitude 5 and it never falls below 7.3, so that it can always be seen with binoculars. It looks rather like a tiny glowing coal.

Another variable, R Cancri, near Beta, is a Mira star with a range from magnitude 6 to 11.8 and a period of 362 days.

This Month's Solar Eclipse. On 20 March 2015, in the morning, the central, dark part of the Moon's shadow will race across the North Atlantic Ocean, just missing the south-eastern corner of Iceland and making landfall at only two places – the Faroe Islands and the Svalbard archipelago. Eclipse chasers wishing to see a total eclipse – one of Nature's greatest spectacles – will need to make their way to one of these two locations to watch from on land, or they can either observe from on board a ship or, perhaps, from an aircraft flying along the path of totality. The greatest duration of totality will be 2m 47s from a location north of the Faroe Islands.

Nowhere in the British Isles will witness totality on 20 March, but a very significant partial eclipse will be seen, ranging from a maximum 85 per cent obscuration in the south-east of England to over 97 per cent in the far north and north-west of Scotland; the whole event lasting well over two hours. For a detailed overview of this event – the most significant solar eclipse in Britain for sixteen years – see the article in Part Two of this Yearbook.

April

Full Moon: 4 April *New Moon:* 18 April

MERCURY is too close to the Sun for observation at first as it passes through superior conjunction on 10 April. However, from about 20 April, it becomes visible in the evenings, low above the west-northwestern horizon, for observers in equatorial and northern latitudes. For observers in northern temperate latitudes this will be the most favourable evening apparition of the year. Figure 9 shows, for observers in latitude 52°N, the changes in azimuth and altitude of Mercury on successive evenings when the Sun is 6° below the horizon. This condition is known as the end-of-evening civil twilight and in this latitude, and at this time of year, occurs about thirty-five minutes after sunset. The changes in the brightness of the planet are indicated by the relative sizes of the circles marking Mercury's position at five-day intervals. Mercury is at its brightest before it reaches greatest eastern elongation on 7 May, fading from magnitude –1.3 on 20 April to –0.4 on the 30th. Mercury (magnitude –1.1) passes less than 1.5 degrees north of the much fainter Mars (+1.4) on 23 April.

VENUS continues to be visible as a magnificent object in the evening sky, magnitude –4.1. Conditions vary quite noticeably as the observer's latitude changes. Those in the latitudes of the British Isles will be able to see it for over four hours after sunset, by the end of the month, whilst those in temperate southern latitudes will enjoy only about two hours. Interestingly, similar conditions will prevail with the morning apparition of Venus in six months' time. Venus passes 2.5 degrees south of the Pleiades open star cluster on 11 April and about 7 degrees north of the Hyades cluster and the red star Aldebaran, in Taurus, on 18–20 April, with the waxing crescent Moon very close to Aldebaran on the 21st. The phase of Venus decreases from 78 per cent to 68 per cent during the month.

MARS, magnitude +1.4, is moving rapidly eastwards in Aries, but is

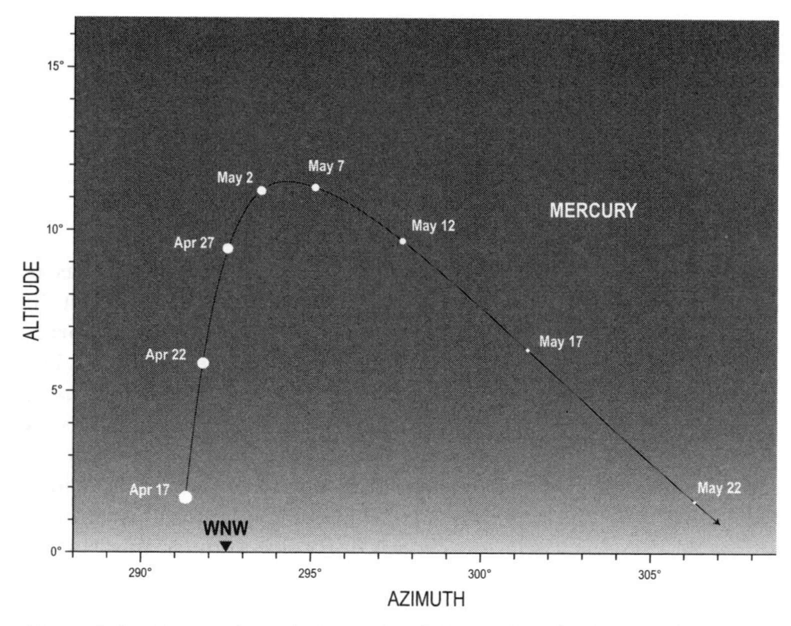

Figure 9. Evening apparition of Mercury from latitude 52°N. The planet reaches greatest eastern elongation on 7 May 2015. It is at its brightest in late April, before elongation.

now drawing ever closer to the Sun in the sky. The planet sets about two hours after the Sun at the beginning of the month, but will be difficult to see against the bright twilight sky. By month end, it will be setting only about an hour after the Sun and to all intents and purposes will be lost in the twilight.

JUPITER reaches a stationary point on 8 April and thereafter resumes its direct (i.e. eastwards) motion once more. The planet will be recognized easily as darkness falls, a brilliant object among the stars of Cancer. Now well past opposition, the planet continues to fade slightly, from magnitude –2.3 to –2.1 during the month. From the latitudes of northern Europe and North America, the planet remains visible until well after midnight at the end of April, but from more southerly locations the period of visibility is much reduced on account of the planet's northerly declination. The waxing gibbous Moon passes about six degrees south of Jupiter on the evening of 26 April.

SATURN will be at opposition in May, so it rises not long after sunset and is observable until dawn, the period of visibility being longer for those in equatorial and Southern Hemisphere latitudes. The planet brightens very slightly from magnitude +0.3 to +0.1 during April. It continues to move retrograde among the fainter stars of Scorpius, the Scorpion.

Thatcher's Comet and the April Lyrids. This month sees the return of the annual Lyrid meteor shower. The position of the radiant at the peak of the shower is RA (Right Ascension) 18h 08m, declination +32°, actually right on the border between Lyra and Hercules, about eight degrees south-west of the brilliant Vega. The shower begins around the 19th, reaches its maximum on the night of 22/23 April, and ends on the 25th. Since the Moon is new on the evening of the 18th, 2015 should be a good Lyrid year in terms of lack of interference by moonlight, so we have to hope that the weather cooperates! The average ZHR (Zenithal Hourly Rate) of the Lyrids at the peak is around 10 meteors per hour, but sometimes, as in 1922 and 1982, this is greatly increased. (The ZHR is the number of naked-eye meteors which would be expected to be seen with the naked eye by an observer under ideal conditions, with the radiant at the zenith. In practice, these conditions are never fulfilled, so that the actual observed rate is always less than the ZHR; moreover, no account is taken of non-shower or sporadic meteors, which may appear from any direction at any moment.)

Most meteor showers, though not all, are associated with known periodic comets, many of which are of short period. The Lyrids, however, are linked with a comet which comes back only once in several centuries, known officially as C/1861 G1 (Thatcher). It was last seen in 1861, and if the estimated period of 415.5 years is anywhere near the truth, we may expect it again around AD 2276. It was discovered on 5 April 1861 by the American astronomer A. E. Thatcher, from New York. It was then described as a tail-less nebulosity about 2 minutes of arc in diameter, with a central condensation; the magnitude was 7.5, below the limit of naked-eye visibility. As it drew inward it brightened, and by the end of the month it was clearly visible without optical aid – indeed, an independent naked-eye discovery was made on 28 April by Carl Wilhelm Baeker in Germany. On 5 May, the comet passed just over 50 million kilometres from the Earth; it was then of magnitude 2.5, with a degree-long tail. Perihelion was passed on 3 June 1861 at a distance of 138 million kilometres from the Sun, slightly inside the Earth's orbit. It faded slowly, but was followed until 7 September, when the magnitude had fallen to 10.

There is some uncertainty in the comet's orbital period, but the orbit is certainly elliptical, so that Thatcher's Comet will return one day. Meanwhile, we can see its debris every year in the form of the April Lyrid meteor shower.

Felix Savary and Xi Ursae Majoris. The French astronomer Felix Savary was born in Paris on 4 October 1797. He was a student at the École Polytechnique, and subsequently taught there, becoming Professor of Astronomy and Geodesy in 1831, and it is where he spent all of his working life. He died on 15 July 1841. He also served as librarian at the Bureau des Longitudes and worked on electromagnetism (some work being carried out with André-Marie Ampère), the rotation of magnets and electrical discharges. Savary accomplished much good work, but is now remembered chiefly as being the first man to apply the laws of gravity to compute the orbits of double stars.

The existence of binary systems had been demonstrated by William Herschel, but no orbits had been worked out, because, of course, for most pairs the changes in separation and position angle are comparatively slow. Savary concentrated upon Xi Ursae Majoris which, in 1780, Herschel had discovered was a visual double star, and which has very similar components – near-twin yellow G-class dwarf stars similar to our Sun, of magnitudes 4.3 and 4.8. We now know that Xi Ursae Majoris is relatively nearby – its distance from us is only 27 light years – though of course star distances would not be determined until 1838, when Friedrich Bessel calculated the distance of 61 Cygni. In 1828, Savary announced that the orbital period of Xi Ursae Majoris was of the order of 60 years, and in this he was correct, since the actual value is 59.878 years. A rather high orbital eccentricity takes the two main components to as far as 29.6 AU (Astronomical Units), 4,430 million kilometres, from each other to as close as 13.4 AU (2,005 million kilometres).

The present angular separation of the two main components is 1.7 arc seconds and needs at least a 7.5-cm telescope to split it. The separation is increasing and by 2035 will be as much as 3.1 arc seconds, but by 2053 it will have decreased again to only 0.8 arc seconds. It is now known that both components are spectroscopic binaries, so that Xi

Ursae Majoris is certainly a quadruple system – and there have even been suggestions that the star may be quintuple, or even sextuple! Indeed, in March 2012, a team of astronomers lead by Edward L. Wright announced that the Wide-field Infrared Survey Explorer (WISE) spacecraft has revealed a T8.5 brown dwarf (WISE J111838.70+312537.9) that exhibits common proper motion with the Xi Ursae Majoris system. The angular separation is 8.5 arc minutes, and the projected physical separation is about 4,000 AU (600,000 million kilometres). The infrared luminosity and colour of the brown dwarf suggests the mass of this companion ranges between 14 and 38 Jupiter masses for system ages of 2,000 million and 8,000 million years, respectively.

Xi (Greek letter ξ) is easy to find, some way south of the famous 'Plough' pattern of Ursa Major (the Great Bear), not far from the borders with Leo and Leo Minor (Figure 10); it makes up a neat little pair with Nu Ursae Majoris, which is slightly the brighter of the two at magnitude 3.5. Xi does have an old proper name, Alula Australis, though this is hardly ever used; Nu has been called Alula Borealis.

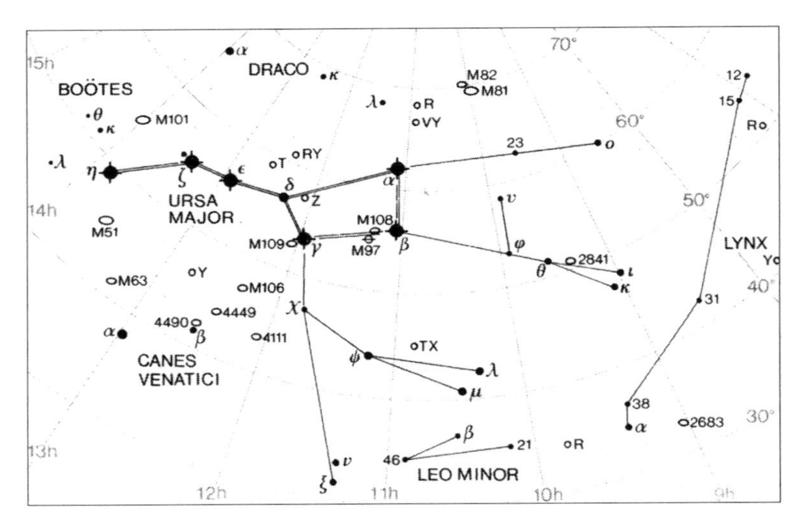

Figure 10. Xi (Greek letter ξ) Ursae Majoris lies quite a long way south of the famous 'Plough' pattern of Ursa Major (the Great Bear), not far from the borders with Leo and Leo Minor. Xi makes up a neat little pair with Nu Ursae Majoris, which is slightly the brighter of the two stars.

May

Full Moon: 4 May *New Moon:* 18 May

MERCURY is at greatest eastern elongation (21°) on 7 May and therefore continues to be visible as an evening object for observers in tropical and northern latitudes for the first two weeks of the month, in the west-north-western twilight sky for a short while around the end of evening civil twilight. The planet is brightest before elongation, fading from magnitude –0.3 to +1.5 during the first two weeks of May. Observers should refer to Figure 9 given with the notes for April. For the rest of the month Mercury is too close to the Sun for observation, passing through inferior conjunction on 30 May. It is not suitably placed for observation by those in more southerly latitudes.

VENUS continues to be visible as a magnificent object dominating the western sky in the evenings, setting over three-and-half hours after the Sun from northern temperate latitudes. The planet will be seen passing south of the Heavenly Twins, Castor and Pollux, in Gemini, at the end of the month. Venus brightens from magnitude –4.1 to –4.3 and the phase decreases from 68 per cent to 53 per cent during May. The waxing crescent Moon will be some way south of Venus in the evening twilight sky on 21 May.

MARS is in conjunction with the Sun in mid-June and is unlikely to be seen this month, being lost in the bright evening twilight sky. The planet is in Taurus, magnitude +1.5.

JUPITER is moving eastwards among the faint stars of Cancer and continues to be a brilliant object in the evenings in the western sky, observable until the early morning hours for those living in the latitudes of the British Isles, although the period of visibility is somewhat shorter for those living further south. Jupiter fades slightly from magnitude –2.1 to –1.9 during May. The waxing crescent Moon will appear about six degrees south of Jupiter on the evening of 23 May.

SATURN, magnitude +0.1, is at opposition in Libra on 23 May, the planet's westwards motion having carried it across the border from neighbouring Scorpius earlier in the month. Figure 11 shows the path of Saturn against the background stars during the year. The planet becomes visible in the south-eastern sky as soon as darkness falls and is observable all night long. Saturn is a lovely sight in even a small telescope, with the rings beautifully displayed at an angle of 24.4° as viewed from the Earth. At opposition the planet is 1,341 million kilometres (834 million miles) from the Earth. The Moon, just past full, will appear fairly close to Saturn on the night of 5/6 May.

Ludwig's Star. One of the most famous stars in the sky is Mizar or Zeta Ursae Majoris, the second star in the 'tail' of the Great Bear – almost overhead as seen from Britain on May evenings. It is of the second magnitude, and close beside it is the fourth-magnitude Alcor (80 Ursae Majoris). The separation is twelve minutes of arc, so that both members of the pair are very easy to see with the naked eye. Together, Mizar and Alcor are sometimes called the 'Horse and Rider'. Telescopically, Mizar itself is seen to be double, with one component decidedly brighter than the other; the two components are magnitudes 2.3 and 4.0, with a separation of 14.4 arc seconds. Mizar was the first double star to be found telescopically and for a long time it was thought that this discovery was due to the seventeenth-century astronomer Giovanni Battista Riccioli (best remembered today for giving names to the craters of the Moon), in about 1650, but it appears that a friend of Galileo, Benedetto Castelli, actually saw it in January 1617! Today we know that each star in the Mizar pair is itself double, making this a quadruple system of two binary stars.

Strangely, the Arabian astronomers of a thousand years ago said that Alcor was difficult to see with the naked eye! Yet the Arabs were nothing if not keen-sighted; light pollution was much less of a problem in those times than it is now, and nowadays anyone with average eyesight can see Alcor with no difficulty at all. It does not seem likely that there has been any real alteration, so can there be another explanation?

Roughly halfway between the Mizar pair and Alcor is an eighth-magnitude star, again first noted by Benedetto Castelli – in November 1616 – and given the curious name of Sidus Ludoviciana more than a century later, when it was observed by Johann Georg Leibknecht in 1722. He believed it to be a new star or even a planet(!) and named it

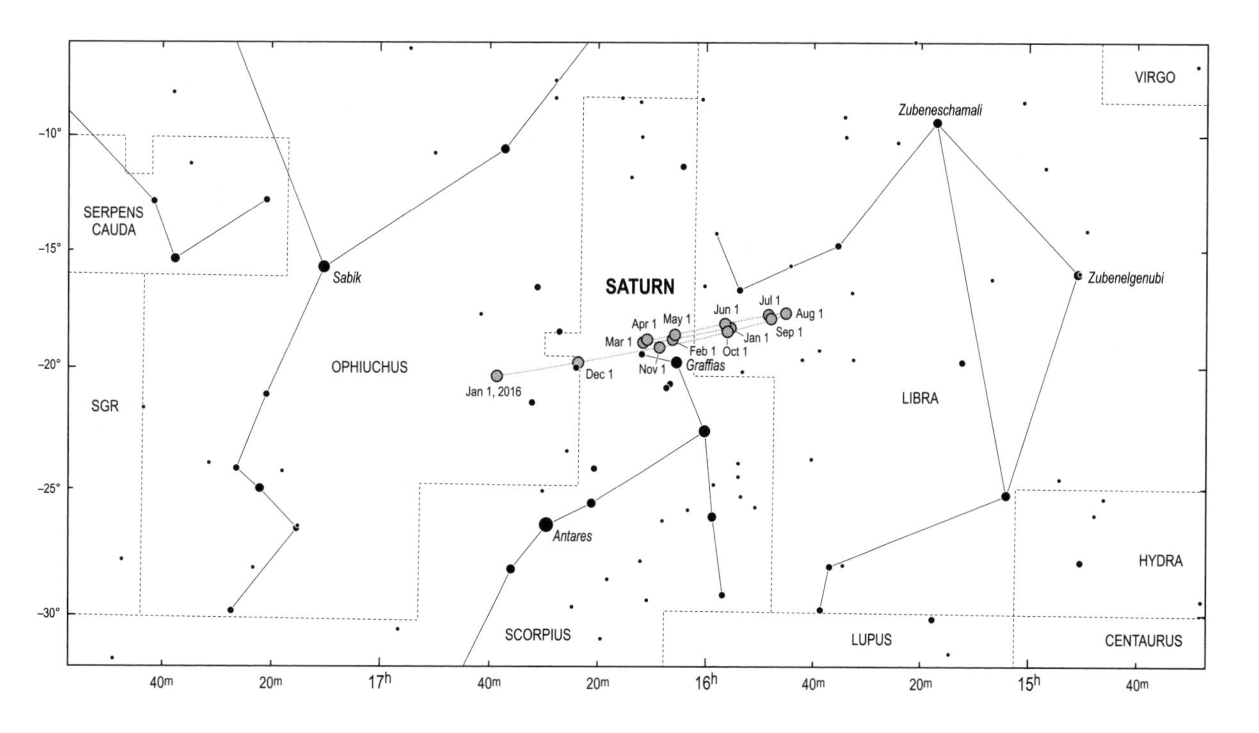

Figure 11. The path of Saturn against the background stars of Libra and Scorpius during 2015.

Sidus Ludoviciana, in honour of his monarch Ludwig, the Landgrave of Hesse-Darmstadt. Could this, not Alcor, be the 'difficult' object of the Arabs? Again it seems most unlikely; Ludwig's Star is not connected with the Mizar group, as it lies well in the background, and it has shown no sign of variability. So the mystery remains – and today it is very clear that Sidus Ludoviciana is well beyond the range of naked-eye visibility.

Incidentally, it was discovered in 2009 that Alcor is itself a binary star and that this pair is most probably gravitationally bound to Mizar, bringing the total number of stars in the entire system to six.

The Largest and Smallest Constellations. During May evenings, the large, faint constellation of Hydra, the Watersnake, sprawls across the southern sky, beginning not far from Procyon in Canis Minor and ending below Spica in Virgo (Figure 12). It has only one bright star, Alphard or Alpha Hydrae (often called 'the Solitary One' because there are no conspicuous stars anywhere near it), which is of second magnitude. The rest of Hydra consists of a straggling line of faint stars below

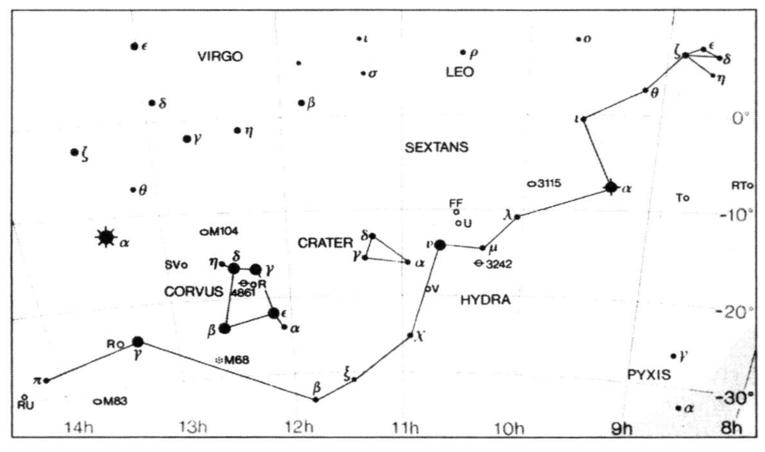

Figure 12. The large, faint constellation of Hydra (the Watersnake) sprawls across the southern sky from northern temperate latitudes on May evenings. There is only one bright star, Alpha (α) Hydrae or Alphard, often called 'the Solitary One' because there are no conspicuous stars anywhere near it. The rest of Hydra consists of a straggling line of faint stars extending below Leo and Virgo.

Leo and Virgo. Extensive though they are, such large and faint constellations attract little notice in comparison with much smaller constellations having brighter stars. There is little order and method in the division of the sky into its various constellations. Of course, some of the constellations were formed in ancient times – Ptolemy, who died in or about AD 180, listed a total of forty-eight – while others have been added much more recently.

Star maps used to include the great sailing ship Argo Navis, which held pride of place with regard to area. In mythology, *Argo* was the vessel which carried Jason and his companions, the Argonauts, in their successful quest to remove the golden fleece of the ram (Aries) from its sacred grove, where it was guarded by a dragon. But Argo was considered so large and unwieldy that, as part of a review of constellation boundaries, the International Astronomical Union divided it into four separate parts, Puppis (the Poop), Carina (the Keel), Vela (the Sails) and Pyxis (the Compass). Canopus, the second brightest star in the sky, which used to be known officially as Alpha Argûs, is now known as Alpha Carinae. Incidentally, the southern constellation of Volans (the Flying Fish) is generally considered to be associated with Argo.

So, in area, Hydra is now the largest of the eighty-eight constellations accepted today. These areas are not difficult to calculate nowadays, because the boundaries of all the constellations are straight lines, or, to be more precise, arcs of great circles on the celestial sphere. Before 1930, the boundaries were left to the fancy of the artists who designed the old pictorial star atlases, and there was little general agreement as to how the boundaries should be drawn to include the outstretched arm of a figure, or the tail of an animal. In 1930, under the direction of Eugene Delporte of the Brussels Observatory, the boundaries were fixed as regular lines, and now it is definitely possible to assign any particular star to one constellation only. Of the eighty-eight constellations, Hydra is the largest, followed by Virgo, Ursa Major and Cetus; the smallest are Sagitta and Equuleus, in the Northern Hemisphere, and the famous Southern Cross, Crux Australis, near the South Pole.

The largest constellations are as follows:

	Name	Area, square degrees
1.	Hydra	1303
2.	Virgo	1294
3.	Ursa Major	1280
4.	Cetus	1231
5.	Hercules	1225
6.	Eridanus	1138
7.	Pegasus	1121
8.	Draco	1083
9.	Centaurus	1060
10.	Aquarius	980

At the other end of the scale come:

1.	Crux Australis	68
2.	Equuleus	72
3.	Sagitta	80
4.	Circinus	93

Remarkably, Orion, the Hunter, has an area of just 594 square degrees, and comes a mere 25th in order of size, despite its exceptional prominence and its collection of bright stars.

June

Full Moon: 2 June *New Moon:* 16 June

Solstice: 21 June

MERCURY attains its greatest western elongation (22°) on 24 June. For observers in the latitudes of the British Isles the long duration of twilight makes observation impossible. But nearer the Equator and in the Southern Hemisphere, Mercury can be seen as a morning object in the east-north-eastern sky at the time of the beginning of morning civil twilight after about the middle of the month. Figure 13 shows, for observers in latitude 35°S, the changes in azimuth and altitude of Mercury on successive mornings when the Sun is 6° below the horizon. This condition is known as the beginning of morning civil twilight and in this latitude and at this time of year occurs about twenty-five minutes before sunrise. The changes in the brightness of the planet are indicated by the relative sizes of the circles marking Mercury's position at five-day intervals. Mercury will be at its brightest after it reaches greatest western elongation, brightening rapidly from magnitude +1.7 on 15 June to –0.1 by the end of the month.

VENUS reaches greatest eastern elongation (45°) on 6 June and continues to be visible as a magnificent object in the western sky in the evenings, brightening very slightly from magnitude –4.3 to –4.4 during the month. The planet's declination north of the celestial equator decreases by almost ten degrees as it moves eastwards from Gemini, through Cancer and into Leo during June. This means that the planet's period of visibility after sunset decreases from three-and-a-half hours to less than two hours for observers in the Northern Hemisphere, but for those living in the Southern Hemisphere the conditions are reversed. The phase of Venus decreases from 53 per cent to 34 per cent during June. Venus will pass less than half a degree (one Moon diameter) south of the somewhat less brilliant Jupiter on 1 July, so the two planets will be in the same part of the sky during the last week of June

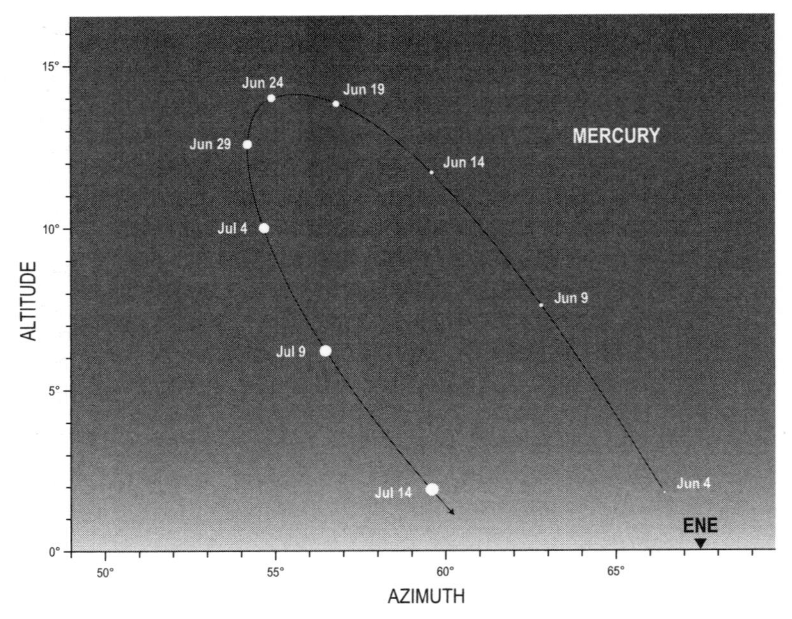

Figure 13. Morning apparition of Mercury from latitude 35°S. The planet reaches greatest western elongation on 24 June 2015. It will be at its brightest in mid-July, several weeks after elongation.

and throughout July. On the evening of 20 June, the waxing crescent Moon will make a beautiful grouping with the two bright planets in the western twilight sky – a wonderful picture opportunity.

MARS is in conjunction with the Sun on 14 June and consequently will be unobservable throughout the month.

JUPITER is still a lovely object in the western sky in the evenings, moving slowly eastwards from Cancer into neighbouring Leo in mid-month, and fading slightly from magnitude –1.9 to –1.8 during June. From northern temperate latitudes the planet will be setting only two hours after the Sun by month end. As mentioned above, Jupiter will be in the same part of the sky as the far more brilliant Venus towards the end of June and during July.

SATURN, just past opposition, is visible in the south-south-east from northern temperate latitudes as darkness falls, and is observable for most of the night. From more southerly locations the planet is situated much higher in the sky. Saturn is moving retrograde in Libra. It fades slightly from magnitude +0.1 to +0.3 during the month. The almost full Moon will appear very close to Saturn on the evening of 1 June and again on the 28th.

Antares and its Neighbours. The constellation of Scorpius, the Scorpion, is one of the most magnificent in the sky (Figure 14). It is well south of the celestial equator, and is always rather low down from British latitudes; the 'sting', which contains the star Shaula (Lambda Scorpii) that is only just below the first magnitude, barely rises even from southern England, and part of the constellation remains permanently below the horizon, though all of it can be fairly well seen from most of the United States. This is the best time of the year for seeing it. From

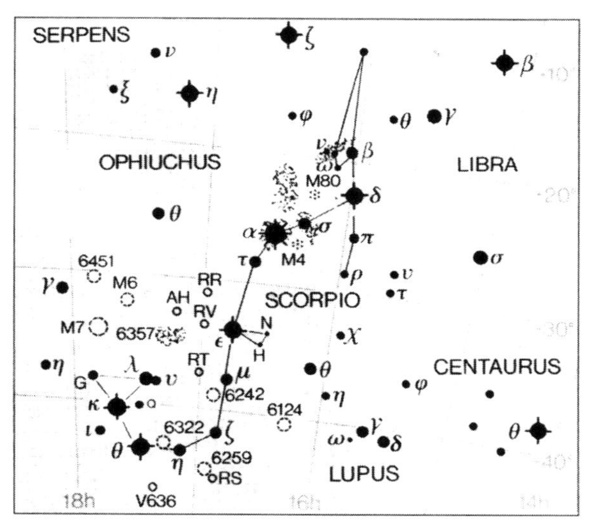

Figure 14. Scorpius (the Scorpion) is visible low down in the southern sky on June evenings from northern temperate latitudes, although the most southerly part of the Scorpion's tail and sting do not rise at all from such locations. The brightest star in the constellation, which marks the Scorpion's heart, is the red supergiant Antares, Alpha (α) Scorpii. On either side of Antares are two much fainter stars, Sigma Scorpii to the north and Tau Scorpii to the south; Tau is slightly the brighter of the two.

countries such as South Africa, Australia and New Zealand, the Scorpion is almost overhead on June evenings, and dominates the night sky.

The brightest star in the constellation, and marking the Scorpion's heart, is the red supergiant Antares, with an apparent visual magnitude of 1.06 (very slightly variable). It has a radius about 700 times that of our Sun, so if it were placed in the centre of our Solar System, its outer atmosphere would extend to between the orbits of Mars and Jupiter. Antares is the reddest of all the brilliant stars (ranked sixteenth brightest in the sky), though it is rivalled by the somewhat brighter Betelgeuse in Orion, and the name 'Ant-Ares' literally means 'the rival of Mars'. Antares is a relatively cool star with a surface temperate of about 3600 K, radiating most of its energy in the infrared part of the spectrum. When this is taken into account, it must be about 60,000 times as luminous as the Sun – although there is some uncertainty here because Antares is embedded in an interstellar dust cloud, so it may be even more luminous still. Antares is some fifteen to eighteen times more massive than our Sun so it will certainly end its days exploding as a brilliant supernova. Based upon parallax measurements, the distance of Antares is about 600 light years.

Antares has a greenish companion of magnitude 5.4, at an apparent separation of 2.6 arc seconds. It is not at all easy to see with small telescopes, particularly from the latitudes of the British Isles, because it is so overpowered by the glare of its primary. The companion may sometimes be observed with a small telescope for a few seconds when Antares is occulted by the Moon; indeed, it was during such an occultation on 31 April 1819 that the companion was discovered by the Austrian mathematician and astronomer Johann Tobias Bürg, although not confirmed for some years. The companion's apparent greenish hue is due largely to contrast.

Flanking Antares to either side are two much fainter stars, Sigma Scorpii to the north and Tau Scorpii to the south; both stars are sometimes known by the traditional Arabic name of Alniyat or Al Niyat, meaning 'the arteries', but Tau is actually a little brighter than Sigma.

Sigma is a spectroscopic binary consisting of a core pair of two hot, white or blue-white stars having a combined apparent magnitude of 2.9, with one of the stars being very slightly variable in light. It belongs to the so-called Beta Cephei or Beta Canis Majoris class; these are pulsating stars, but the magnitude range is very slight – only about a tenth of a magnitude in the case of Sigma Scorpii, and there are multiple periods ranging from a few hours to several days. Orbiting the main pair of stars in a period of over a hundred years is a fifth-magnitude companion and there is also a distant attendant of the ninth magnitude, which shares their motion through space. Recent estimates place the distance of the Sigma Scorpii system at about 570 light years.

Tau Scorpii, with an apparent magnitude of 2.8, is a single, massive star – about fifteen times the mass of our Sun – lying some 470 light years distant; like Sigma, it is bluish-white. Tau Scorpii is a magnetic star whose surface magnetic field has been mapped by a technique known as Zeeman–Doppler imaging. Tau Scorpii is rotating relatively slowly with a period of forty-one days.

Also in this region is the fine globular cluster M4 (Figure 15), which is easily seen in binoculars and is on the fringe of naked-eye visibility.

Figure 15. The fine globular cluster Messier 4 (M4), located about one degree west of the red star Antares in Scorpius, is well shown in this CCD image taken with the historic 0.9-metre telescope at the National Science Foundation's Kitt Peak National Observatory in March 1995. M4 is one of the smallest globular clusters and is somewhat less condensed than most globulars. In 2001, the WIYN Consortium, led by the University of Wisconsin-Madison and Indiana University, assumed operational responsibility for the 0.9-meter telescope. As part of the acquisition, the consortium chose to upgrade the control system and install new motors and encoders in all axes, for a state-of-the-art observing facility. (Image courtesy of T2KA, KPNO 0.9-m Telescope, NOAO, AURA, NSF.)

Its distance is believed to be in the region of 7,000 light years, but with considerable uncertainty because of the presence of major obscuring interstellar dust clouds near it. It is certainly one of the closest of the globular clusters – much nearer than, for example, Omega Centauri, 47 Tucanae or M13 in Hercules – and it is easy to find because it lies about one degree west of Antares, only slightly away from a line joining Antares to Sigma, and is in the same binocular field with them. M4 was discovered by De Cheseaux in 1746 and is one of the smallest and sparsest of the globular clusters and rather less condensed than most globulars. Even when low down, as it always is from Britain, it is well worth finding, and from southern latitudes it is one of the most impressive globular clusters in the sky.

The Southern Triangle. Very few of the constellations bear the slightest resemblance to the objects they are supposed to represent. It takes a lively effort of the imagination to make a Bull out of Taurus, for example, and most of the 'modern' constellations are even worse; how does one twist the dim stars of Microscopium into a figure of a microscope? But there are a few exceptions, and one of these is Triangulum Australe, the Southern Triangle (Figure 16).

It lies in the far south, and is therefore never visible from the latitudes of Europe or North America. There are three stars above the third magnitude, and they really do make a well-formed triangle:

Star	R. A.	Declination	Magnitude	Spectrum
		° ′ ″		
Alpha	16h 48m 40s	–69 01 40	1.9	K2
Beta	15h 55m 09s	–63 25 50	2.8	F2
Gamma	15h 18m 55s	–68 40 46	2.9	A1

The other stars above the fifth magnitude are Delta (3.9), Epsilon (4.1) and Zeta (4.9), but these do not destroy the main pattern; Epsilon lies midway between Beta and Gamma, while Delta is close to Beta.

There is no difficulty in locating the Southern Triangle, because it lies not far from Alpha and Beta Centauri, the brilliant Pointers to the Southern Cross. Moreover, Alpha Trianguli Australis, with its K2-type spectrum, is distinctly orange. There is not much else of real note in the constellation, apart from the open cluster NGC 6025, which lies near

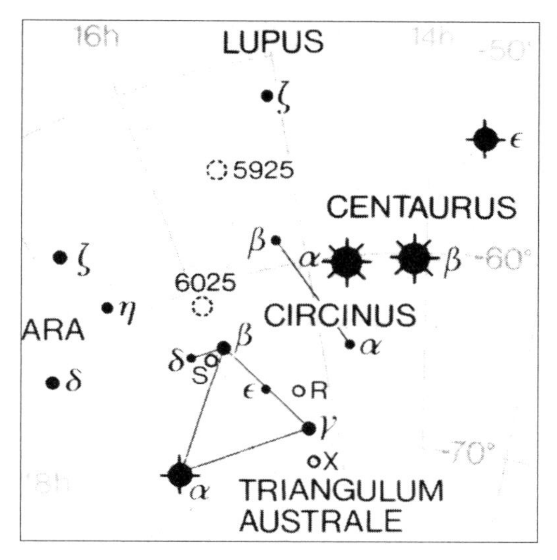

Figure 16. Triangulum Australe (the Southern Triangle) lies not far from Alpha (α) and Beta (β) Centauri, the brilliant Pointers to the Southern Cross. The little constellation of Circinus (the Compasses) lies between Triangulum Australe and Alpha and Beta Centauri. Alpha Circini is exactly midway between Beta Centauri and Gamma (γ) Trianguli Australis.

the border with neighbouring Norma. It has an integrated magnitude of about 5 and contains at least sixty stars. Its distance is about 2,700 light years and the age of the cluster has been estimated as 100 million years. It was discovered by Abbé Nicolas Louis de Lacaille during his South African tour in 1752.

The variable star R Trianguli Australis is a Cepheid with a range of magnitude 6.3 to 7.0 and a period of 3.389 days, while S Trianguli Australis is also a Cepheid, varying from 6.0 to 6.8 in a period of 6.323 days.

The little constellation of Circinus, the Compasses, lies between Triangulum Australe and the Pointers, Alpha and Beta Centauri. The star Alpha Circini (magnitude 3.2) lies exactly midway between Beta Centauri and Gamma Trianguli Australis.

July

Full Moon: 2 and 31 July *New Moon:* 16 July

EARTH is at aphelion (furthest from the Sun) on 6 July at a distance of 152 million kilometres (94.5 million miles).

MERCURY, for observers in equatorial and southern latitudes, may be seen in the morning twilight sky about twenty-five minutes before sunrise, towards the east-north-east, for the first ten days of the month, as shown in Figure 13 given with the notes for June. During this period its magnitude brightens from –0.1 to –0.9. Thereafter, it is too close to the Sun for observation as it passes through superior conjunction on 23 July.

VENUS continues to be visible as a brilliant object in the western sky in the evenings, magnitude –4.5. However, for those in northern temperate latitudes, the period available for observation is shortening noticeably and soon after the middle of the month the planet will be lost in the glare of sunset. In the tropics and the Southern Hemisphere, the planet is visible right through to the end of July. Venus attains its greatest brilliancy on 12 July, with a magnitude of –4.5, but the phase decreases from 33 per cent to just 8 per cent (a slim crescent) during the month. Venus passes less than half a degree south of Jupiter on 1 July, Venus magnitude –4.4, Jupiter –1.8, and they will remain in fairly close proximity to one another throughout July as they both gradually become lost in the evening twilight. The waxing crescent Moon will make a lovely grouping with the two planets on the evening of 18 July, but this event will only be visible to those living in more southerly climes.

MARS was in conjunction with the Sun in mid-June and is unlikely to be seen at all this month, being lost in the bright dawn twilight sky.

JUPITER, magnitude –1.8, continues to be visible as an evening object, though only visible for a short while low above the west-north-western horizon. Observers in the latitudes of the British Isles and suffering from the long evening twilight are unlikely to be able to see it after the first two weeks of the month. As mentioned above, Jupiter will lie less than half a degree north of the more brilliant Venus on 1 July.

SATURN is an evening object in Libra, magnitude +0.3, setting soon after midnight from northern temperate latitudes by the end of the month; a couple of hours later from locations much further south.

PLUTO, officially a dwarf planet, reaches opposition on 6 July, in the constellation of Sagittarius, at a distance of 4,770 million kilometres (2,964 million miles). It is visible only with a moderate-sized telescope since its magnitude is +14.

Solar Superstorm – A Near Miss! Our Sun varies in roughly an eleven-year cycle of activity as the number of sunspots and spot groups on the visible surface of the Sun (the photosphere) rises and falls. At sunspot minimum, the Sun may be entirely spotless for many days, even weeks (the last minimum was in 2008–9), but at sunspot maximum the Sun usually displays many spot groups and large, active sunspots (Figure 17); there may be violent solar eruptions and brilliant displays of the aurora as electrified solar particles cascade down into Earth's upper atmosphere.

The current sunspot cycle (Cycle 24) has been the weakest for over a century and it has exhibited a double peak – the first in February/March 2012 (when the sunspot number reached 67) and the second extending from late 2013 into early 2014 when the sunspot number reached 76. Many cycles are double-peaked – the last two maxima around 1989 and 2001 had not one but two peaks, where solar activity goes up, then dips slightly, and then goes up for a second time over a period of a couple of years. But Cycle 24 is the first in which the second peak in sunspot number was greater than the first – the 2012 peak being due primarily to sunspot activity in the Sun's Northern Hemisphere and the later peak to activity in the Southern Hemisphere. In spite of the late surge, Cycle 24 is still the weakest sunspot cycle since Cycle 14 in February 1906, which had a sunspot number of just 64.

In April 2014, scientists, government officials and emergency planners travelled to Boulder in Colorado for the NOAA Space Weather Workshop, a meeting held every year to discuss the Sun's

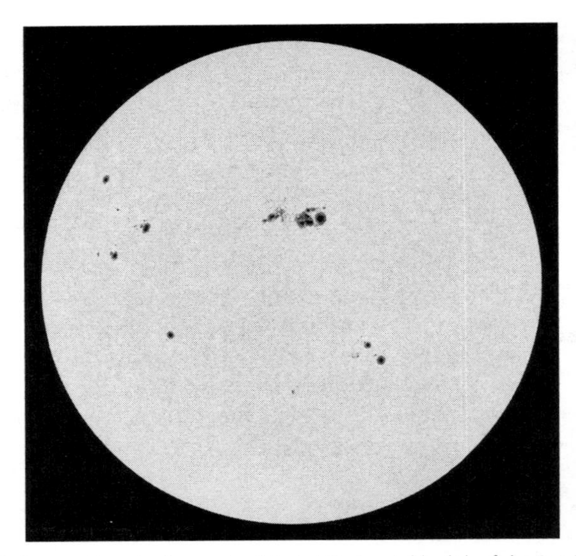

Figure 17. Approaching the first peak of Cycle 24, the visible disk of the Sun is peppered with sunspots. Caused by intense magnetic fields emerging from the interior, sunspots appear to be dark only when contrasted against the rest of the solar surface, because they are slightly cooler than their surroundings. This image was acquired by the HDI instrument of the Solar Dynamics Observatory spacecraft on 8 November 2011 at 1300 UT. Just five days earlier, active region 1339 (near the centre of the disk here), one of the largest sunspots in years – 40,000 kilometres wide and at least twice that in length – unleashed an X-class solar flare. Waves of ionization in the upper atmosphere created a strong radio blackout. The related coronal mass ejection (CME) was not directed towards the Earth. (Image courtesy of Solar Dynamics Observatory/NASA Goddard Space Flight Center.)

activity and in particular major solar storms, their effects on the Earth and how best to deal with them. Given the current relatively low level of overall solar activity, one might have expected the scientists' discussions to be similarly low-key but there was, in fact, enormous excitement about a massive solar eruption that had occurred in 2012 and had only narrowly missed the Earth.

The solar superstorm had occurred on 23 July 2012. An enormous plume of electrified gas and charged particles – a coronal mass ejection or CME – had erupted from the Sun and hurtled outwards at 3,000 kilometres per second, about four times faster than a typical CME. The

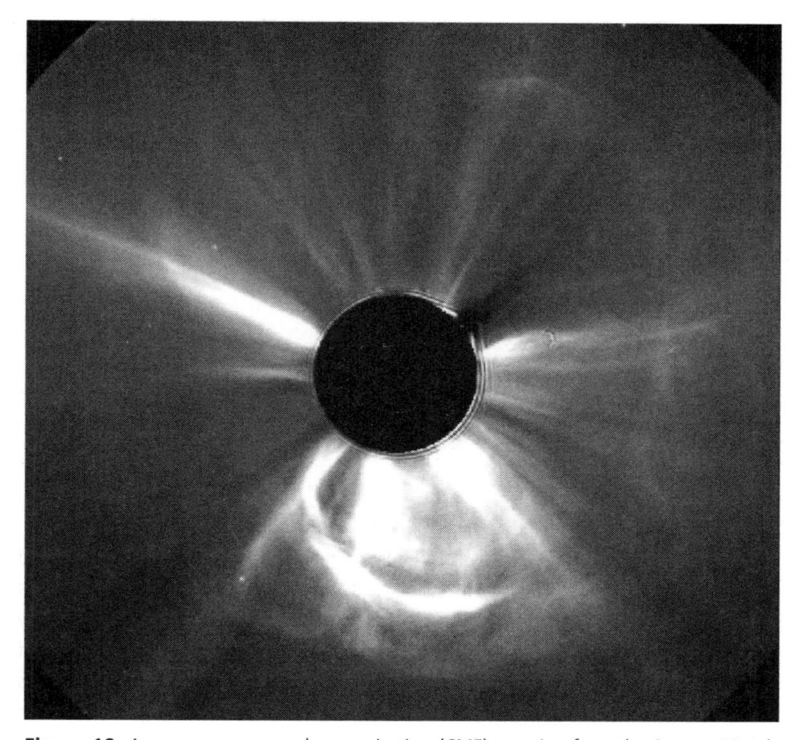

Figure 18. An enormous coronal mass ejection (CME) erupting from the Sun on 23 July 2012, as captured by NASA's Solar TErrestrial RElations Observatory - Ahead (STEREO-A). Because the CME headed in STEREO-A's direction, it appeared like a giant halo around the Sun (whose brilliant face is blocked by a dark disk in this image). Using the STEREO data, scientists at NASA's Goddard Space Flight Center in Greenbelt, Maryland clocked this CME travelling at between 2,900 and 3,500 kilometres per second as it left the Sun. (Image courtesy STEREO/NASA Goddard Space Flight Center.)

storm tore through Earth's orbit but, as luck would have it, the Earth wasn't there; instead the storm hit the STEREO-A spacecraft, one of two almost identical space-based observatories – one ahead of the Earth in its orbit, and the other trailing behind (Figure 18). These twin spacecraft have provided the first-ever stereoscopic measurements to study the Sun and the nature of its coronal mass ejections (CMEs).

The July 2012 event was actually two CMEs separated by only ten to fifteen minutes. The cloud of charged particles from this double eruption travelled far more quickly than normal through interplanetary space because that space had been cleared out by another CME only four days earlier, and there was no deceleration as they travelled through the interplanetary medium.

The scientists who analyzed the STEREO spacecraft data concluded that the 2012 storm was one of the most powerful in recorded history – possibly even stronger than the Carrington Event of 1 September 1859 – a series of powerful CMEs that hit Earth head-on, sparking displays of the aurora seen down to the tropics and intense geomagnetic storms which caused global telegraph lines to spark, setting fire to telegraph offices and disabling communications.

A similar storm today could have a catastrophic and lasting effect on modern power grids, global telecommunication networks and sensitive technologies such as GPS, satellite communications and the Internet. Had the July 2012 eruption occurred just one week earlier, the site of the eruptions would have been facing Earth, rather than off to one side, so it looks as though we had a relatively narrow escape!

First Close-up Views of Pluto. Pluto is at opposition in Sagittarius on 6 July. Eight days later, at 11h 50m UT on 14 July 2015, after a journey of nine-and-half years and about 5,000 million kilometres, the New Horizons spacecraft will make a dramatic high-speed flyby past this icy dwarf planet and its moons (Figure 19). It is a moment that many people (including the editor of this *Yearbook*) have waited for their entire lives – the first close-up look at this enigmatic little world on the outskirts of our Solar System. New Horizons is only the fifth space probe to traverse interplanetary space so far from the Sun – and the first ever to travel to Pluto.

Although New Horizons briefly encountered an asteroid and then proceeded to make a flyby of Jupiter (which provided a gravitational assist increasing the probe's speed by 14,000 kilometres per hour), the spacecraft was launched primarily to study the dwarf planet Pluto and its moons. Hopefully, it will go on to visit one or two more distant Kuiper Belt objects, depending on which may be in a suitable position to be explored. The Hubble Space Telescope is carrying out an in-depth survey of the outer Solar System in a bid to discover potential target objects in the right location to be visited by New Horizons after the Pluto encounter.

Figure 19. Artist's impression of the New Horizons spacecraft passing Pluto in July 2015. (Image courtesy of NSSDC/NASA Goddard Space Flight Center.)

The New Horizons mission was approved in 2001 following the cancellation of two previous planned probes – the Pluto Fast Flyby and Pluto Kuiper Express. The mission profile was proposed by a team led by principal investigator Alan Stern of the Southwest Research Institute. After several delays on the launch site, New Horizons blasted off on 19 January 2006, setting the record for the highest velocity of a human-made object from Earth – 58,540 kilometres per hour. For a more detailed look at the New Horizons mission, see the main article by David Harland elsewhere in this Yearbook.

Pluto is almost totally unknown. It is so far away that even the Hubble Space Telescope can see little in the way of detail (Figure 20); its best images reveal Pluto's shape (spherical) and colour (reddish). Over the years, variations in its surface reflectivity hint at changes, but no one knows what they are. By late April 2015, New Horizons will be close enough to Pluto to take pictures rivalling those of Hubble – and it will get better from there.

At closest approach in July 2015, New Horizons will be a scant 10,000 kilometres above the surface of Pluto and nobody knows what it will discover; there may be rings, icy geyser eruptions, towering mountains, deep valleys and even frozen lakes. What an exciting prospect!

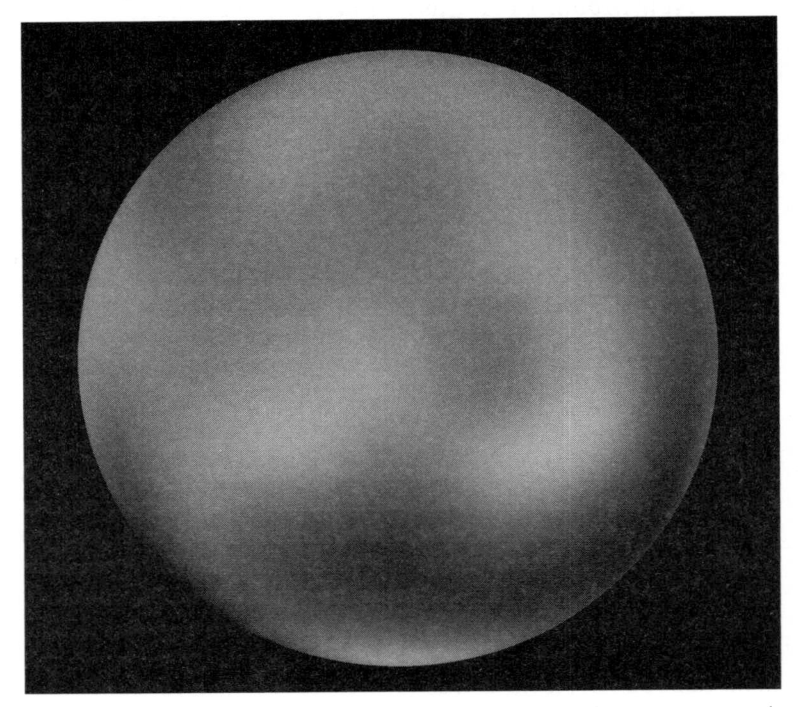

Figure 20. An image of Pluto acquired by the Advanced Camera for Surveys on NASA's Hubble Space Telescope and released in February 2010 showing an icy, mottled, dark molasses-coloured world undergoing seasonal surface colour and brightness changes. The images showed that between 2000 and 2002, Pluto had become significantly redder, while its illuminated northern hemisphere was brightening. These changes are most likely consequences of surface ice melting on the sunlit pole and then refreezing on the other pole, as the dwarf planet heads into the next phase of its 248-year-long seasonal cycle. Hubble images like this will remain the sharpest views of Pluto until NASA's New Horizons probe is within six months of its flyby during 2015. The Hubble images are invaluable for picking the planet's most interesting hemisphere for imaging by the New Horizons probe because the craft will pass by Pluto so quickly that only one hemisphere will be photographed in detail. (Image courtesy of NASA, ESA, and M. Buie (Southwest Research Institute.)

August

New Moon: 14 August *Full Moon:* 29 August

MERCURY reaches its greatest eastern elongation (27°) on 4 September and, apart from the first few days of the month, is visible throughout August to observers in equatorial and southern latitudes, for whom this is the best evening apparition of the year. Unfortunately, Mercury will not be visible to observers in northern temperate latitudes at all this month. Figure 21 shows, for observers in latitude 35°S, the changes in azimuth and altitude of Mercury on successive evenings when the Sun is 6° below the horizon. This condition is known as the end of evening civil twilight and in this latitude and at this time of year occurs about twenty-five minutes after sunset. The changes in the brightness of the planet are indicated by the relative sizes of the circles marking Mercury's position at five-day intervals. It will be noticed that Mercury is at its brightest well before it reaches greatest eastern elongation. During its period of observation Mercury's magnitude fades from –0.9 to +0.2. Mercury passes a little over half a degree (one Moon diameter) north of the more brilliant Jupiter (Mercury magnitude –0.6, Jupiter –1.7) early on 7 August and the brighter planet may be a useful guide to picking up Mercury on the evenings of 6–8 August. The even brighter Venus will be in the same part of the sky, but about six degrees further south.

VENUS, although lost in the bright evening twilight sky for observers in the latitudes of the British Isles, further south may still be seen low above the western horizon after sunset, for the first 10–12 days of the month, fading from magnitude –4.3 to –4.0 during this period. Telescopes will reveal the planet as a very thin crescent. Venus passes rapidly through inferior conjunction on 15 August when it is 7.4° south of the ecliptic and 43.2 million kilometres (26.8 million miles) from the Earth. A few days later, as it emerges from the glare of the rising Sun, the planet becomes observable again to Southern Hemisphere observers, low in the eastern sky shortly before sunrise. In the last week of August it should be visible to those living further north. By the end

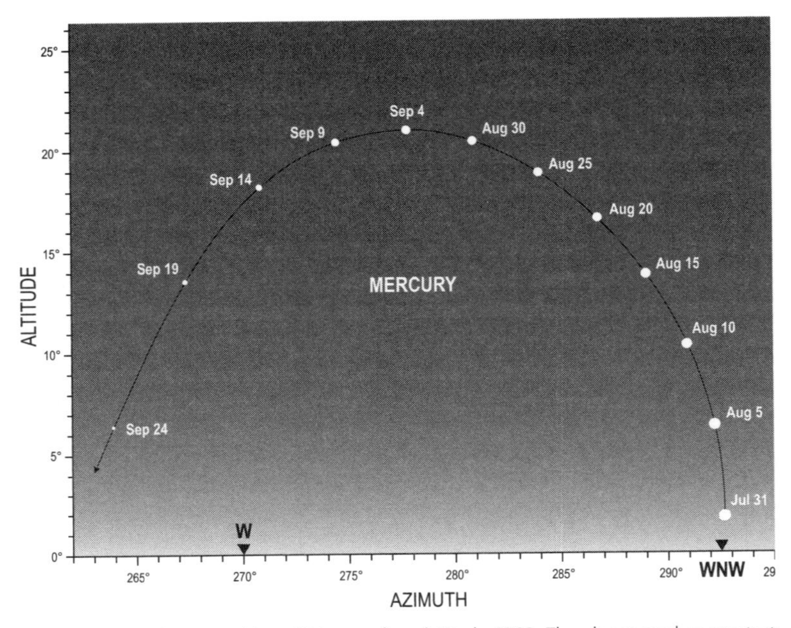

Figure 21. Evening apparition of Mercury from latitude 35°S. The planet reaches greatest eastern elongation on 4 September 2015. It was at its brightest in mid-August, several weeks before elongation.

of the month, Venus will be rising an hour and a half before the Sun for observers in the Southern Hemisphere, slightly less for those in the latitudes of the British Isles, magnitude –4.3.

MARS will be rising over two hours before the Sun by the end of the month from northern temperate latitudes, but only about half this time from the Southern Hemisphere. The planet is moving direct through Cancer and is not particularly prominent at magnitude +1.8, so it will be difficult to pick out in the dawn twilight sky. Figure 22 shows the path of Mars from August until the end of the year.

JUPITER, magnitude –1.7, is only likely to be glimpsed by observers in southern and equatorial latitudes for the first ten days of the month, low in the west-north-western sky shortly after sunset. As mentioned above, Jupiter will lie close to the elusive Mercury early on 7 August.

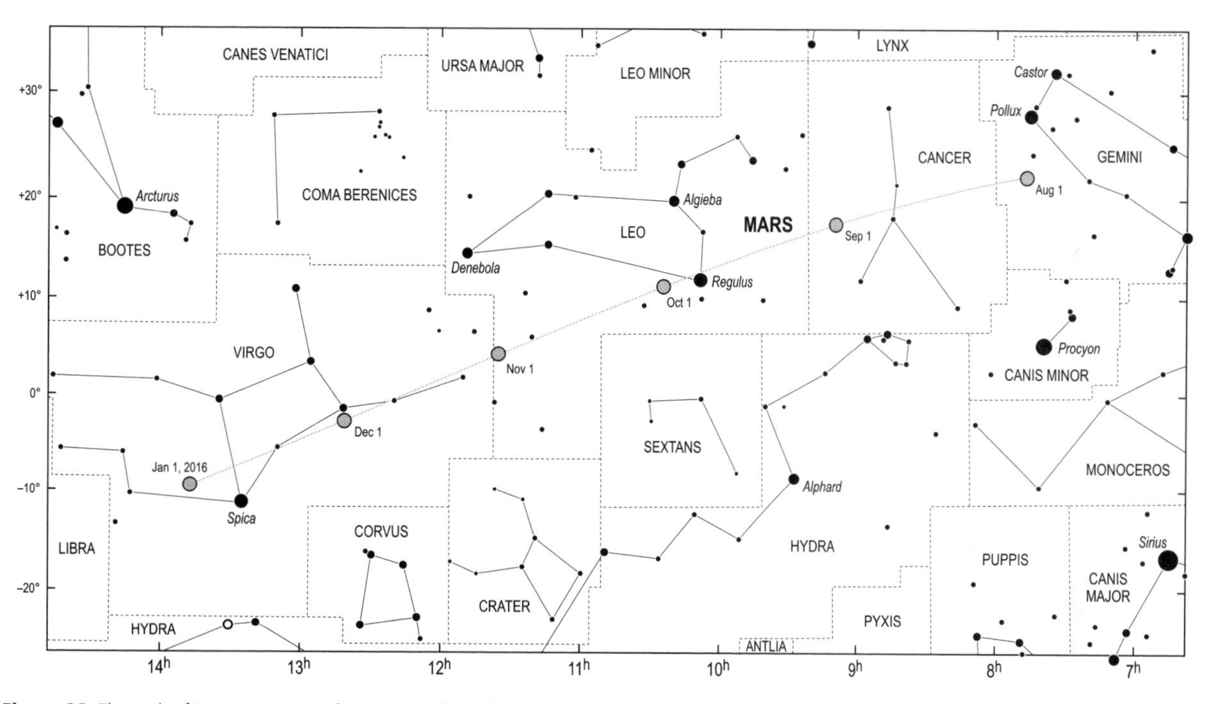

Figure 22. The path of Mars as it moves from Cancer through Leo and into Virgo between 1 August and the end of the year. The motion of the planet is direct (i.e. eastwards) throughout this period.

Thereafter, Jupiter passes through conjunction on 26 August and will not be visible until it emerges from the morning twilight next month.

SATURN reaches its second stationary point on 2 August and thereafter resumes its direct motion once more. The planet is in Libra and is visible as an evening object in the south-western sky throughout August, still setting about three hours after the Sun from northern temperate latitudes by month end; some two hours longer from further south. Its brightness decreases from magnitude +0.4 to +0.6 during the month. The waxing half-Moon will lie just north of Saturn on the evening of 22 August.

Chi Cygni. During August evenings, Cygnus, the Swan, is at its best, though part of it is always circumpolar from Britain. It is a large and imposing constellation, containing one star of the first magnitude (Deneb) and several more of the second and third; it is crossed by the Milky Way, and there are many rich starfields.

Cygnus is also rich in variable stars. Of these, one of the most interesting is Chi Cygni, which should be visible with the naked eye this month. Its variability was discovered by the German astronomer Gottfried Kirch – one of the earliest systematic observers of variable stars – as long ago as 1686. Chi Cygni is a rather rare spectral class S (S6) star, which means that visually it is deep red in colour. Lying about three hundred light years distant, it is a red giant star that has expanded to about three hundred times the diameter of our Sun, so if it were placed in the centre of the Solar System it would extend out to the orbit of Mars. Such bloated stars pulsate slowly, causing the variations in brightness, and there is a steady loss of mass from the star's extended atmosphere.

Chi Cygni belongs to the Mira class of long-period variables; the current period is around 408 days, though of course this is subject to significant fluctuations from one cycle to the next. It also seems likely that the star's average period has increased by about four days since its discovery more than three centuries ago. What makes Chi Cygni notable is its tremendous magnitude range – one of the greatest known. Occasionally, it can brighten to magnitude 4, but at minimum it can sink to below 14, so that it disappears altogether as viewed through small telescopes.

Chi Cygni is easy to locate, because it forms part of the pattern of

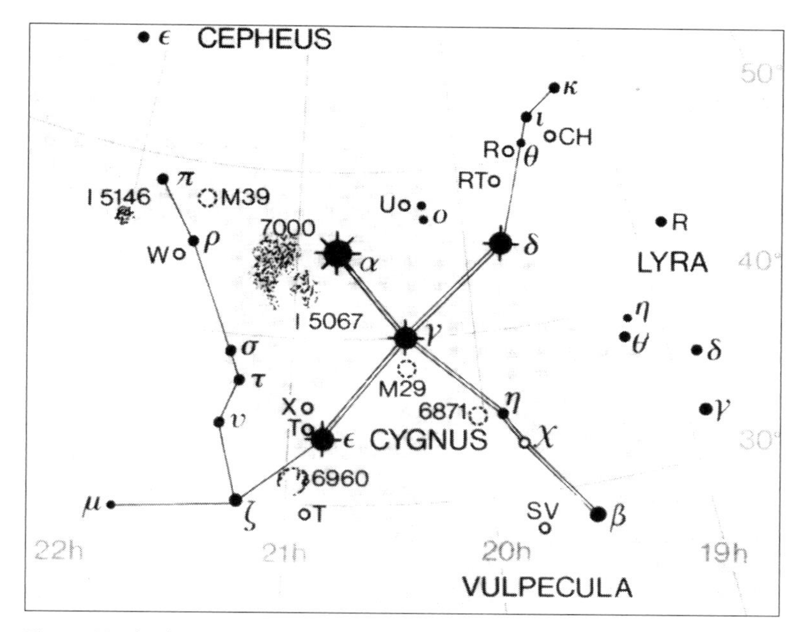

Figure 23. The deep red, long-period variable star Chi (χ) Cygni is easy to find because it forms part of the pattern of the 'Cross' of Cygnus (the Swan). It is in the neck of the Swan lying between Gamma (γ) and the double star Beta (β) or Albireo. Two other stars in the Swan's neck, Eta (η) Cygni (between Gamma and Beta) and Phi Cygni (between Chi and Beta), are handy comparison stars when Chi Cygni is bright.

the 'Cross of Cygnus' lying between the centre star (Gamma) and the lovely coloured double Beta or Albireo (Figure 23). Not far from Chi, and also in the neck of the Swan, are the stars Eta Cygni (magnitude 4.0, roughly midway between Gamma and Beta) and Phi Cygni (magnitude 4.8, between Chi and Beta), which make excellent comparisons when the variable is bright.

It is useful to note the maximum magnitude. On rare occasions the star has reached magnitude 3, but certainly most maxima are much below this, and there are some years in which the star remains a rather faint magnitude-5 object at maximum, just discernible with the naked eye. When observed with a telescope or with binoculars, when the star is near maximum, the deep red colour is very obvious indeed.

By the autumn, Chi Cygni will have dropped below naked-eye visibility, and by the beginning of March next year, it will be approaching minimum, so that it cannot then be identified except with a powerful telescope and an adequate set of detailed charts.

Sagittarius and the Lettering of Stars. During August evenings, from northern latitudes, the constellation of Sagittarius, the Archer, may be seen low in the southern sky (Figure 24). From London or New York it is never visible to advantage, but from parts of the southern United States, such as California and Arizona, it is very prominent. It has no distinctive shape, but it contains various bright stars; eight of Sagittarius's brightest stars form a figure widely recognized as a

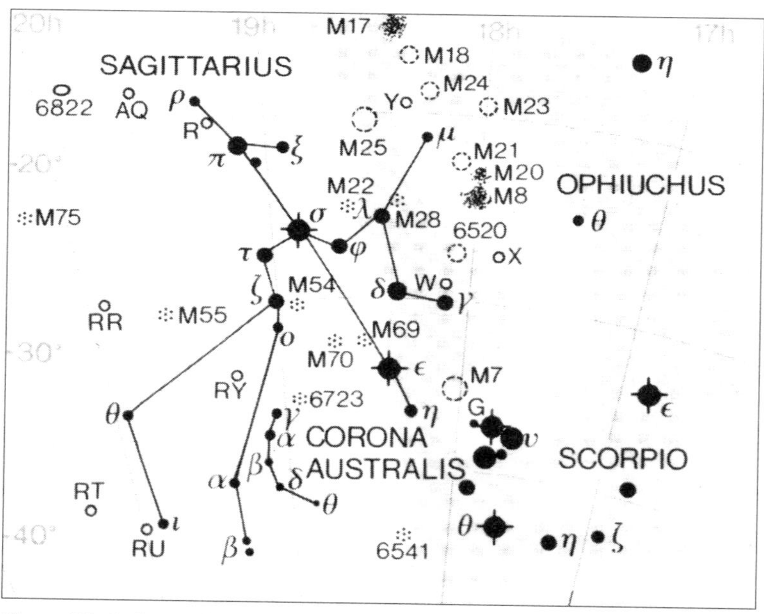

Figure 24. Sagittarius (the Archer) may be seen low down in the southern sky early on August evenings from northern temperate latitudes, although its most southerly stars remain below the horizon. With Sagittarius, the usual system of lettering the brightest star in a constellation Alpha, the second Beta, and so on down to Omega, has not been followed at all. The brightest stars are Epsilon, Sigma, Zeta, Delta, Lambda, Pi and Gamma, with both Alpha and Beta much fainter still.

'teapot', with its handle to the east and its spout to the west. The Milky Way is very rich-flowing through Sagittarius and, visually at least, the dense star clouds mask our view towards the centre of the Galaxy. The first-magnitude stars Deneb and Altair in the 'Summer Triangle' make almost a direct line to Sagittarius, with Altair in the middle position; this is probably the easiest way to find the constellation. Sagittarius is in the Zodiac, though it contains no planets at the present time; however, Pluto (now a dwarf planet) is moving slowly through the constellation.

The system of allotting Greek letters to stars was introduced by Johann Bayer in 1603. The principle is to letter the brightest star in a constellation Alpha, the second Beta, and so on down to Omega, the last letter in the Greek alphabet. Bayer's system is convenient, and is still in use. The Greek alphabet is as follows:

Alpha α	Epsilon ε	Iota ι	Nu ν	Rho ρ	Phi φ
Beta β	Zeta ζ	Kappa κ	Xi ξ	Sigma σ	Chi χ
Gamma γ	Eta η	Lambda λ	Omicron ο	Tau τ	Psi ψ
Delta δ	Theta θ	Mu μ	Pi π	Upsilon υ	Omega ω

Perhaps unfortunately, the strict brightness sequence is not always followed; thus in Orion, Beta (Rigel) is brighter than Alpha (Betelgeuse). With Sagittarius, the alphabetical order can only be described as chaotic! The brightest stars are Epsilon (magnitude 1.9), Sigma (2.0), Zeta (2.6), Delta (2.7), Lambda (2.8), Pi (2.9) and Gamma (3.0), with both Alpha and Beta considerably fainter. Incidentally, both Alpha and Beta (which is a wide naked-eye double star) are too far south to rise in Britain or the northern United States.

September

New Moon: 13 September *Full Moon:* 28 September

Equinox: 23 September

MERCURY reaches greatest eastern elongation (27°) on 4 September and passes through inferior conjunction on 30 September, but remains unsuitably placed for observation by those in the latitudes of the British Isles. Further south the planet continues to be visible as an evening object in the western sky for the first two to three weeks of the month, and reference should be made to Figure 21, given with the notes for August. Mercury's brightness fades from magnitude +0.2 on 1 September to +1.7 by the 21st.

VENUS is drawing rapidly out from the glare of the Sun and becomes a magnificent object completely dominating the eastern sky before dawn. It attains its greatest brilliancy on 20 September with a magnitude of –4.5. By the end of September, Venus will be rising nearly four hours before the Sun from the latitudes of northern Europe and North America, but only about half this time from the tropics and the Southern Hemisphere. The phase of Venus increases from 10 per cent to 34 per cent during the month. At the beginning of September, Venus lies about 8.5° south of Mars and may serve as a guide to locating the considerably fainter planet. Indeed, by the end of the month, Venus, Mars and Jupiter will be strung out in a line across the eastern sky before dawn (Venus the highest and Jupiter the lowest), with the first-magnitude star Regulus in Leo also visible just to the west of Mars. The waning crescent Moon will make a beautiful pairing with Venus on the morning of 10 September.

MARS is an early morning object, and is still rather faint and inconspicuous (magnitude +1.8), but by the end of the month it may be seen in a dark sky rising more than three hours before the Sun from northern temperate latitudes, although its period of visibility is much reduced from locations further south. During September, Mars moves from Cancer into neighbouring Leo, passing just 0.7° north of the star Regulus (Alpha Leonis, magnitude +1.3) on 24 September.

As mentioned earlier, Mars will be part of a line of three planets in the eastern early morning sky at the end of September, with the much brighter Venus above and Jupiter lower down. The waning crescent Moon will join the line of planets on the mornings of 10 and 11 September, being equidistant from Mars and Regulus on the 11th.

JUPITER, magnitude –1.7, emerges from the morning twilight in mid-September for observers in northern temperate latitudes and the tropics, becoming observable above the eastern horizon shortly before dawn. However, it does not become visible to observers in the Southern Hemisphere until the end of the month. Jupiter will be the lowest object in a line of three planets strung out in the early morning twilight sky at the end of September.

SATURN, magnitude +0.6, is visible in the south-western sky as soon as darkness falls. From northern temperate latitudes, Saturn will be getting rather low in the twilight at dusk by month end, although rather easier for observers further south to see. The planet continues to move eastwards in Libra. The waxing crescent Moon will be slightly west of Saturn on the evening of 18 September.

NEPTUNE is at opposition on 1 September, in the constellation of Aquarius. It is not visible with the naked eye since its magnitude is +7.8. At opposition Neptune is 4,331 million kilometres (2,691 million miles) from the Earth. Figure 25 shows the path of Neptune against the background stars during the year.

Synodic Periods of the Planets. The planet Neptune is at opposition this month; Uranus is at opposition next month. During 2015 there are oppositions of Jupiter, Saturn, Uranus, Neptune and Pluto – as indeed there are in almost every year. With Mars, the situation is different. Mars was last at opposition on 8 April 2014 and is next at opposition on 22 May 2016; there is no opposition in 2015. This is because the synodic period of Mars – 780 days – is much longer than that of any other planet.

The synodic period is the mean interval between successive

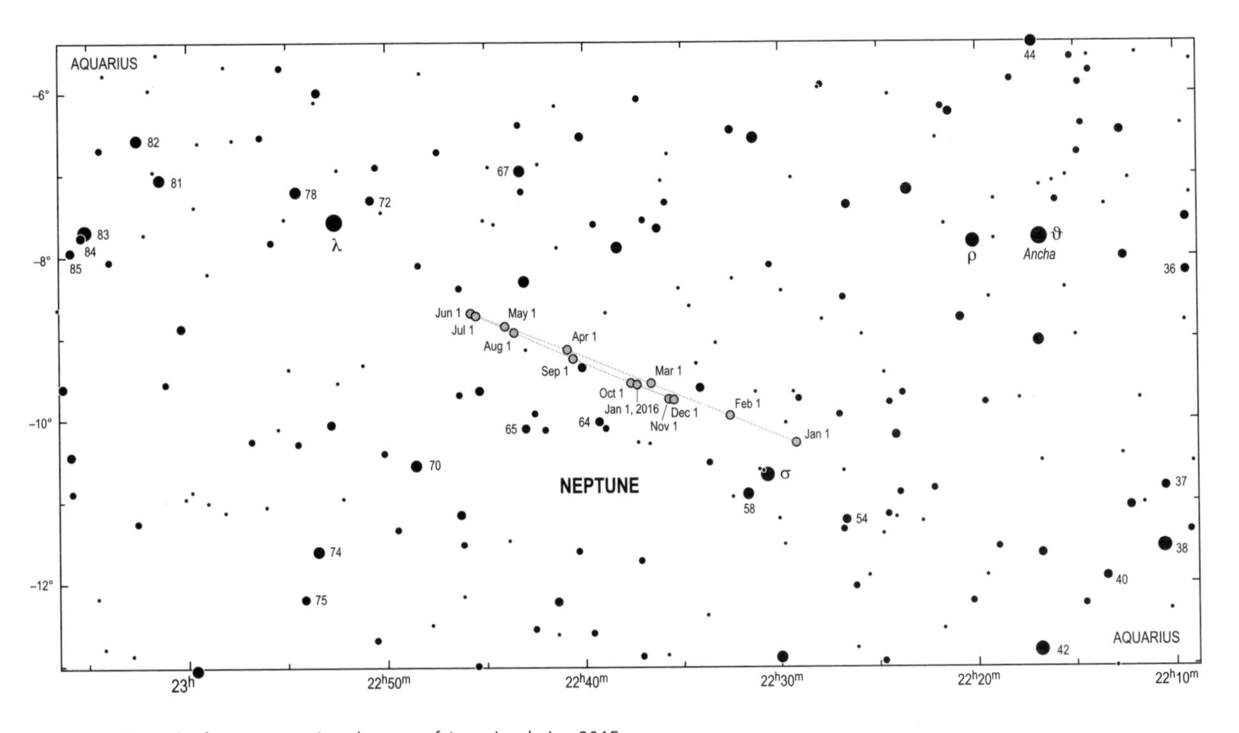

Figure 25. The path of Neptune against the stars of Aquarius during 2015.

oppositions of a planet. To understand what is meant, let us first consider Pluto, which moves round the Sun at a mean distance of more than 5,900 million kilometres, though its orbit is unusually eccentric, and at its closest it can come within the orbit of Neptune. As well as having a large orbit, Pluto is slow-moving. In one year, the time taken for the Earth to go once round the Sun, Pluto covers only a tiny fraction of its orbit – so that the Sun, Earth and Pluto are lined up every 366.73 days. Having been right round the Sun in 365.25 days, the Earth takes only an extra day and a half to catch up with Pluto.

With the closer planets, more time is required; the mean synodic periods for Neptune, Uranus, Saturn and Jupiter are 367.49, 369.66, 378.09 and 398.88 days, respectively – so that, for instance, oppositions of Jupiter occur approximately 33 days later in each year; the opposition of 6 February 2015 follows that of 5 January 2014, and there will also be oppositions in March 2016, April 2017, May 2018 and so on.

Mars is a special case. Its orbital velocity is comparable with that of the Earth, and the mean synodic period is therefore a great deal longer at 780 days. Consequently, oppositions of Mars do not occur every year; there were oppositions in March 2012 and April 2014, and there is one in May 2016, but there are none in 2015 – and, following next year's opposition, 2017 will be another 'blank year' so far as Mars is concerned.

Migrating Stars. During northern summer evenings, and well into early autumn, the evening sky is dominated by three bright stars – Vega in Lyra (the Lyre or Harp), Altair in Aquila (the Eagle), and Deneb in Cygnus (the Swan) – which form a large triangle, an asterism popularly nicknamed the 'Summer Triangle', although this label is inappropriate in the Southern Hemisphere, where it really should be renamed the 'Winter Triangle'.

Altair (ranked twelfth brightest star in the sky) is flanked on either side by two fainter stars, Gamma Aquilae (Tarazed) to the north-west, which is orange, and the white Beta Aquilae (Alshain) to the south-east of Altair. Overall, Aquila (Figure 26) is a well-formed constellation, yet a strange thing happened to it in December 1992. Aquila lost one of its stars, Rho Aquilae, which migrated eastwards across the border into the neighbouring constellation, Delphinus (the Dolphin).

Stellar proper motions are very slight, and incidences of stars changing constellations are rare. Moreover, the constellations themselves are

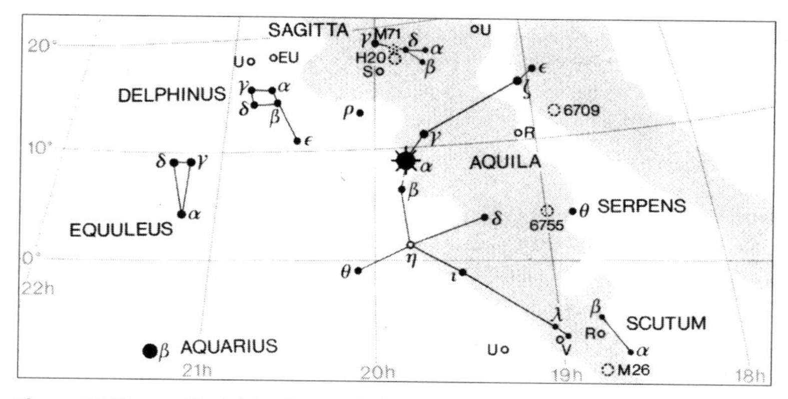

Figure 26. The star Rho (ρ) Aquilae may be found above the centre of this star map, roughly midway between Altair (α) and the small but distinctive pattern of Delphinus (the Dolphin). Rho Aquilae is an example of a migrating star, having crossed the border from Aquila into neighbouring Delphinus in 1992, on account of its very slight eastwards proper motion.

quite arbitrary, since the stars in them are not associated with each other. Our current system is largely based on that of the Greeks, though it has been drastically modified, and the present constellation boundaries have been laid down by the International Astronomical Union. Rho Aquilae had been right on the border between the Eagle and the Dolphin; but in 1992 it 'crossed the floor'.

It is interesting to look at other cases of naked-eye stars which will also change constellations in the foreseeable future:

Year	Star		Constellation
2400	Gamma Caeli		Columba
2640	Epsilon Indi		Tucana
2920	Epsilon Sculptoris		Fornax
3200	Lambda Hydri	*will*	Tucana
4500	Mu[1] Cygni	*move*	Pegasus
5200	Chi Pegasi	*into*	Pisces
5200	Mu Cassiopeiae		Perseus
6300	Eta Sagittarii		Corona Australis
6400	Zeta Doradûs		Pictor

Careful examination of any star map will reveal that all of the above stars, bar one, appear right on the border, or very close to the border, of the neighbouring constellation into which they will eventually migrate. The exception is Mu Cassiopeiae, a binary system, which today appears some way from the border with neighbouring Perseus. It turns out that compared with other nearby stars, this pair of stars is travelling at the relatively high speed of 167 kilometres per second through the Galaxy and consequently they are moving quite rapidly towards the border with Perseus, crossing over in about AD 5200.

Obviously, these stellar migrations are of academic interest only – and who can tell whether the constellations in use in the year 6400 will be the same as those which we use today?

October

New Moon: 13 October   *Full Moon:* 27 October

Summer Time in the United Kingdom ends on 25 October.

MERCURY reaches greatest western elongation (18°) on 16 October and is visible as an early morning object in the east-south-eastern sky from about 8 October until the last few days of the month. For observers in northern temperate latitudes this will be the most favourable morning apparition of the year. Figure 27 shows, for observers in latitude 52°N, the changes in azimuth and altitude of Mercury on successive mornings when the Sun is 6° below the horizon. This condition is known as the beginning of morning civil twilight and in this latitude and at this time of year occurs about thirty-five minutes before sunrise. The changes in the brightness of the planet are indicated by the relative sizes of the circles marking Mercury's position at five-day intervals. During its period of visibility, Mercury brightens rapidly from magnitude +1.3 to –1.0, thus Mercury is at its brightest after it reaches greatest western elongation. Although the planet will be visible from the tropics around the time of greatest elongation, it will be unsuitably placed for observation from the Southern Hemisphere.

VENUS reaches greatest western elongation (46°) on 26 October and continues to be visible as a magnificent morning object, visible in the eastern sky for several hours before sunrise. From northern temperate latitudes the planet rises well over four hours before the Sun at the end of October, but is visible for only about half this time from the Southern Hemisphere. Its brightness decreases slightly from magnitude –4.5 to –4.3 and the phase increases from 35 per cent to 53 per cent during October. Venus passes 2.5° south of the first-magnitude star Regulus in Leo on 9 October and that same morning Venus, the waning crescent Moon, Mars and Jupiter will all lie close together in the pre-dawn sky – with the much fainter Mercury possibly glimpsed low down shortly before dawn. Venus passes just a degree (two Moon

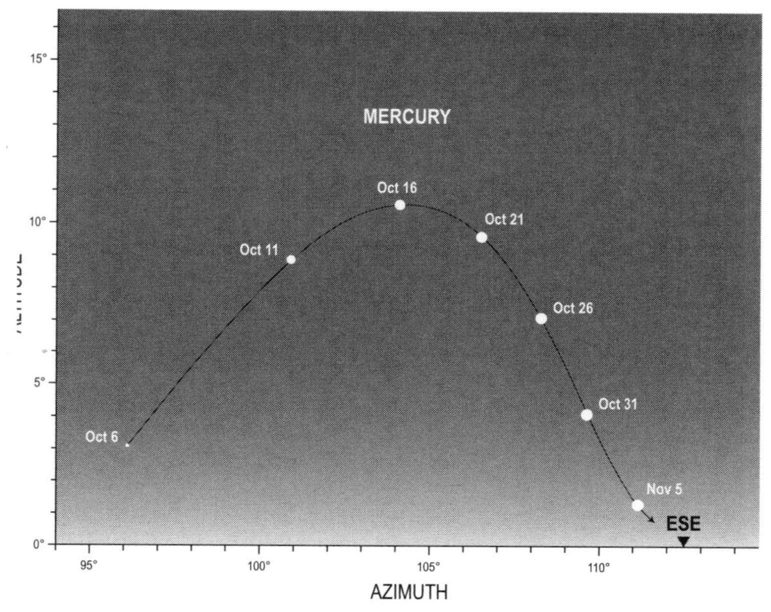

Figure 27. Morning apparition of Mercury from latitude 52°N. The planet reaches greatest western elongation on 16 October 2015. It will be at its brightest in early November, some weeks after elongation.

diameters) south of Jupiter early on 26 October, and the two brilliant planets will be a magnificent sight in the pre-dawn sky during the month. With the much fainter Mars also close by, there will be many lovely picture opportunities.

MARS is an early morning object in Leo, rising over four hours before the Sun from the latitudes of the British Isles by month end, but less well placed for observers further south. Its magnitude is only +1.8, but the planet's proximity to the much brighter Venus and Jupiter will make it easy to locate. Mars will lie only about half a degree (one Moon diameter) north of Jupiter on the morning of 17 October and so may be fairly easily located for several days before and after this date.

JUPITER, magnitude –1.8, continues to be visible as a brilliant morning object in the eastern sky. Jupiter is in the constellation of Leo and its proximity to the even brighter Venus during the month will make for a lovely spectacle in the early mornings. As mentioned above, Jupiter passes less than half a degree south of Mars on 17 October. Jupiter's four Galilean satellites are readily observable with a small telescope or even a good pair of binoculars, provided that they are held rigidly.

SATURN, magnitude +0.6, is visible in the west-south-western sky as soon as darkness falls. From northern temperate latitudes, Saturn will be inconveniently low in the twilight at dusk by month end, although will be somewhat easier to observe for those further south. The planet's eastwards motion carries it from Libra into neighbouring Scorpius during the month.

URANUS is at opposition on 12 October in the constellation of Pisces. Uranus is barely visible to the naked eye as its magnitude is +5.7, but it is easily located in binoculars. Figure 28 shows the path of Uranus against the background stars during the year. At opposition Uranus is 2,840 million kilometres (1,765 million miles) from the Earth.

The 'Hot Poles' of Mercury. No Earth-based telescope will show much detail on the surface of Mercury. The planet is always inconveniently close to the Sun in the sky; it is small, and a long way away. Therefore, most of our detailed knowledge of it comes from the MESSENGER spacecraft which flew past the planet three times in 2008–9, went into orbit around it in March 2011, and had completed 3,000 orbits of Mercury by April 2014, sending back over 240,000 images of the planet while in orbit – and its work continued daily as this Yearbook went to press.

Up to the 1960s it was thought that Mercury had a captured or synchronous rotation, in which case the 'day' would have been equal to the 'year' of eighty-eight Earth-days, and Mercury would have kept the same face turned sunward all the time (just as the Moon always keeps the same face turned towards the Earth). In fact, this is not so. Mercury's axial rotation period is 58.6 days, or two-thirds of a Mercurian year. Because the eccentricity of Mercury's orbit is the highest of any of the major planets, its distance from the Sun at perihelion is 45.9 million kilometres and at aphelion as much as 69.7 million kilometres, so the apparent diameter of the Sun in the sky changes, and so does its apparent motion in the Mercurian sky.

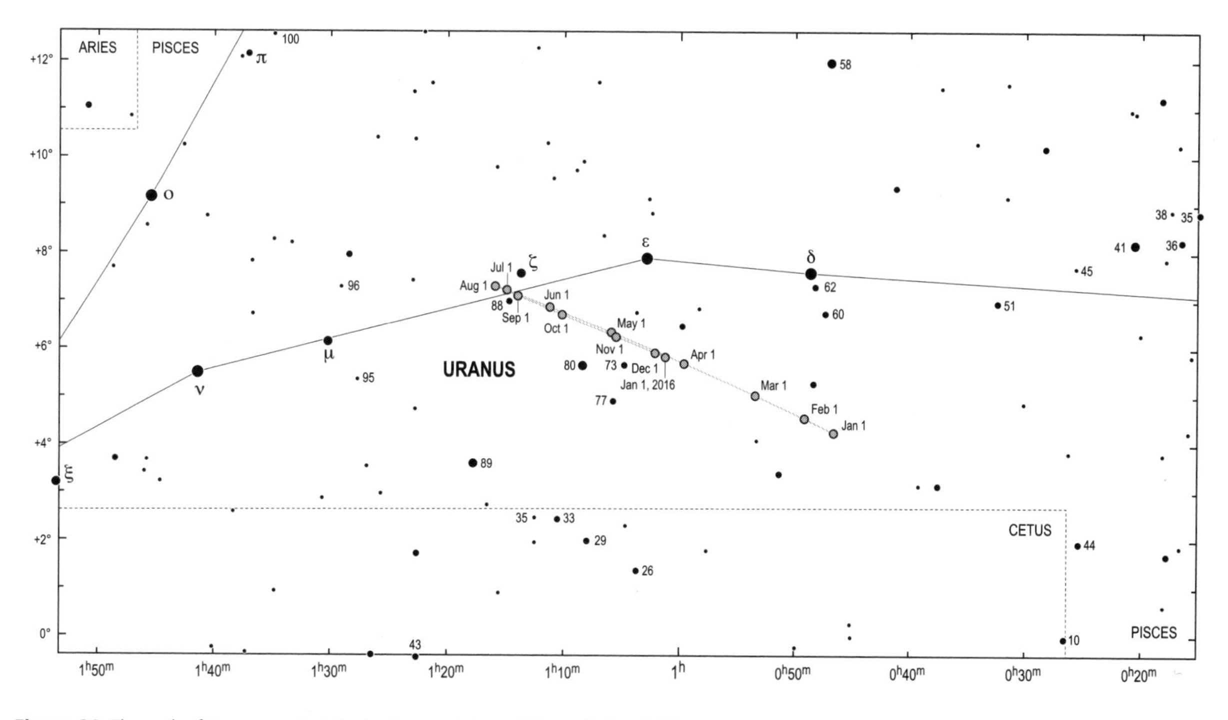

Figure 28. The path of Uranus against the background stars of Pisces during 2015.

Mercury has two so-called 'hot poles', where the Sun is overhead at the time of perihelion, and the surface temperature reaches around 430 °C. One of these hot poles is at zero degrees latitude and 180° longitude on Mercury's surface. The other is at 0° latitude and 0° longitude. Mercury's eccentric orbit and spin-orbit coupling combine so that one or the other of these 'hot poles' always points towards the Sun when Mercury is at perihelion, and the Sun is overhead longer at these points

Figure 29. A portion of a global mosaic basemap acquired by the Mercury Dual Imaging System (MDIS) during MESSENGER's first year in orbit. The large crater at centre right (Atget crater) is 100 kilometres across. Caloris Basin, one of two so-called 'hot poles' and the largest young impact basin on Mercury, dominates the scene. With an east-west diameter of 1,550 kilometres, Caloris hosts a wide variety of tectonic features and shows widespread evidence of both extension and compression, an uncommon tectonic combination on Mercury. Since different processes produce extensional and contractional land forms, the Caloris Basin has clearly had a complex and detailed geological history. Understanding how these structures developed will yield an insight into tectonism in Mercury's largest basin, and large impact craters in general. (Image courtesy of NASA/Johns Hopkins University Applied Physics Laboratory/Carnegie Institution of Washington.)

than anywhere else on the planet. Consequently, the 'hot poles' receive about two and a half times more solar radiation than locations near longitudes 90° and 270°, and so get much hotter. Since Mercury's rotation axis has essentially no tilt relative to its orbital plane, Mercury does not have true seasons. But, depending on their location, portions of the surface will experience hotter periods.

One of the hot poles is marked by the great Caloris Basin, which is 1,550 kilometres across and is bounded by rings of mountains up to two kilometres high (Figure 29, opposite). At sunrise over Caloris, the Sun would be at its smallest; as it climbed towards the zenith it would swell, but before reaching the zenith it would 'backtrack' for a period before resuming its normal motion. As it dropped towards the horizon, it would shrink, finally setting eighty-eight Earth-days after having risen. The interval between one Mercurian sunrise and the next is 176 Earth-days or two Mercurian years.

To an observer 90° in longitude away from Caloris, the Sun would be largest when rising and setting, and there would be a 'false sunrise' and a 'false sunset' as the Sun bobbed above and below the horizon. The reason for this strange behaviour is that when Mercury is near perihelion its orbital speed is very great, whereas the rate of axial spin remains constant. Because Mercury is the closest planet to the Sun it might be expected to be the hottest, but this is not so; the surface temperature of Venus is appreciably higher – because Venus has a dense, carbon dioxide-rich atmosphere which traps the Sun's heat, causing a runaway greenhouse effect, whereas the atmosphere of Mercury is negligible. We must admit that neither of these inner planets is suited for human exploration, at least in the foreseeable future.

Faint Alphas. Below the Square of Pegasus lies Pisces, the Fishes – officially the last constellation of the Zodiac, though since it now contains the vernal equinox it really ought to be the first. Pisces (Figure 30), which is well placed late on October evenings, is entirely unremarkable, and is one of the more obscure of the Zodiacal groups. It is, however, ancient, and there is even a mythological legend attached to it. It is said that Venus and her son Cupid once escaped from the fire-breathing, hissing monster Typhon (who could live in flames and fire but not in water) by throwing themselves into the River Euphrates and changing themselves temporarily into fishes. Apparently, they tied themselves together with a cord so as not to lose each other in the dark waters of

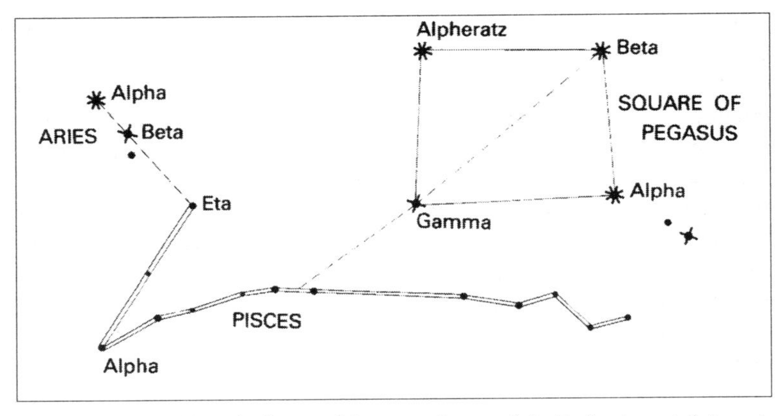

Figure 30. Pisces (the Fishes), one of the more obscure of the Zodiacal constellations, is well placed during October evenings and may be easily found using the stars of the Great Square of Pegasus. Eta (η) Piscium is the brightest star in the constellation, followed by Gamma (γ), with Alpha only the third brightest.

the river. Subsequently, two fishes tied together by their tails were placed in the sky in memory of this rather curious episode!

There are few notable objects in Pisces, though Alpha, sometimes known by its old proper name of Al Rischa, is a fairly easy double, with two components of magnitudes 4.2 and 5.1, separation 1.8 seconds of arc, position angle 262 degrees. Actually, it is not the brightest star in the constellation; this distinction goes to Eta, or Alpherg, which is slightly more conspicuous, even though it is still well below the third magnitude.

When Johann Bayer allotted Greek letters to the stars in each constellation, in his star maps of 1603, he usually followed the system of making the brightest star Alpha, the second brightest Beta, the third brightest Gamma, and so on. There are, however, exceptions to the rule – Pisces being one. The most glaring examples are Corvus and Sagittarius. In Corvus there are four stars of magnitude 3 or above, making up a conspicuous little quadrilateral; these are Gamma, Beta, Delta and Epsilon, while Alpha Corvi is a full magnitude fainter. The other faint Alpha is in Sagittarius, where the two brightest stars are Epsilon (magnitude 1.9) and Sigma (2.0), while Alpha is a modest 4.0.

Of course, in many cases where Alpha is not the leader of a constellation, the difference is so slight that naked-eye estimates might lead to an error; thus in Ursa Major Epsilon (Alioth) is 1.77 and Alpha (Dubhe) fractionally fainter at 1.79 – although in this case it seems that Bayer lettered the principal stars in order of right ascension rather than brilliancy. There are also cases of revisions of constellation boundaries. In Puppis the brightest star is Zeta; but Puppis is part of the dismembered constellation of Argo Navis, and Alpha, Beta, Gamma, Delta and Epsilon of Argo are now in different parts. (Alpha is, of course, Canopus.)

The following are cases of 'faint Alphas', where the star lettered Alpha is at least 0.4 magnitude fainter than the brightest star in the constellation.

Constellation	Brightest Star	Magnitude of Alpha
Cancer	Beta, 3.5	4.3
Capricornus	Delta, 2.9	3.5 (a double star)
Cetus	Beta, 2.0	2.5
Corvus	Gamma, 2.6	4.0
Draco	Gamma, 2.2	3.6
Gemini	Beta (Pollux), 1.2	1.6 (Castor)
Sagitta	Gamma, 3.5	4.4
Sagittarius	Epsilon, 1.8	4.0

The Alphas of Andromeda, Aquarius, Camelopardalis, Delphinus, Hydrus, Libra, Microscopium, Monoceros, Pegasus, Pisces and Ursa Major are slightly fainter than the leader of each constellation. We also have two variable Alphas, in Hercules and Orion; Alpha Herculis (Rasalgethi) is a semi-regular red giant ranging between magnitudes 3 and 4, whereas the two leaders, Beta and Zeta, are both ranked as 2.8; and, of course, Alpha Orionis (Betelgeuse) is not quite equal to Rigel, though when it is at maximum, the two are more or less comparable. (Herschel, on one occasion, even made Betelgeuse the brighter, though this was a naked-eye estimate and cannot be regarded as precise.) Finally, there are suspicions that Castor and Pollux are less equal than they used to be in ancient times, though the evidence is very uncertain.

November

New Moon: **11 November** *Full Moon:* **25 November**

MERCURY passes through superior conjunction on 17 November and thus is unsuitably placed for observation throughout the month.

VENUS, magnitude –4.3, continues to be visible as a magnificent object in Virgo, completely dominating the eastern sky for several hours before dawn. The planet will still be rising over four hours before the Sun from northern temperate latitudes at the end of November, but is visible for only about half of this time from the Southern Hemisphere. The phase of Venus increases from 54 to 67 per cent during the month. Venus passes just 0.7° south of Mars on 3 November and will be a useful guide to locating the much fainter planet. The waning crescent Moon will be close to both Venus and Mars on the morning of 7 November, with Jupiter lying about seven degrees further west. On 28 November, Venus passes four and a half degrees north of the first-magnitude star Spica in Virgo.

MARS is an early morning object, brightening slightly from magnitude +1.7 to +1.5 during November. From northern temperate latitudes the planet rises some five and a half hours before the Sun at the end of November, but the period of visibility is considerably less from locations further south. The planet's eastwards motion carries it from Leo into neighbouring Virgo early in the month. The planet will be at aphelion (greatest distance from the Sun) on 20 November.

JUPITER continues to be visible in the eastern sky as a brilliant morning object in the constellation of Leo. Its period of visibility is gradually increasing, but it is still rising shortly after midnight at the end of the month from northern temperate latitudes and slightly later from more southerly locations. The brightness of Jupiter increases slightly from magnitude –1.8 to –2.0 during November. The waning crescent Moon will make a nice pairing with Jupiter on the morning of 6 November.

SATURN is in conjunction with the Sun on 30 November and consequently will not be visible at all this month for observers in the latitudes of the British Isles. For those in the tropics and the Southern Hemisphere, the planet may be glimpsed low in the west-southwestern sky at dusk for the first few days of the month. The planet is in Scorpius, magnitude +0.5.

Finding the South Pole Star. Northern navigators have always been glad to have a reasonably bright pole star – Polaris in Ursa Minor (the Little Bear), which is of the second magnitude and is very easy to find. Things are less easy in the south, where Polaris remains below the horizon; there is no bright South Pole star, and the nearest which is reasonably prominent (Beta Hydri, magnitude 2.8) is well over twelve degrees from the polar point. So we have to make do with the obscure Sigma Octantis, magnitude 5.4 – which is none too easy to identify even with a clear sky, while the slightest mist or haze will hide it.

Octans, the Octant, is a very barren constellation. The brightest star in it, Nu Octantis, is only of magnitude 3.8. Neither is there any definite pattern. The pole lies about midway between the Southern Cross and the brilliant Achernar in Eridanus, and a slightly curved line extended from the long arm of the Southern Cross (from Gamma Crucis through Alpha Crucis) to Achernar will pass fairly close to the position of the pole, but to identify Sigma Octantis something more detailed is required.

One method is to start with Alpha Centauri, the brighter of the two Pointers to the Southern Cross. Quite close to it is Alpha Circini, magnitude 3.2. These show the way to the little constellation of Apus; Alpha Apodis is of magnitude 3.8. In the same binocular field as Alpha Apodis are two dim stars, Epsilon Apodis (5.2) and Eta Apodis (5.0); these point straight to the orange Delta Octantis (4.3).

Delta can also be identified because of the two faint stars close beside it, Pi^1 and Pi^2 Octantis. Now put Delta Octantis in the edge of the field, and continue the line from Apus. Chi Octantis (5.2) will be on the far side of the field; centre it, and then you will see three more stars of about the same brightness, Sigma, Tau and Upsilon. Together with Chi, these three stars make up a distinctive trapezium or small 'half hexagon' shape, in the same field with, say, 7×50 binoculars. Sigma is at one corner of the trapezium, nearest to Chi.

This may sound rather complicated, but it is actually easier to do than it is to describe! Moreover, Sigma is almost one degree from the polar point. It is an F-class subgiant star, lying at a distance of 270 light years.

The Crazy Tilt of Uranus. Uranus was at opposition last month, and it is well placed in Pisces (magnitude +5.7, on the fringe of naked-eye visibility), but surface details are by no means easy to make out and, even with large instruments, the pale greenish disk has a decidedly bland appearance.

For reasons which are still rather unclear, the rotation axis of Uranus is tilted to its orbital plane by 98 degrees – more than a right angle. So, for a planet where the axial tilt is so extreme, how do you decide which is the north pole, and which is the south? The International Astronomical Union (IAU), the controlling body of world astronomy, is quite emphatic about this. All poles above the ecliptic (i.e. the plane of Earth's orbit around the Sun) are north poles, while all poles below the ecliptic are south poles. By this definition, it was Uranus's south pole which was in sunlight during the Voyager 2 spacecraft flyby in January 1986.

The extreme axial tilt also gives the planet a weird calendar. Uranus takes eighty-four years to complete a solar orbit, so each of its four seasons lasts twenty-one years, and near the time of the solstices, one pole is turned continuously towards the Sun while the other pole is turned away. At such times, near the equator on the side nearest to the sunward pole, the Sun will be very low down on the horizon. Just over the other side of the equator it will be dark. Forty-two years later, on the opposite side of Uranus's orbit, the orientation of the poles towards and away from the Sun will be reversed.

Near the time of the equinoxes, the Sun shines down onto the equatorial regions of Uranus, producing more evenly distributed sunlight and giving a period of day-and-night cycles similar to those seen on other planets. Uranus reached its most recent equinox in December 2007. Consequently, spring (in the northern hemisphere of Uranus) and autumn (in the southern hemisphere) started in December 2007. For the next 42 years – until 2049 – the north pole will be sunny all of the time, while the south pole will be in darkness.

When the Voyager 2 spacecraft flew by in 1986, it was the time of the southern winter solstice, and Uranus appeared virtually featureless. The northern hemisphere of Uranus is now just coming out of the grip of its decades-long winter. As the sunlight reaches some latitudes for the first time in many decades, it will warm the atmosphere and trigger gigantic springtime storms comparable in size to North America. The last time this happened there were no instruments that could resolve any features on the distant planet, but now with some of the very large ground-based telescopes that are available and, of course, the Hubble Space Telescope, it is possible to see what is happening (Figure 31).

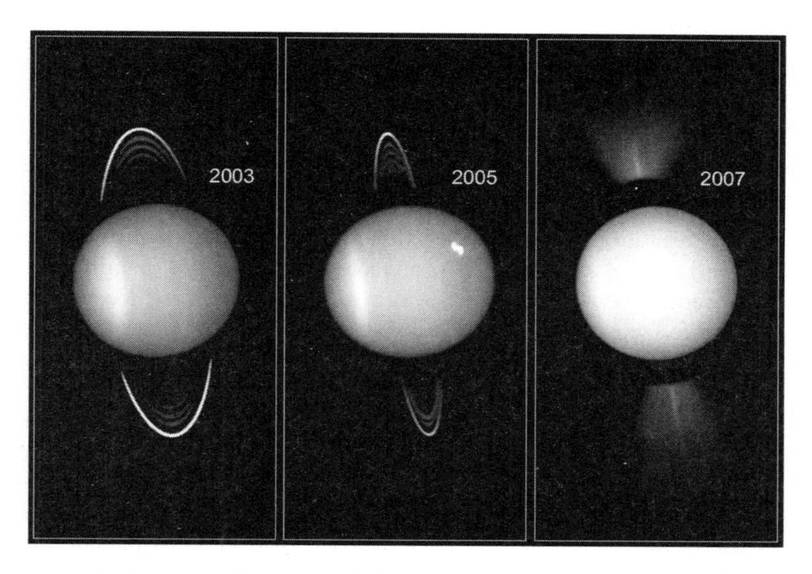

Figure 31. This series of images from NASA's Hubble Space Telescope shows how the ring system around the distant planet Uranus appears at ever more oblique (shallower) tilts as viewed from Earth – culminating in the rings being seen edge-on in three observing opportunities in 2007. The best of these events appears in the far-right image taken with Hubble's Wide Field Planetary Camera 2 on 14 August 2007. Some atmospheric cloud features are visible in the images acquired in 2003 and 2005. (Images courtesy of NASA, ESA and M. Showalter (SETI Institute).)

December

New Moon: 11 December *Full Moon:* 25 December

Solstice: 22 December

MERCURY reaches greatest eastern elongation (20°) on 29 December. For observers in tropical and southern latitudes it is visible as an evening object low in the west-south-western sky at the end of evening civil twilight, about twenty-five minutes after sunset, for all but the first few days of the month. Observers in the latitudes of the British Isles will find observation much more difficult and it is only during the last few evenings of the month that they stand any chance of locating the planet, low in the south-western sky about thirty-five minutes after sunset. During the month, the magnitude of Mercury fades slowly from –0.8 to –0.3, so it will be at its brightest before it reaches greatest eastern elongation.

VENUS is a beautiful object in the early mornings, magnitude –4.2. At the end of the year it is still rising above the east-south-eastern horizon over three hours before the Sun, for observers in the latitudes of the British Isles; about half an hour less for those in temperate southern latitudes. The planet's rapid eastwards motion carries it from Virgo into Libra in December. The phase of Venus increases from 67 per cent to 77 per cent during the month. The waning crescent Moon will make a nice pairing with Venus on the mornings of 7 and 8 December.

MARS is in Virgo throughout December, and from northern temperate latitudes it rises soon after 02h all month long on account of the fact that it is moving both eastwards and further south as the month goes on. Consequently, for observers in the tropics and the Southern Hemisphere, the planet starts the month rising soon after 02h, but rises steadily earlier as the month progresses, coming up soon after midnight by the end of the year. Mars brightens slightly from magnitude +1.5 to +1.3 during December. Mars passes about 1.5° south of the star Porrima

(Gamma Virginis) on 1 December, and the waning crescent Moon will lie close to Mars early on 6 December. Indeed, on the mornings of 4–8 December, the Moon will move down the line of three planets – Jupiter, Mars and Venus – which are strung out in a line across eastern Leo and Virgo. There was no opposition of Mars in 2015; it will next come to opposition on 22 May 2016 when the planet will be in Scorpius.

JUPITER, magnitude −2.1, is a brilliant morning object moving direct in Leo, rising in the east before midnight by the end of the month. From northern temperate latitudes, the planet will be found in the south-eastern sky during the early morning hours; from the Southern Hemisphere it will be towards the north-east. The waning crescent Moon will lie close to Jupiter on the early morning of 4 December.

SATURN was in conjunction with the Sun at the end of November and becomes visible low in the south-eastern sky shortly before dawn at the end of December. The planet is in Ophiuchus, magnitude +0.5.

Sir John Herschel and Gamma Virginis. In early December, Mars passes just south of the star Gamma Virginis, one of the most interesting binary stars in the sky. This star has several proper names: Arich, Porrima, and Postvarta. Since it lies only one and a half degrees south of the celestial equator it can be seen from every inhabited country, although in December it will only be on view in the pre-dawn hours. Virgo is best seen during evenings in May.

Gamma Virginis is one of our nearer neighbours, at a distance of thirty-eight light years. There are two virtually identical components, each of spectral type FO and each about four times as luminous as the Sun. Many observers call them yellowish; but others will see no colour in them. The revolution period is 168.9 years, and the orbit is decidedly eccentric, so that, as seen from Earth, the angular separation changes considerably; at its greatest it is 6.2 seconds of arc, as in 1920 and again 2089, while at its least the components are so close together that in any ordinary telescope they appear as one star.

Gamma Virginis was the second binary system to have its orbit worked out (the first was Xi Ursae Majoris, whose orbit was computed in 1828 by Felix Savary). With Gamma Virginis, the calculations were undertaken by Sir John Herschel. In 1832, he published the following statement with regard to his measurements:

> If they be correct, the latter end of the year 1833, or the beginning of the year 1834, will witness one of the most striking phenomena which sidereal astronomy has yet afforded, viz. the perihelion passage of one star round another, with the immense angular velocity of between 60 and 70 degrees per annum, that is to say, of a degree in 5 days. As the two stars will then, however, be within little more than half a second of each other, and as they are both large, and nearly equal, none but the very finest telescopes will have any chance of showing this magnificent phenomenon.

The closest approach actually occurred in 1836. Three observers were following it: Admiral W. H. Smyth at Bedford in England, F. G. W. Struve at Dorpat in Estonia, and Sir John Herschel himself, who had then taken his twenty-foot reflector to the Cape of Good Hope and was busy undertaking the first really detailed survey of the far-southern stars which can never be seen from Europe. Smyth wrote in January 1836 that 'instead of the appulse which a careful projection had led me to expect, I was astonished to find it a single star! In fact, whether the real disks were over each other or not, my whole powers, patiently worked from ×240 to ×1200, could only make the object round.' At Dorpat, Struve was able to use the best telescope in the world at that time – the Fraunhofer refractor – and using a magnification of ×848 he could see that Gamma Virginis was elongated, though he could not split it. From the Cape, Herschel recorded on 27 February 1836 that:

> Gamma Virginis, at this time, is to all appearance a single star. I have tormented it, under favourable circumstances, with the highest powers I can apply to my telescopes, consistently with seeing a well-defined disk, till my patience has been exhausted; and that lately, on several occasions, whenever the definition of stars generally, in that quarter of the heavens, would allow of observing with any chance of success, but I have not been able to procure any decisive symptom of its consisting of two individuals.

Within a few months the two components could be separated again. The events of 1836 did not, of course, indicate that the components had really closed up; everything depends upon the angle from which we view them. In the true orbit of the two stars the distance between them varies from about 700 million kilometres to 12,270 million kilometres. Greatest separation occurred in 1920. Thereafter, the two components closed once more and in April 2005 the star again appeared single, except in the largest telescopes. The binary is now widening quickly and in 2015 the separation is expected to be about 2.2 seconds of arc; they are a beautiful pair in a 10-centimetre telescope.

Both components have been suspected of being slightly variable; thus Struve wrote that between 1825 and 1831 the star now officially classed as the secondary appeared the brighter, but that in 1851 the converse was true. The reality of these changes is very doubtful, but observers may care to check whether there really are any relative changes. All in all, the two appear as perfect twins.

The Belt of Orion. During late evenings in December, Orion (Figure 32) dominates the scene; it is crossed by the celestial equator, which

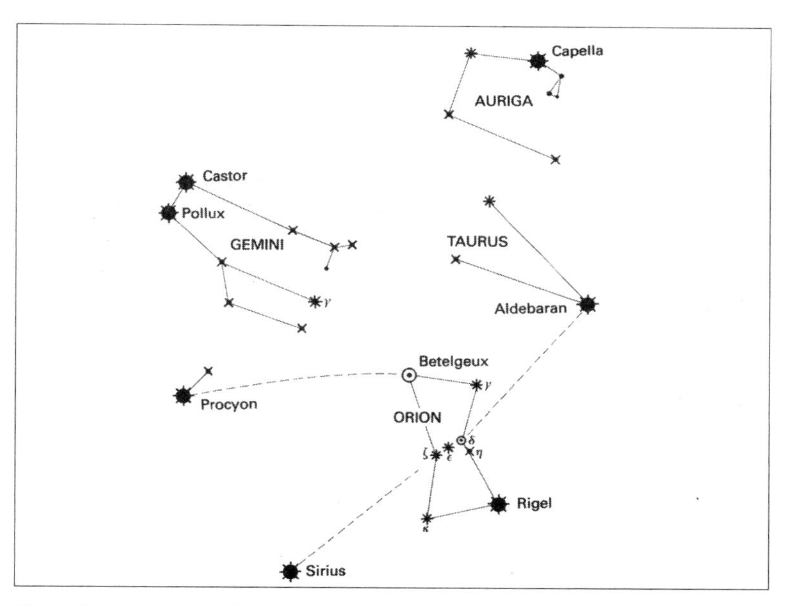

Figure 32. During December evenings, the brilliant winter constellations have again come into view. Of these, Orion, the Hunter, dominates the southern aspect of the sky, surrounded by its retinue of Aldebaran in Taurus (the Bull), Capella in Auriga (the Charioteer), Castor and Pollux in Gemini (the Twins), Procyon in Canis Minor (the Little Dog) and the brilliant Sirius in Canis Major (the Great Dog). The line marked by the three stars of Orion's Belt can be extended north and west to find Aldebaran, and south and east to locate Sirius.

passes very close to Mintaka (Delta Orionis) in the Hunter's Belt, so that the constellation is visible from every inhabited continent. Orion's two leading stars, the orange-red Betelgeuse (ranked tenth brightest in the sky) and the brilliant blue-white Rigel (ranked seventh), cannot be mistaken. If the star's output in the ultraviolet region of the spectrum is taken into account, Rigel is over 80,000 times as luminous as the Sun. It is interesting to note that Saiph or Kappa Orionis (magnitude 2.06), the other star marking the southern end of the main pattern, is almost as powerful as Rigel, but is hotter and radiates more of its energy in the ultraviolet and so appears fainter to the human eye even though they lie at similar distances from us.

The three stars of Orion's Belt, which are all of the second magnitude, were lettered by the German astronomer Johann Bayer in order from west to east, rather than in order of brightness, so Mintaka (Delta) is actually the faintest of the three. The Arabian astronomers, rather fittingly, referred to the stars of Orion's Belt as 'the string of pearls'. The three stars are:

Star	Apparent Magnitude	Spectral type	Distance, light years
Delta (Mintaka)	2.21v	O9.5 + B0.5	915
Epsilon (Alnilam)	1.70	B0.1	1340
Zeta (Alnitak)	1.75	O9.7, O9 + B0	800

All are very hot, very luminous and bluish-white in colour. Since all are very remote, their distances are naturally somewhat uncertain, but the values given above are of the right order.

Mintaka, the westernmost star of the Belt, is an eclipsing binary with a small range of less than 0.2 magnitude, and a period of 5.73 days. The components are both giant stars, roughly equal in luminosity, about 90,000 times that of the Sun (after correction for interstellar dust absorption), with masses about 20 times that of the Sun and destined to explode as supernovae later in life. There is also a seventh-magnitude companion at a separation of almost a minute of arc, which equates to a distance of about half a light year. In 1834, Sir John Herschel announced that Mintaka is variable, with a range of from magnitude 2¼ to 2¾; subsequently, the German astronomer Arthur von Auwers gave a period of 16 days, while Eduard Schonfeld could find no regular period. Since the magnitude range is in fact too small to be noticed with the naked eye, these estimates were almost certainly erroneous. Mintaka was the first star known to show stationary spectral lines, due to the presence of very tenuous gas and dust in the interstellar medium; the discovery was made by Johannes Hartmann, at Potsdam Observatory, in 1904.

Alnilam, the central star of the Belt and visually the brightest of the three, is a highly luminous supergiant lying at a distance of over 1,300 light years (roughly one and a half times the distance of Mintaka and Alnitak). Indeed, Alnilam is the most luminous of the three Belt stars –

Figure 33. The spectacular star-forming region known as the Flame Nebula (NGC 2024) and its surroundings, including the star Alnitak in Orion's Belt (upper right). This image was the first to be released publicly from VISTA, the world's largest survey telescope. In views of this evocative object in visible light the core of the nebula is completely hidden behind obscuring dust, but in this VISTA view, taken in infrared light, the cluster of very young stars at the object's heart is revealed. The wide-field VISTA view also includes the glow of the reflection nebula NGC 2023, just below centre, and the ghostly outline of the Horsehead Nebula (Barnard 33) towards the lower right. (Image courtesy of ESO/J. Emerson/VISTA and Cambridge Astronomical Survey Unit.)

about 370,000 times the luminosity of the Sun, when its ultraviolet radiation is taken into account – and it is surrounded by a faint reflection nebula, NGC 1990.

Alnitak, the easternmost star of the Belt, and lying at about eight hundred light years distant, is actually a triple-star system. The primary is another very hot, highly luminous supergiant star, with a luminosity perhaps one hundred thousand times that of the Sun when its ultraviolet emissions are considered – the brightest O-type star in the night sky. In 1819, the German amateur astronomer Georg Kunowsky first reported that Alnitak was a double star; the magnitudes are 1.9 and 4.0, the separation is 2.2 arc seconds, and the revolution period must be around 1,500 years. Then, in 1998, a team at the Lowell Observatory discovered that the bright primary itself had a close blue companion, again of the fourth magnitude.

The region around Alnitak is quite beautiful, being bathed in the nebulosity of the bright emission nebula IC434, with the so-called Flame Nebula immediately to the east, and the famous Horsehead dark nebula to the south of Alnitak, which is spectacular when imaged, but remarkably difficult to observe visually (Figure 33).

Although the difference in magnitude between Alnilam and Alnitak is only five hundredths of a magnitude, many observers say that they find it quite easy to see that Alnilam is visually the brighter of the two – a comment upon the sensitivity of the human eye.

Eclipses in 2015

MARTIN MOBBERLEY

During 2015 there will be four eclipses: two total lunar eclipses, a total eclipse of the Sun, and a partial eclipse of the Sun. All four eclipses occur in two specific periods, namely late March/early April and September.

1. *A total eclipse of the Sun* on 20 March will be visible along a track stretching from the waters off Newfoundland at sunrise, crossing the Faroe Islands and then passing over the Norwegian islands of Svalbard, in particular the main island Spitsbergen. The umbral track first hits the Earth's surface at 09h 10m UT and leaves it at 10h 21m UT. A partial solar eclipse will occur over the North Atlantic, European, Russian and North African regions of the Northern Hemisphere. From London the partial eclipse will peak at 09h 31m UT with almost 87 per cent of the solar disc being covered and the Sun positioned at 28 degrees altitude in the south-east. From the Faroe Islands capital of Tórshavn, totality will last for 2 minutes 2 seconds, centered on 09h 42m UT with the Sun at an altitude of 20 degrees. Some 300 kilometres north of the Faroe Islands, in the North Atlantic, the greatest duration of totality occurs: some 2 minutes 47 seconds, centered on 09h 46m UT. From Longyearbyen, on Spitsbergen, totality lasts 2 minutes 27 seconds, centered on 10h 12m UT, but the Sun will only be at an altitude of 11 degrees at an azimuth of 167 degrees, so not having any obstructions to the south will be a major consideration for the traveller. Due to the glancing angle at which the lunar shadow strikes the surface of the Earth the north-south umbral footprint is quite large, up to 462 kilometres, in fact. Eclipse chasers heading to this eclipse should bear in mind a number of factors, namely that the likelihood of cloud is very high, the daytime temperature on Spitsbergen may well be around –15 °C (even without wind chill) and the North Atlantic will be at its coldest and quite choppy too. In addition, those planning on seeing aurorae will probably be disappointed, as late

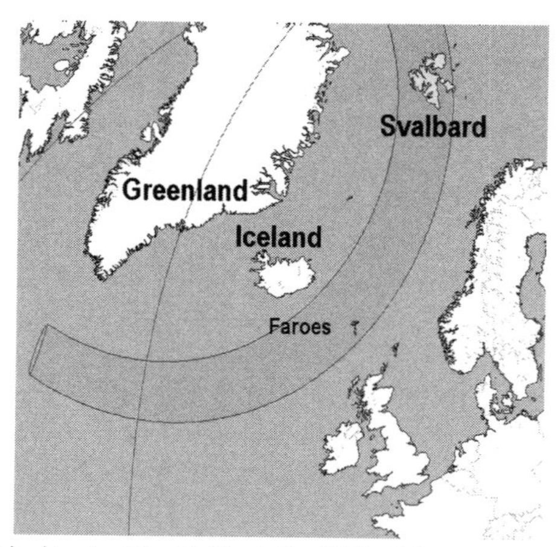

Figure 1a. The umbral track of the 20 March Total Solar Eclipse curves east and north as it moves between Scotland and Iceland, passing over the Faroe Isles and, eventually, Svalbard.

Figure 1b. A close-up of the umbral track as it crosses over the island of Spitsbergen and its main town Longyearbyen.

March marks the end of the Aurora season as the Sun goes no lower than twelve degrees below the horizon, so the night sky will never be truly dark. The danger from wandering polar bears is also something to bear in mind! Nevertheless, Spitsbergen features a spectacular frozen and rocky landscape and the sight of a totally eclipsed Sun above that scenery, for almost 2 minutes and 30 seconds, would be awesome.

2. *A total eclipse of the Moon* on 4 April will be visible in its entirety from the central Pacific Ocean, but, sadly, will be totally unobservable from regions close to the Greenwich meridian, such as the entire United Kingdom, Scandinavia and Africa. As the Moon enters the umbra the partial phases prior to totality will be visible across the eastern US with the eastern seaboard witnessing umbral entry just prior to sunset and the western seaboard witnessing the entire eclipse. From the opposite side of the Pacific Ocean Australia will witness totality and all of the umbral phases, with moonrise occurring as the Moon enters the umbra from the extreme western coastline of the Australian continent. From the longitude of India moonrise occurs just after totality. It should be stressed that totality itself, when the Moon is immersed in the umbra and so not receiving any direct sunlight, is very brief for this lunar eclipse, lasting just four minutes and forty-three seconds, centred on 12h 00m UT. This is because the Moon passes through the extreme northern part of the umbral shadow, so even during totality the

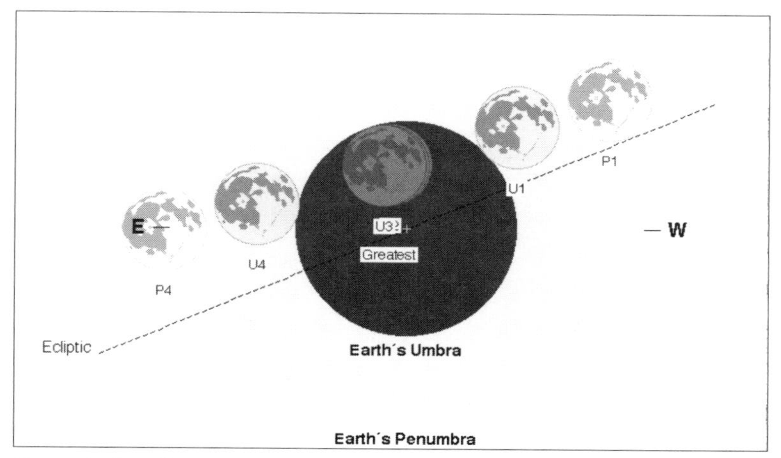

Figure 2. The Full Moon passes through the extreme northern part of the Earth's umbral shadow on 4 April, but only immerses its whole disc for less than five minutes at totality. The UT times are as follows: P1 = 09h 01m 27s, U1 = 10h 15m 45s, U2 = 11h 57m 54s, U3 = 12h 02m 37s, U4 = 13h 44m 46s, P4 = 14h 58m 58s. Diagram by NASA/Fred Espenak.

northern lunar limb will surely still appear quite noticeable, due to the refracted light bending around the Earth's atmosphere. At this time the ghostly Moon will be drifting through the stars of Virgo, ten degrees to the north-west of the brilliant first-magnitude star Spica. Specific umbral timings, supplied by NASA, are as follows: First umbral contact: 10h 15m 45s UT. Start of Totality: 11h 57m 54s UT. End of Totality: 12h 02m 37s UT. Last umbral contact: 13h 44m 46s UT.

3. *A partial eclipse of the Sun* on 13 September will be visible from the extreme Southern Hemisphere of the Earth, peaking on the Antarctic continent at 06h 54m UT when the Moon ingresses 78 per cent across the solar disc, but with the eclipse occurring very low on the Antarctic horizon. The southern African countries of South Africa, Namibia, Botswana, Zimbabwe and Mozambique, along with southern Madagascar, are the only other areas of land to experience a partial solar eclipse, with 40 per cent of the solar disc encroached upon as viewed from the southern tip of South Africa. The partial phase first becomes visible at sunrise along the western Atlantic coast of southern

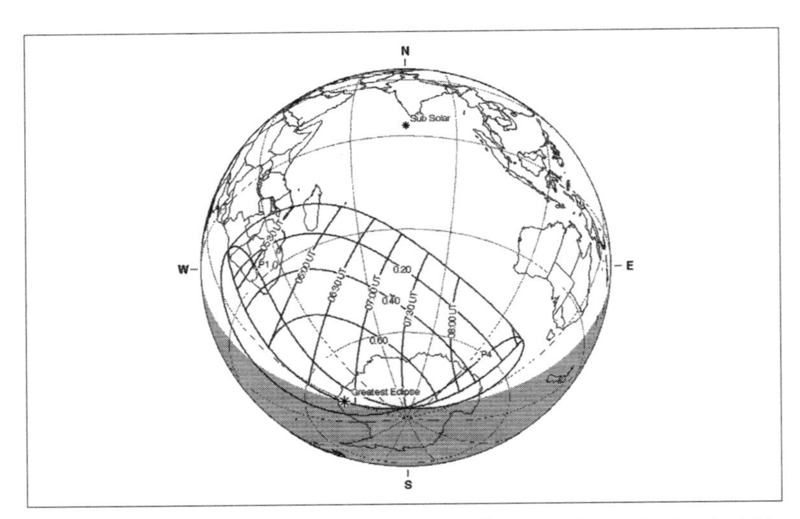

Figure 3. As described in the text, the partial solar eclipse of 13 September is only visible from the far Southern Hemisphere. The vertical contours indicate UT times of 05h 30m to 08h 00m UT in half-hour intervals and the horizontal contours represent zero, 20, 40 and 60 per cent of the solar disc being covered. Diagram by NASA/Fred Espenak.

Africa at 04h 42m UT and finally disappears at sunset in the ocean between Antarctica and Australia at 09h 06m UT.

4. *A total eclipse of the Moon* on 28 September will be visible in its entirety from the Atlantic Ocean, the UK and Ireland, France, Spain, western Africa, South America and the eastern US, with totality visible from the entire US, albeit very low in the eastern sky from western US states. None of the phases of the eclipse will be visible from Australia, Japan, Asia or the western Pacific Ocean. Weather permitting, this will be the best total lunar eclipse visible from the UK for many years, although it will be at an unfavourable time of night, in the early hours, and at a fairly low altitude. Totality lasts for a healthy 1 hour 12 minutes with the Moon passing through the Southern Hemisphere of the Earth's shadow, so it is likely that the Moon's southern limb will appear relatively bright during totality, unless this turns out to be a very dark eclipse overall. The precise darkness is impossible to predict and depends on the clarity of the Earth's atmosphere during totality, which can be highly influenced by any recent volcanic eruptions, which leave dust in the Earth's upper atmosphere. From London the Moon will be

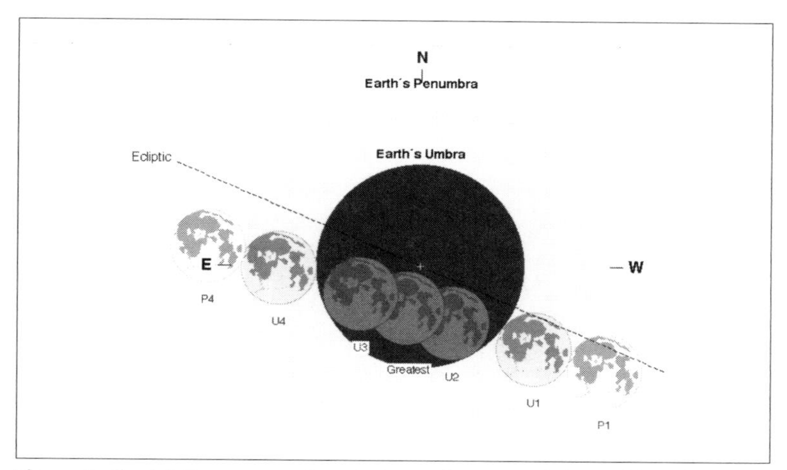

Figure 4. The full Moon passes through the southern part of the Earth's umbra on 28 September. The UT times corresponding to its passage through the penumbra and umbra are as follows: P1 = 00h 11m 47s UT, U1 = 01h 07m 11s UT, U2 = 02h 11m 10s UT, U3 = 03h 23m 05s UT, U4 = 04h 27m 03s UT, P4 = 05h 22m 27s UT.

thirty-five degrees above the south-south-west horizon when it first enters the umbra at 01h 07m UT (02h 07m a.m. British Summer Time). When totality starts, at 02h 11m UT, the Moon will have dropped to an altitude of 30 degrees. Mid-totality occurs at 02h 48m UT and totality ends at 03h 23m UT with the Moon just 23 degrees above the London skyline. By the time the Moon leaves the umbra, at 04h 27m UT, it will be a mere 13 degrees above the western London horizon and dawn twilight will be breaking on the opposite horizon. From the US, the eclipse takes place at a far more sociable time with Eastern states most favoured with totality occurring in the late evening. For example, from New York, mid-totality occurs just before 11 p.m. EDT (Eastern Daylight Time) with the Moon 42 degrees above the south-east horizon.

Occultations in 2015

NICK JAMES

The Moon makes one circuit around the Earth in just over twenty-seven days and as it moves across the sky it can temporarily hide, or occult, objects that are further away, such as planets or stars. The Moon's orbit is inclined to the ecliptic by around 5.1° and its path with respect to the background stars is defined by the longitude at which it crosses the ecliptic passing from south to north. This is known as the longitude of the ascending node. After passing the node the Moon moves eastwards relative to the stars, reaching 5.1° north of the ecliptic after a week. Two weeks after crossing the ascending node it crosses the ecliptic moving south and then it reaches 5.1° south of the ecliptic after three weeks. Finally, it arrives back at the ascending node a week later and the cycle begins again.

The apparent diameter of the Moon depends on its distance from the Earth but at its closest it appears almost 0.6° across. In addition, the apparent position of the Moon on the sky at any given time shifts depending on where you are on the surface of the Earth. This effect, called parallax, can move the apparent position of the Moon by just over one degree. The combined effect of parallax and the apparent diameter of the Moon mean that if an object passes within 1.3° of the apparent centre of the Moon, as seen from the centre of the Earth, it will be occulted from somewhere on the surface of our planet. For the occultation to be visible the Moon would have to be some distance from the Sun in the sky and, depending on the object being occulted, it would have to be twilight or dark.

For various reasons, mainly the Earth's equatorial bulge, the nodes of the Moon's orbit move westwards at a rate of around 19° per year, taking 18.6 years to do a full circuit. This means that, whilst the Moon follows approximately the same path from month to month, this path gradually shifts with time. Over the full 18.6-year period, all of the stars that lie within 6.4° of the ecliptic will be occulted.

Only 4 first-magnitude stars lie within 6.4° of the ecliptic. These are

Aldebaran (5.4°), Regulus (0.5°), Spica (2.1°) and Antares (4.6°). As the nodes precess through the 18.6-year cycle there will be a monthly series of occultations of each star followed by a period when the star is not occulted. In 2015, there are thirteen occultations of Aldebaran, one every lunar month.

In 2015, there will be twenty occultations of bright planets. Two each of Mercury and Mars, three of Venus and thirteen of Uranus. Sixteen of these events take place at a solar elongation of greater than thirty degrees. The table lists occultations of the brighter planets which are potentially visible from somewhere on the Earth and where the solar elongation exceeds thirty degrees. More detailed predictions for your location can often be found in magazines or the *Handbook of the British Astronomical Association*.

Object	**Date and time of Minimum Distance (UT)**		**Minimum Distance**	**Elongation**	**Best Visibility**
Uranus	25 Jan 2015	11:30	0.6	68	China
Aldebaran	29 Jan 2015	17:15	1.2	121	Arctic Canada
Uranus	21 Feb 2015	22:03	0.3	41	Eastern US
Aldebaran	25 Feb 2015	23:15	1.0	93	Northern Canada
Aldebaran	25 Mar 2015	07:10	0.9	66	Alaska
Aldebaran	21 Apr 2015	16:46	0.9	39	Northern Russia
Uranus	15 May 2015	11:58	0.2	−36	Western South America
Uranus	11 June 2015	20:37	0.5	−61	Australia
Uranus	9 July 2015	03:06	0.7	−86	South Africa
Aldebaran	12 July 2015	18:13	0.9	−40	Alaska
Venus	19 July 2015	00:56	0.4	34	Pacific
Uranus	5 Aug 2015	09:20	1.0	−112	South America
Aldebaran	8 Aug 2015	23:44	0.7	−66	Central Russia
Uranus	1 Sept 2015	16:32	1.0	−139	South Pacific
Aldebaran	5 Sept 2015	05:30	0.5	−92	Eastern Canada
Uranus	29 Sept 2015	01:25	1.0	−167	South Africa
Aldebaran	2 Oct 2015	13:10	0.5	−119	Western US
Venus	8 Oct 2015	20:08	0.7	−45	South Pacific
Mercury	11 Oct 2015	11:15	0.9	−17	Southern South America

Uranus	26 Oct 2015	10:46	0.9	165	South Pacific
Aldebaran	29 Oct 2015	23:01	0.6	−146	Europe
Uranus	22 Nov 2015	19:10	0.9	137	South Pole
Aldebaran	26 Nov 2015	09:49	0.7	−174	Western Canada
Mars	6 Dec 2015	02:37	0.1	−60	Middle East
Venus	7 Dec 2015	17:15	0.6	−42	North America
Uranus	20 Dec 2015	01:25	1.1	109	South Pole
Aldebaran	23 Dec 2015	19:25	0.6	158	Western Russia

Comets in 2015

MARTIN MOBBERLEY

More than sixty short-period comets should be observed approaching perihelion in 2015, although four of these, D/Brooks (1886 K1), D/Barnard (1884 O1), 34D/Gale and D/Helfenzrieder (1766 G1), have been declared defunct and so have a D/ prefix. Essentially, a D/ prefix means that these comets have either run out of volatile material or disintegrated. However, with so many deep surveys now scouring the night sky they may yet be rediscovered. Indeed, in last year's Yearbook I stated that I thought there was a reasonable chance of 72D/Denning-Fujikawa being recovered, and it was indeed found by Sato last June! All of the comets expected to return in 2015 orbit the Sun with periods of between 4 and 64 years and many are too faint for amateur visual observation, even with a large telescope. Bright or spectacular comets have much longer orbital periods and, apart from a few notable exceptions like 1P/Halley, 109P/Swift-Tuttle and 153P/Ikeya-Zhang, the best performers usually have orbital periods of many thousands of years and are often discovered less than a year before they come within amateur range. For this reason it is important to regularly check the best comet websites for news of bright comets that may be discovered well after this Yearbook is finalized. Some recommended sites are:

British Astronomical Association Comet Section: www.ast.cam.ac.uk/~jds/

Seiichi Yoshida's bright comet page: www.aerith.net/comet/weekly/current.html

CBAT/MPC comets site: www.minorplanetcenter.net/iau/Ephemerides/Comets/

Yahoo Comet Images group: http://tech.groups.yahoo.com/group/Comet-Images/

Yahoo Comet Mailing list: http://tech.groups.yahoo.com/group/comets-ml/

The CBAT/MPC web page above also gives accurate ephemerides of comet positions in right ascension and declination.

As many as thirty periodic comets might reach a magnitude of fourteen or brighter during 2015 and so they should all be observable with large amateur telescopes equipped with amateur CCD imaging systems, in a reasonably dark sky, from the Northern or Southern Hemispheres. Sadly, only a few comets are, at the time of writing, predicted to be tenth magnitude or brighter and so visual targets for large amateur telescopes will be few and far between unless some bright prospects are discovered in 2014 or 2015, namely after this Yearbook has gone to press. The current cometary highlight for the coming year is arguably the return of 67P/Churyumov-Gerasimenko, as it will be the rendezvous target for the Rosetta probe with the comet arriving at perihelion during August. However, this is only likely to be a space probe highlight as the comet will be poorly placed in September, a month after perihelion, and although 67P could reach tenth magnitude, it may become no brighter than magnitude 12. Nevertheless, for all comet fans this spacecraft rendezvous should be fascinating. The Rosetta probe began its final approach towards 67P/Churyumov-Gerasimenko in June 2014, around the time this Yearbook was being finalized. Hopefully, just as this Yearbook is published, in November 2014, Rosetta will release the lander, called Philae, which will anchor itself to the surface of the comet. The lander will then send back data from the surface of the comet for the whole of 2015 as 67P reaches perihelion and, hopefully, for a further four months after this date. The mission is scheduled to end in December 2015. Exciting stuff!

Two visual highlights in this rather barren cometary year may be 10P/Tempel, which could achieve tenth magnitude for Southern Hemisphere observers in October, and 88P/Howell, which might reach tenth or eleventh magnitude in the southern skies during April. A more recent discovery, 2012 K1 PanSTARRS, may start the year as a ninth- or tenth-magnitude object in southern skies. Perhaps a more promising prospect, right at the end of the year, is comet 2013 US10 Catalina, which may become a pre-dawn binocular object.

It should perhaps be explained that the distances of comets from the Sun and the Earth are often quoted in Astronomical Units (AU) where 1 AU is the average Earth–Sun distance of 149.6 million kilometres or 93 million miles.

The brightest cometary prospects for 2015 are listed below in the order they reach best visibility, which often coincides with reaching perihelion. The dual status asteroid/comet (596) Scheila and the comet 29P/Schwassmann-Wachmann are nowhere near perihelion during the year but are best placed in January and June respectively. However, Scheila will be a fading evening object in Taurus as 2015 starts and 29P will be far south of the celestial equator during its peak April to August months, hovering around the adjoining border regions of north-eastern Scorpius, western Sagittarius and southern Ophiuchus. This enigmatic comet will often be too faint for visual detection even in a large amateur telescope, but it can rise to eleventh magnitude when in outburst. Comet 19P/Borrelly returns to perihelion in May and would be a tenth-magnitude target except for the fact that its tiny solar elongation will make observation impossible for months either side of the perihelion date.

Comet	Period (years)	Perihelion	Peak Magnitude
(596) Scheila	5.0	May 2017	13 in outburst
P/1998 U3 Jäger	15.2	12 Mar 2014	14 in Jan
C/2012 K1 (PanSTARRS)	Long	27 Aug 2014	9 in Jan
289P/Blanpain	5.3	28 Aug 2014	14 in Jan
32P/Comas Sola	9.6	17 Oct 2014	14 in Jan
108P/Ciffreo	7.2	18 Oct 2014	14 in Jan
C/2013 A1 (Siding Spring)	Long	25 Oct 2014	10 in Jan
110P/Hartley	6.9	17 Dec 2014	14 in Jan
15P/Finlay	6.5	27 Dec 2014	12 in Jan
201P/LONEOS	6.4	14 Jan 2015	14 in Jan
D/Brooks (1886 K1)	5.7	21 Jan 2015	11 if found
7P/Pons-Winnecke	6.3	30 Jan 2015	13 in Feb
6P/d'Arrest	6.6	2 Mar 2015	14 in Mar
88P/Howell	5.5	6 Apr 2015	11 in Apr
C/2012 F3 (PanSTARRS)	Long	7 Apr 2015	14 in Apr
19P/Borrelly	6.8	28 May 2015	10 in May
29P/Schwassmann-Wachmann 1	14.7	12 Apr 2019	11 in outburst
51P/Harrington	7.2	12 Aug 2015	13 in Sept
67P/Churyumov-Gerasimenko	6.5	12 Aug 2015	12 in Sept
141P/Machholz	5.3	24 Aug 2015	11 in Sept
34D/Gale	11.2	8 Sept 2015	13 if found
61P/Shajn-Schaldach	7.1	1 Oct 2015	13 in Oct

Comet	Period (years)	Perihelion	Peak Magnitude
22P/Kopff	6.4	26 Oct 2015	10 in Oct
10P/Tempel	5.4	15 Nov 2015	10 in Nov
230P/LINEAR	6.3	18 Nov 2015	13 in Nov
C/2013 US10 (Catalina)	Long	15 Nov 2015	8 in Dec
204P/LINEAR-NEAT	7.0	11 Dec 2015	13 in Dec
116P/Wild	6.5	12 Jan 2016	14 in Dec
P/Arend (50P)	8.3	7 Feb 2016	14 in Dec
P/2010 V1 (Ikeya-Murakami)	5.4	10 Mar 2016	10 in Dec

WHAT TO EXPECT

Comet 2012 K1 (PanSTARRS) may be the best prospect as 2015 starts. This object has been around for three years now and peaked in August 2014. Nevertheless, it may still be a ninth-magnitude target in the Southern Hemisphere constellation of Sculptor throughout January. Sadly, its elongation from the Sun becomes poor very rapidly during February and by May, when it will be better placed for Southern

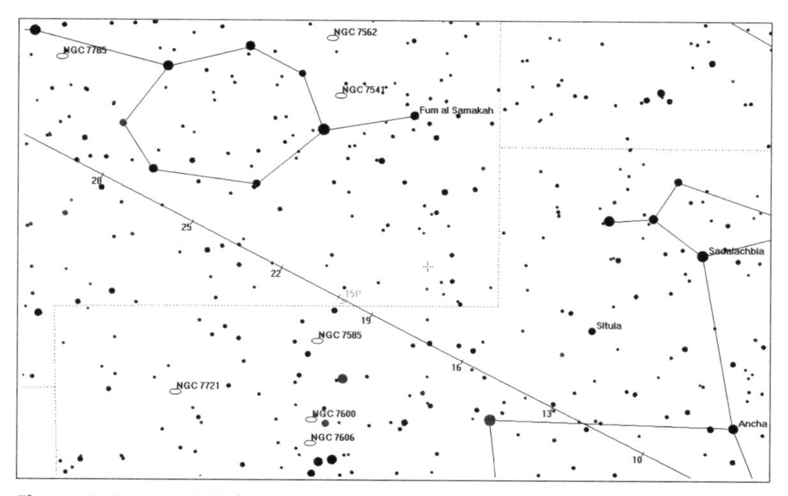

Figure 1. Comet 15P/Finlay moves north-west from Aquarius to Pisces during January, crossing the constellation boundary around January 19/20.

Hemisphere observers, while in Cetus, it will probably have faded to thirteenth magnitude.

Comet 15P/Finlay reached a perihelion of 0.97 AU from the Sun on 27 December 2014, and may be a twelfth-magnitude object low in the early evening sky, in Aquarius, as January starts. However, it will be fading and its solar elongation of 45 degrees is far from ideal.

C/2013 A1 (Siding Spring) reached perihelion in October 2014 but as 2015 starts it has a very poor elongation from the Sun as it is located in Ophiuchus, close to the Serpens border. However, by late January, the comet is clearing the morning twilight as it heads towards Hercules at tenth or eleventh magnitude. So for those brave Northern Hemisphere souls who can face a freezing pre-dawn observing session in February, C/2013 A1 may be the best target around.

88P/Howell was discovered by Ellen Howell on photographic plates taken with the 0.46-m Schmidt telescope at Mount Palomar Observatory on 29 August 1981. It currently has an orbital period of 5.5 years. The comet has a perihelion distance of 1.36 AU from the Sun, which it reaches on 6 April while close to the Aquarius/Capricornus border at a declination of –15 degrees, so this comet will primarily be a target for

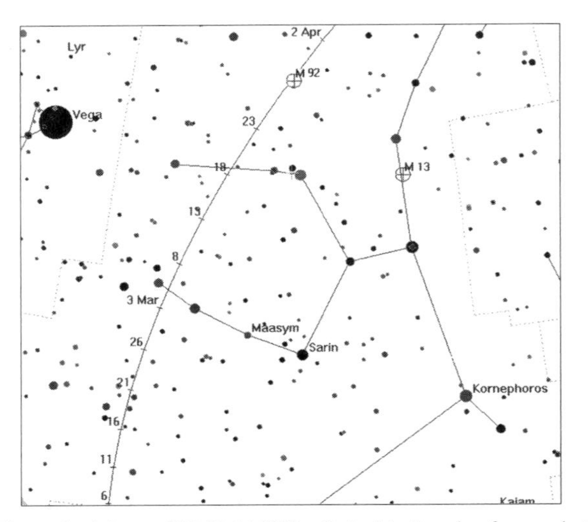

Figure 2. The track of Comet C/2013 A1 (Siding Spring) in Hercules, from early February to late March 2015. On the morning of 29 March, the comet will pass within ten arc minutes of the globular cluster M92, but will probably have faded to twelfth magnitude by that time.

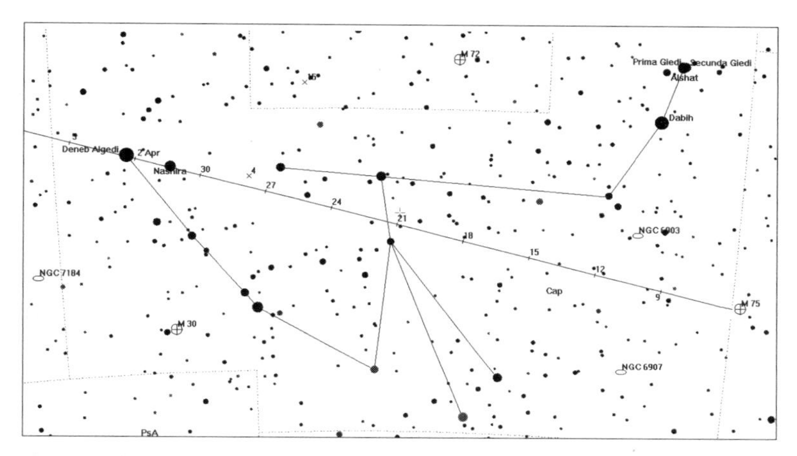

Figure 3. The track of 88P/Howell through Capricornus during March 2015. As April starts, the comet will be midway between Nashira and Deneb Algedi. On the night of 5/6 March, the comet passes just five arc minutes from the globular cluster M75.

Southern Hemisphere observers. Around perihelion, the comet will be 1.8 AU from Earth and it is unlikely to become brighter than tenth or eleventh magnitude, although CCD images will probably show it with a one-arc-minute diameter coma and a short tail around this time.

As I have already mentioned, the space probe highlight of 2015 will be the data that returns from the 67P/Churyumov-Gerasimenko lander as the comet moves towards perihelion. Despite the fact that the comet will be a borderline visual object even in the largest amateur telescopes, and less than impressive even with CCDs, many amateurs will want to observe or image the comet with so much likely media coverage. The comet reaches its perihelion distance from the Sun of 1.24 AU on 12 August, at which time it will be 1.77 AU from the Earth. Throughout July and August it stays at a remarkably constant 43 degrees elongation from the Sun as it moves through Taurus and Gemini, low down in the morning twilight sky and probably no brighter than twelfth magnitude. As the solar declination drops, with the comet's declination holding steady at around +24 degrees, it will slowly clear the morning twilight and should be visible in a reasonably dark sky by late August. However, it is unlikely that even at its best, a week or two after perihelion, it will even achieve magnitude 11, but this will not dissuade those dedicated

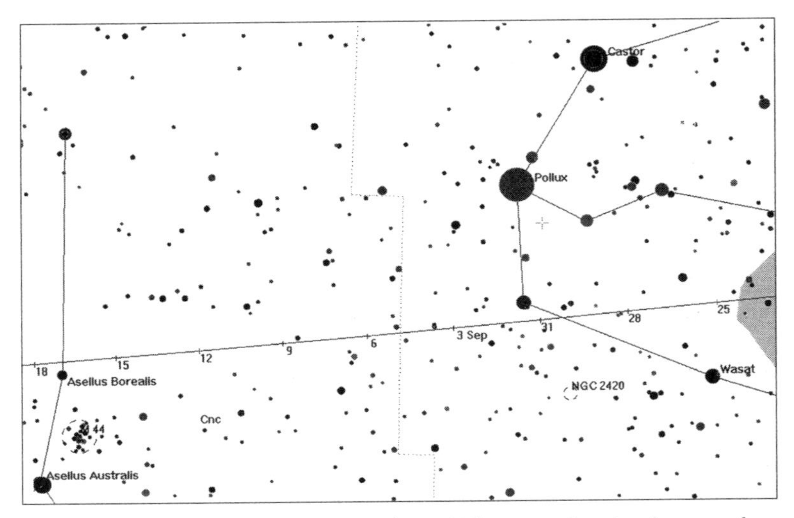

Figure 4a. From late August to mid-September 67P/Churyumov-Gerasimenko passes from eastern Gemini to western Cancer.

Figure 4b. The Philae lander at work on Comet 67P/Churyumov-Gerasimenko. With luck, just as this Yearbook is published, this scene will become reality, with the lander hopefully functioning perfectly during 2015. Copyright ESA/AOES Medialab.

comet observers who wish to capture the comet during this historic Rosetta/Philae monitoring period.

Comet 141P/Machholz was discovered twenty-one years ago by the tireless Californian amateur astronomer Don Machholz, who has now discovered eleven comets visually. Soon after discovery the Austrian comet imaging maestro Michael Jaeger detected that the comet appeared to be splitting into multiple components. With a period of 5.2 years, the comet has already returned to perihelion (0.75 AU from the Sun) three times since the year of discovery, but the last return was unfavourable and dismal, indicating that maybe its magnitude in 1994 was unusually bright. Even so, with such a small perihelion distance it may surprise us. In 2015, perihelion occurs on 24 August, but the closest approach to the Earth is almost six weeks earlier, on 16 July, when comet 141P passes within 0.689 AU of us on 16 July. However, it is the perihelion date that is the most significant and the main cometary component should peak in brightness in late August when it is a morning object in Gemini. From mid-July to mid-September, comet Machholz travels from Triangulum, through Perseus, then through Auriga, rapidly accelerating into Gemini and in early September moving into Cancer. Just what the peak magnitude of this unpredictable comet (or its largest surviving chunk) will be in late August remains to be seen. In outburst it could conceivably reach eighth magnitude, but this is highly unlikely and it may only achieve magnitude 12. We can only wait and see.

Comet 22P/Kopff returns to perihelion every 6.4 years and 25 October marks its perihelion date at this return. On that day the comet will pass within 1.57 AU of the Sun. Sadly, this is not a favourable return for Northern Hemisphere observers because just as it becomes interesting it plunges into the Southern Hemisphere. However, observers south of the Equator will enjoy a better performance. It should reach tenth magnitude during October, when it travels through the Libra, Scorpius and Ophiuchus border regions. On 3/4 October, it travels within six arc minutes of Saturn as seen from our perspective, but in reality the comet is more than five times closer to the Earth, at a distance of 2.0 AU. Despite its southerly declination even Southern Hemisphere comet observers will find its decreasing solar elongation means it will stay low in the evening sky.

10P/Tempel was discovered as long ago as 1873 and this will be its twenty-third return to perihelion since that time. Like many comets

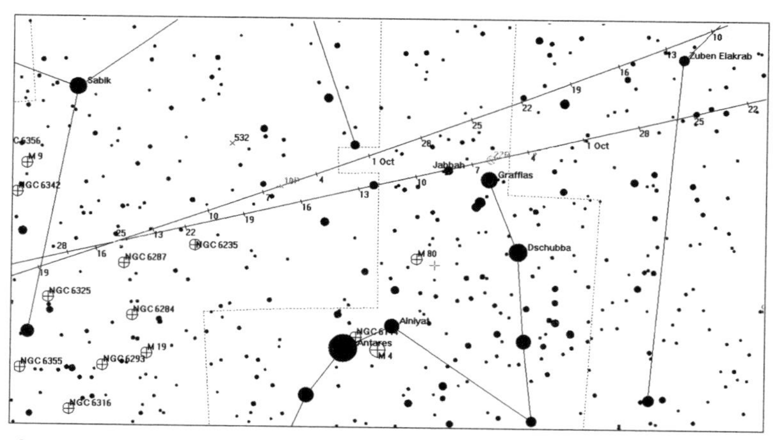

Figure 5. Periodic comets 10P/Tempel and 22P/Kopff follow very similar tracks from late September to late October as they both head east and slightly south as they pass above Antares and the head of the Scorpion and into Ophiuchus. The upper track is that of 10P, while the lower is that of 22P.

returning in 2015 it will be at its brightest when in southern skies. Perihelion occurs on 14 November when the comet should peak, near the 'Teapot handle' in Sagittarius, at around magnitude 10 or 11. Sadly, UK observers' last view of it will probably be in July, as it brightens to twelfth magnitude in Virgo. Comet 10P follows a similar course along the ecliptic as 22P/Kopff during the Autumn, often leading (east) of that other comet by some ten degrees or so.

Although Comet P/2010 V1 (Ikeya-Murakami) does not reach perihelion until March 2016, it will be well observable and well placed from late evening onwards in UK December skies at the end of 2015. The comet tracks slowly eastwards during the month, until by year end it is some six degrees above the top of Leo's 'Sickle'. It glides less than a degree below the constellation boundary set at +33 degrees declination, which separates Cancer from Lynx and, beyond 9h 22m, Leo from Leo Minor. After 10 Dec the comet crosses from Cancer into Leo, slowly brightening, at around tenth magnitude.

Finally, an object discovered on 31 October 2013 was initially thought to be a very large near-Earth asteroid, but was quickly re-designated as comet 2013 US10 (Catalina). This comet will be a pre-dawn December 2015 object for UK observers, passing very close to Arcturus as 2015 changes to 2016. It reaches perihelion on 15 November 2015, but continues to approach the Earth as 2015 ends (closest in mid-January 2016) so it should be a binocular object for the entire December 2015/January 2016 period. Hopefully it will be at least magnitude 8 at this time, but it may be a bit brighter, and with a perihelion distance of 0.82 AU it could already have a nice tail by early December. Southern Hemisphere observers with telescopes will have been able to follow the comet brightening throughout the summer and autumn months.

Of course it goes without saying that the phase of the Moon and its proximity to the field where a comet lies have a huge bearing on the object's visibility. Even when a comet is high up at midnight, a full Moon ten or twenty degrees away will trash any prospects of observing it visually, so potential comet spotters need to be very aware of the position and brightness of the Moon when planning any comet observing. CCDs are much more tolerant of a bright sky, though, and image processing can work wonders in compensating.

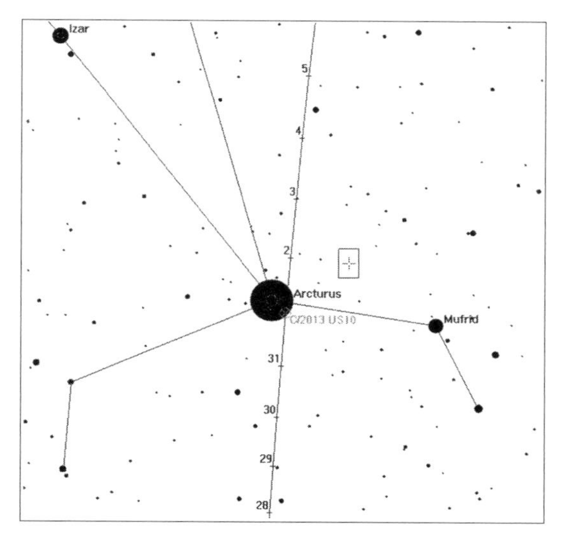

Figure 6. As the final days of December 2015 tick past and the year changes to 2016, comet C/2013 US10 (Catalina) passes just half a degree west of blazing Arcturus on January 1, travelling north at almost 2 degrees per hour.

THE DEFUNCT COMETS

By their very nature comets are highly unpredictable performers. Their nuclei come in a variety of shapes and sizes, are covered with patches of icy material on the surface or under the surface, and they spin at various rates too. The smallest ones are highly prone to disintegration when they come close to the Sun and others can have their orbit altered drastically, now and again, by the giant planet Jupiter. Every apparition is different from the Earth-based observer's viewpoint. The comet may arrive at perihelion with the Earth poorly placed, such as when the comet is on the opposite side of the Sun, or, conversely, the comet may pass within tens of millions of kilometres of our home planet. In addition, short-period comets lose material every time they reach perihelion and after thousands of orbits a comet may simply have no more material to discharge. Some comets slowly fragment on each passage with more components being detected on every return.

It is therefore hardly surprising that some comets are simply lost and declared 'Defunct', implying they have simply disappeared or broken up and will never be seen again. However, with deep CCD surveys of the night sky becoming ever more powerful, some of these formerly missing comets are occasionally recovered, and modern computing power often enables an object with a slightly different orbit to be revealed as an old comet, not seen for many decades.

In 2015, four of these D/ category comets are predicted to return to perihelion. In all probability they will not be recovered, but a recovery of one or two of them cannot be ruled out. The 2015 D/ comets are D/Brooks (1886 K1), D/Barnard (1884 O1), 34D/Gale and D/Helfenzrieder (1766 G1). If their orbits are unaltered from their last appearance these comets would reach perihelion on 21 January, 13 March, 8 September and 4 November respectively. Their orbital periods when last seen are 5.7, 5.4, 11.2 and 4.5 years. In addition, if they are as active as when they were last observed, they could be between eleventh and fifteenth magnitude at this return. Perhaps the most likely returning comet to escape its defunct status will be 34D/Gale which, by virtue of it having returned on a second perihelion after discovery, is listed as periodic comet number 34. The comet was discovered by Walter Frederick Gale in Sydney Australia on 7 June 1927 and recovered on its next return by Leland Cunningham in 1938. However, in 1949 and at every eleven-year interval since, it has not been observed. It could reach magnitude 13 if found in 2015. Comet D/Brooks (1886 K1) could be the brightest prospect if recovered, as it might reach magnitude 11. Its discoverer, W. R. Brooks of Phelps, New York, discovered the comet in May 1886, but it has never been seen again. The other two defunct comets, D/Barnard (1884 O1) and D/Helfenzrieder (1766 G1), are even less likely to be picked up in 2015 but are worth mentioning just in case a deep patrol system like PanSTARRS actually does find these long-lost comets.

EPHEMERIDES

The tables below list the right ascensions and declinations of, arguably, the ten most promising comets in 2015, at their peak, as well as the distances, in AU, from the Earth and the Sun. The elongation, in degrees from the Sun, is also tabulated along with the estimated visual magnitude, which can only ever be a rough guess as comets are, without doubt, a law unto themselves!

C/2012 K1 (PanSTARRS)

Date	RA (2000)			Dec.			Distance from Earth	Distance from Sun	Elongation from Sun	Mag.
	h	m	s	°	'	"	AU	AU	°	
Jan 1	00	09	19.9	−33	39	15	2.320	2.200	70.7	9.8
Jan 6	00	07	48.3	−32	03	52	2.464	2.259	66.5	10.0
Jan 11	00	07	07.7	−30	36	58	2.605	2.318	62.3	10.2
Jan 16	00	07	07.5	−29	17	28	2.742	2.377	58.3	10.5
Jan 21	00	07	40.0	−28	04	27	2.876	2.436	54.2	10.7
Jan 26	00	08	38.7	−26	57	11	3.005	2.494	50.3	10.9

15P/Finlay

Date	RA (2000)			Dec.			Distance from Earth	Distance from Sun	Elong-ation from Sun	Mag.
	h	m	s	°	'	"	AU	AU	°	
Jan 1	21	45	16.8	−14	01	45	1.412	0.979	43.8	12.7
Jan 6	22	07	54.9	−11	26	50	1.401	0.987	44.8	12.7
Jan 11	22	30	25.9	−08	43	16	1.394	1.000	45.8	12.7
Jan 16	22	52	46.8	−05	53	38	1.393	1.019	47.0	12.8
Jan 21	23	14	55.1	−03	00	39	1.399	1.042	48.1	12.9
Jan 26	23	36	48.5	−00	07	13	1.410	1.069	49.2	13.0

C/2013 A1 (Siding Spring)

Date	RA (2000)			Dec.			Distance from Earth	Distance from Sun	Elong-ation from Sun	Mag.
	h	m	s	°	'	"	AU	AU	°	
Feb 1	18	05	34.3	+12	12	04	2.427	1.970	51.4	10.9
Feb 6	18	05	38.6	+14	22	50	2.395	2.017	56.0	10.9
Feb 11	18	05	14.3	+16	41	24	2.361	2.064	60.6	11.0
Feb 16	18	04	16.1	+19	08	09	2.326	2.113	65.3	11.1
Feb 21	18	02	37.8	+21	43	17	2.291	2.162	70.0	11.1
Feb 26	18	00	12.3	+24	26	38	2.256	2.211	74.7	11.2

88P/Howell

Date	RA (2000)			Dec.			Distance from Earth	Distance from Sun	Elong-ation from Sun	Mag.
	h	m	s	°	'	"	AU	AU	°	
Apr 1	21	41	09.4	−16	38	02	1.792	1.360	48.8	11.3
Apr 6	21	58	20.3	−15	19	28	1.779	1.359	49.3	11.2
Apr 11	22	15	08.0	−13	57	23	1.767	1.360	50.0	11.2
Apr 16	22	31	31.3	−12	32	40	1.757	1.363	50.7	11.2
Apr 21	22	47	29.2	−11	06	09	1.748	1.368	51.4	11.3
Apr 26	23	03	00.8	−09	38	39	1.741	1.376	52.2	11.3

67P/Churyumov-Gerasimenko

Date	RA (2000)			Dec.			Distance from Earth	Distance from Sun	Elongation from Sun	Mag.
	h	m	s	°	'	"	AU	AU	°	
July 1	03	38	29.1	+17	22	24	1.916	1.350	42.3	12.7
July 11	04	16	01.1	+19	52	48	1.861	1.308	42.6	12.5
July 21	04	55	20.0	+21	55	38	1.820	1.275	42.7	12.4
July 31	05	35	51.3	+23	24	10	1.791	1.254	42.8	12.2
Aug 10	06	16	50.0	+24	14	07	1.775	1.244	43.0	12.2
Aug 20	06	57	22.5	+24	24	33	1.768	1.246	43.5	12.2

141P/Machholz

Date	RA (2000)			Dec.			Distance from Earth	Distance from Sun	Elongation from Sun	Mag.
	h	m	s	°	'	"	AU	AU	°	
July 20	02	57	59.0	+35	42	31	0.692	0.965	65.6	11.7
July 30	04	23	11.9	+36	45	52	0.731	0.876	57.5	10.6
Aug 09	05	39	31.1	+34	29	28	0.807	0.807	51.1	9.7
Aug 19	06	42	27.4	+30	18	13	0.909	0.768	46.7	9.3
Aug 29	07	33	56.6	+25	21	39	1.024	0.764	44.1	9.5
Sept 08	08	16	58.0	+20	18	29	1.142	0.798	43.0	10.3
Sept 18	08	53	33.1	+15	27	42	1.254	0.862	43.1	11.6

22P/Kopff

Date	RA (2000)			Dec.			Distance from Earth	Distance from Sun	Elongation from Sun	Mag.
	h	m	s	°	'	"	AU	AU	°	
Oct 1	15	48	38.9	−17	58	11	1.989	1.577	51.7	9.6
Oct 6	16	03	19.2	−18	54	05	2.007	1.570	50.3	9.6
Oct 11	16	18	23.2	−19	45	47	2.025	1.565	49.1	9.6
Oct 16	16	33	49.3	−20	32	46	2.045	1.561	47.8	9.6
Oct 21	16	49	35.3	−21	14	30	2.065	1.559	46.6	9.6
Oct 26	17	05	38.6	−21	50	31	2.087	1.558	45.4	9.6

10P/Tempel

Date	RA (2000)			Dec.			Distance from Earth	Distance from Sun	Elongation from Sun	Mag.
	h	m	s	°	'	"	AU	AU	°	
Nov 01	18	01	00.3	−24	25	24	1.797	1.425	52.2	10.1
Nov 06	18	18	32.3	−24	55	33	1.813	1.420	51.2	10.1
Nov 11	18	36	25.1	−25	17	19	1.830	1.418	50.2	10.1
Nov 16	18	54	34.1	−25	30	19	1.849	1.418	49.3	10.1
Nov 21	19	12	54.2	−25	34	18	1.869	1.419	48.4	10.2
Nov 26	19	31	19.8	−25	29	10	1.890	1.423	47.5	10.2

P/2010 V1 (Ikeya-Murakami)

Date	RA (2000)			Dec.			Distance from Earth	Distance from Sun	Elongation from Sun	Mag.
	h	m	s	°	'	"	AU	AU	°	
Dec 01	09	08	46.2	+31	57	12	1.155	1.829	117.1	10.9
Dec 06	09	16	21.9	+31	59	34	1.095	1.806	120.4	10.8
Dec 11	09	23	23.4	+32	03	21	1.039	1.785	123.7	10.6
Dec 16	09	29	46.2	+32	08	39	0.985	1.764	127.2	10.4
Dec 21	09	35	25.6	+32	15	23	0.935	1.744	130.7	10.3
Dec 26	09	40	17.4	+32	23	20	0.887	1.725	134.4	10.1
Dec 31	09	44	16.3	+32	32	13	0.843	1.707	138.1	10.0

C/2013 US10 (Catalina)

Date	RA (2000)			Dec.			Distance from Earth	Distance from Sun	Elongation from Sun	Mag.
	h	m	s	°	'	"	AU	AU	°	
Dec 01	14	19	02.4	−11	28	58	1.535	0.872	32.0	8.2
Dec 06	14	18	27.2	−08	28	25	1.442	0.907	38.4	8.3
Dec 11	14	17	59.8	−05	03	08	1.341	0.950	45.0	8.3
Dec 16	14	17	34.0	−01	04	01	1.234	0.998	52.0	8.3
Dec 21	14	17	00.3	+03	41	23	1.126	1.051	59.3	8.3
Dec 26	14	16	04.8	+09	29	30	1.018	1.108	67.2	8.2
Dec 31	14	14	25.9	+16	40	47	0.918	1.168	75.7	8.1

Minor Planets in 2015

MARTIN MOBBERLEY

Some 650,000 minor planets (also referred to as asteroids) are known. They range in size from small planetoids hundreds of kilometres in diameter to boulders tens of metres across. More than 400,000 of these now have such good orbits that they possess a numbered designation and almost 19,000 have been named after mythological gods, famous people, scientists, astronomers and institutions. Most of these objects live between Mars and Jupiter but some 10,000 have been discovered between the Sun and Mars, and more than 1,400 of these are classed as potentially hazardous asteroids (PHAs) due to their ability to pass within eight million kilometres of the Earth while also having a diameter greater than roughly 200 metres. The first four asteroids to be discovered were (1) Ceres, now regarded as a dwarf planet, (2) Pallas, (3) Juno and (4) Vesta, which are all easy binocular objects when at their peak, due to them having diameters of hundreds of kilometres.

In 2015, most of the first ten numbered asteroids reach opposition during the year. However, with orbital periods of, typically, four or five years, it is often the case that the Earth cannot 'overtake' an asteroid on the inside track, so to speak, in a specific twelve-month period. This is the case when opposition has occurred late in the previous year. In these instances the asteroid will not be transiting the meridian at midnight during any month in 2015 and so will not be seen at its brightest and closest during the year. So, this year the asteroid (5) Astrae will not quite achieve opposition and neither will (6) Hebe; these two minor planets will have to wait until February and March 2016 respectively before they transit the meridian at local midnight.

As far as the remaining top ten asteroids are concerned, the tenth is the first to reach opposition, literally as the year begins, as (10) Hygiea is best placed as 2014 ends and 2015 starts. In January it can be found in western Gemini and on 17 January, it is precisely one degree due north of the bright magnitude 2.9 star mu Geminorum, known as Tejat Posterior. At magnitude 10.5 you will need large tripod-mounted binoculars or a small telescope to easily identify it. Nevertheless, Hygiea's westerly motion of 11 arc minutes per day should enable the patient observer to pin it down given a few clear nights and a bit of determination. The minor planet (3) Juno is also best placed in January, reaching opposition at the end of the month near the head of Hydra, just five degrees below the Hydra/Cancer border. On 21/22 January, it passes just one degree below magnitude 4.4 sigma, also known as Al Minliar al Shuja. At magnitude 8.2 Juno should be rather easier to identify than Hygiea.

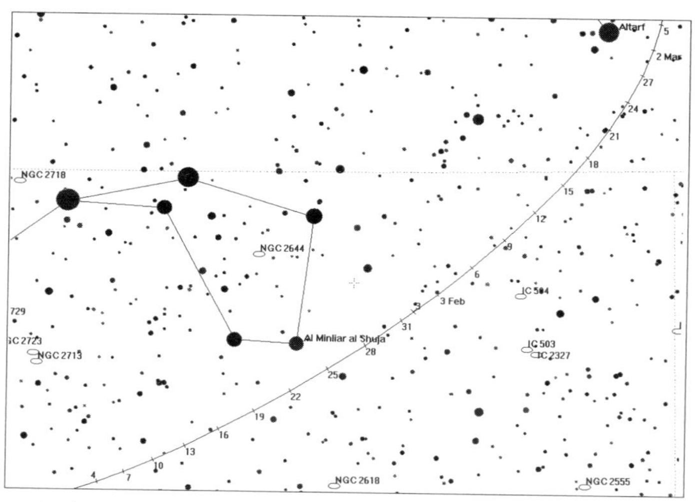

Figure 1. The track of the minor planet Juno as it passes from the head of Hydra into Cancer during January and February 2015.

The next minor planet in the top ten to reach opposition will be (8) Flora, which peaks around 20 February near to the unmistakeable 'Sickle' or backwards question mark that denotes the head of Leo the Lion. Flora will be around magnitude 9.1 in late February, roughly seven degrees north of the bright star Regulus and travelling north-west at a rate of roughly a quarter of a degree per day. By the start of April, the asteroid will have faded to magnitude 10.1 but will lie just ten arc minutes south of the centre of galaxy NGC 2903, which could be a photo opportunity for astro-imagers.

During the first week of March, the minor planet (7) Iris reaches opposition in southern Leo, very close to the Sextans border, some twenty degrees south-east of the bright star Leo. Iris should be a healthy magnitude 8.9 around this time and, as with Flora, its motion will be towards the north-west at a quarter of a degree per day. On 7 March, the minor planet crosses the aforementioned border and enters Sextans.

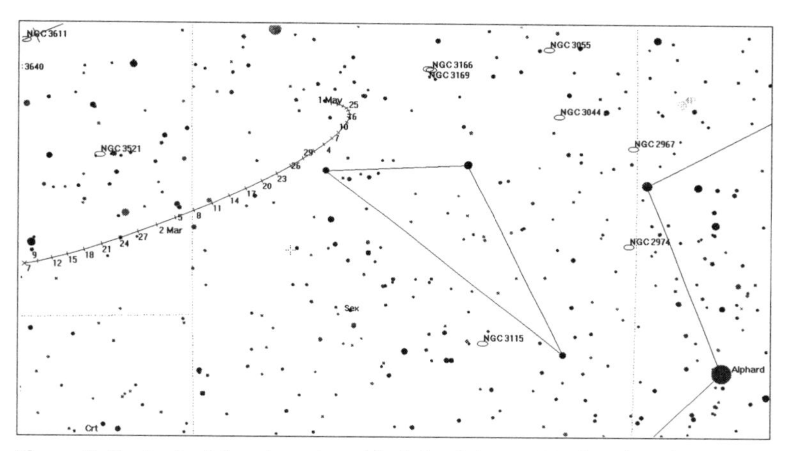

Figure 2. The track of the minor planet Iris during February, March and April 2015 as it passes to the East of the faint triangle of Sextans.

During June and July, the first two asteroids to be numbered, namely (1) Ceres and (2) Pallas, reach opposition, although strictly speaking Ceres is now regarded as a dwarf planet due to its large diameter. The first to reach opposition is Pallas, which peaks around mid-June at magnitude 9.4. It will be very well placed from European and North American latitudes as the asteroid will be travelling slowly through Hercules, a few degrees East of magnitude 3 delta Herculis, also known as Sarin.

Ceres itself reaches opposition at the end of July, peaking at a healthy magnitude 7.5. However, unlike Pallas, the dwarf planet will not be well placed for Northern Hemisphere observers, as it will lie in eastern Sagittarius, close to a declination of –30 degrees and not far from the border with Microscopium, which it crossed on 24/25 July. However, the Ceres media peak and the asteroid highlight of 2015 is scheduled to occur much earlier in the year, because the NASA Dawn spacecraft, which left Vesta on 5 September 2012, is due to arrive at Ceres in February.

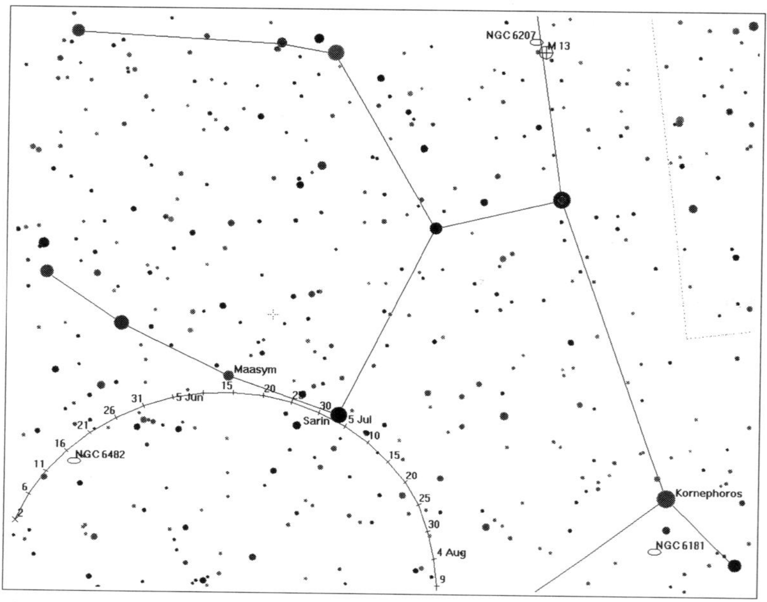

Figure 3. The track of the minor planet (12) Pallas curves beneath the star delta Herculis (Sarin) during May, June and July 2015.

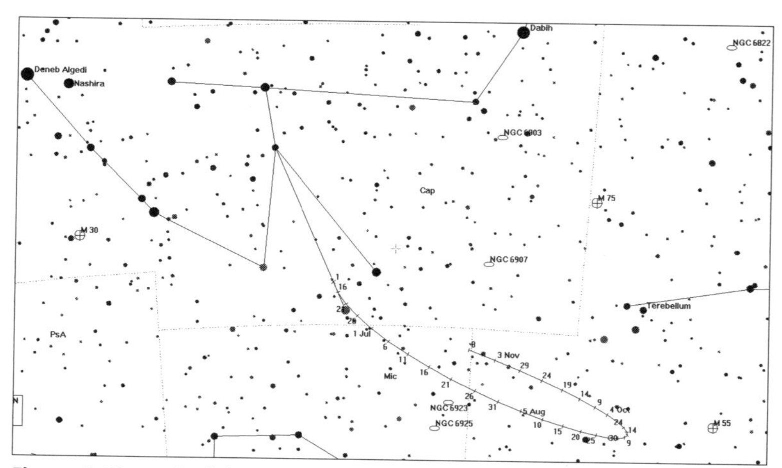

Figure 4. The track of the minor planet Ceres as it loops through Capricornus, Microscopium and Sagittarius, from mid-June to the start of November 2015.

Finally, the month of September will see (9) Metis and (4) Vesta reach opposition. Metis is in Aquarius, close to a declination of –16 degrees and at a respectable magnitude 9.2. At its peak, on 8/9 September, it is located just a few degrees east of magnitude 3.3 delta Aquarii (Skat). The brightest asteroid opposition occurs at the end of the month when (4) Vesta peaks at a splendid magnitude 6.2 in the constellation of Cetus on 30 September. It will only be at a declination of –9 degrees, but will still be a binocular object from UK latitudes, some four degrees east of magnitude 3.5 iota Ceti, also known as Deneb Kaitos Shemali.

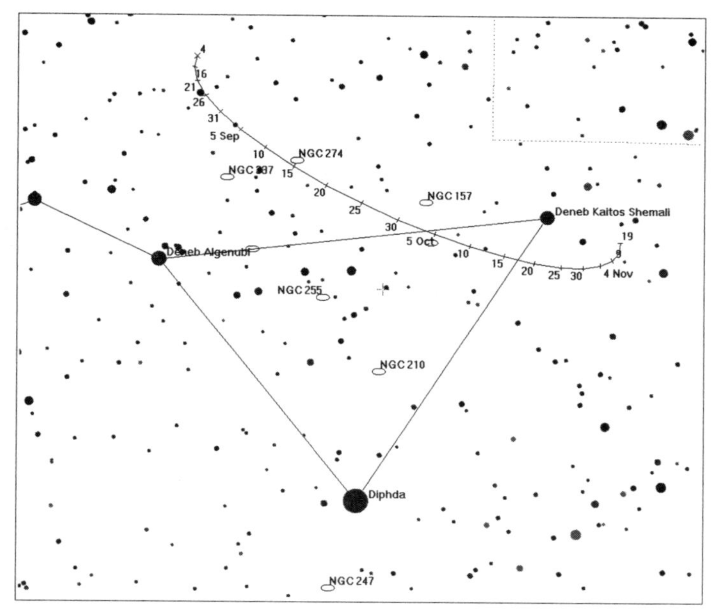

Figure 5. The track of the minor planet Vesta as it curves through the constellation of Cetus from August to November 2015.

Ephemerides for the best-placed bright minor planets at, or approaching, opposition in 2015, listed in calendar order, appear in the following tables.

(10) Hygiea

Date	RA (2000)			Dec.			Distance from Earth	Distance from Sun	Elongation from Sun	Mag.
	h	m	s	°	'	"	AU	AU	°	
Jan 01	06	37	42.1	+23	34	21	2.399	3.382	178.5	10.1
Jan 06	06	33	15.5	+23	33	46	2.401	3.378	172.5	10.2
Jan 11	06	28	56.2	+23	32	35	2.411	3.374	166.4	10.3
Jan 16	06	24	50.4	+23	30	49	2.428	3.370	160.4	10.4
Jan 21	06	21	04.0	+23	28	32	2.452	3.366	154.4	10.5
Jan 26	06	17	42.3	+23	25	48	2.483	3.362	148.6	10.6

(3) Juno

Date	RA (2000)			Dec.			Distance from Earth	Distance from Sun	Elongation from Sun	Mag.
	h	m	s	°	'	"	AU	AU	°	
Jan 15	08	45	04.1	+01	37	09	1.329	2.261	155.5	8.2
Jan 20	08	40	52.6	+02	12	01	1.324	2.274	160.0	8.2
Jan 25	08	36	29.8	+02	52	36	1.326	2.287	163.6	8.2
Jan 30	08	32	06.0	+03	37	54	1.334	2.300	165.3	8.1
Feb 04	08	27	51.3	+04	26	44	1.349	2.313	164.4	8.2
Feb 09	08	23	54.7	+05	17	51	1.370	2.327	161.5	8.3

(8) Flora

Date	RA (2000)			Dec.			Distance from Earth	Distance from Sun	Elongation from Sun	Mag.
	h	m	s	°	'	"	AU	AU	°	
Feb 05	10	13	05.2	+16	58	00	1.283	2.251	165.4	9.2
Feb 10	10	08	04.6	+17	41	22	1.280	2.260	170.9	9.1
Feb 15	10	02	49.6	+18	23	18	1.284	2.268	174.0	9.0
Feb 20	09	57	31.8	+19	02	29	1.295	2.277	171.5	9.1
Feb 25	09	52	23.2	+19	37	46	1.312	2.286	166.2	9.3
Mar 02	09	47	35.0	+20	08	20	1.336	2.294	160.4	9.4

(7) Iris

Date	RA (2000)			Dec.			Distance from Earth	Distance from Sun	Elong-ation from Sun	Mag.
	h	m	s	°	'	"	AU	AU	°	
Feb 20	11	06	31.5	−03	49	43	1.574	2.524	159.7	9.1
Feb 25	11	01	47.5	−03	26	17	1.567	2.536	165.0	9.0
Mar 02	10	56	54.6	−02	59	14	1.567	2.548	169.4	8.9
Mar 07	10	52	01.4	−02	29	27	1.574	2.559	171.0	8.9
Mar 12	10	47	16.6	−01	57	51	1.589	2.570	168.7	8.9
Mar 17	10	42	48.2	−01	25	26	1.610	2.582	164.2	9.1

(2) Pallas

Date	RA (2000)			Dec.			Distance from Earth	Distance from Sun	Elong-ation from Sun	Mag.
	h	m	s	°	'	"	AU	AU	°	
June 01	17	41	45.0	+25	05	57	2.399	3.149	130.1	9.4
June 06	17	37	38.1	+25	22	50	2.399	3.158	130.9	9.4
June 11	17	33	23.0	+25	31	57	2.403	3.166	131.2	9.4
June 16	17	29	05.4	+25	33	10	2.413	3.174	131.1	9.4
June 21	17	24	51.5	+25	26	28	2.427	3.182	130.5	9.4
June 26	17	20	47.3	+25	12	07	2.445	3.190	129.5	9.4

(1) Ceres

Date	RA (2000)			Dec.			Distance from Earth	Distance from Sun	Elong-ation from Sun	Mag.
	h	m	s	°	'	"	AU	AU	°	
July 15	20	36	46.6	−29	12	11	1.949	2.940	164.2	7.6
July 20	20	32	23.7	−29	42	48	1.941	2.942	167.8	7.5
July 25	20	27	48.7	−30	11	09	1.940	2.944	169.2	7.5
July 30	20	23	09.3	−30	36	39	1.946	2.946	167.8	7.5
Aug 04	20	18	33.0	−30	58	49	1.959	2.948	164.2	7.6
Aug 09	20	14	07.3	−31	17	20	1.978	2.950	159.6	7.7

(9) Metis

Date	RA (2000)			Dec.			Distance from Earth	Distance from Sun	Elong-ation from Sun	Mag.
	h	m	s	°	'	"	AU	AU	°	
Aug 25	23	23	15.8	−14	20	29	1.393	2.377	162.6	9.4
Aug 30	23	19	04.7	−14	52	47	1.375	2.370	167.1	9.3
Sept 04	23	14	32.8	−15	23	56	1.363	2.363	170.0	9.2
Sept 09	23	09	48.4	−15	52	49	1.358	2.356	169.7	9.2
Sept 14	23	05	00.9	−16	18	22	1.360	2.349	166.4	9.3
Sept 19	23	00	20.6	−16	39	39	1.367	2.343	161.7	9.3

(4) Vesta

Date	RA (2000)			Dec.			Distance from Earth	Distance from Sun	Elong-ation from Sun	Mag.
	h	m	s	°	'	"	AU	AU	°	
Sept 15	00	51	30.1	−07	13	27	1.438	2.405	159.2	6.4
Sept 20	00	47	24.3	−07	49	49	1.429	2.410	163.7	6.3
Sept 25	00	42	58.9	−08	24	31	1.427	2.415	167.0	6.2
Sept 30	00	38	22.4	−08	56	28	1.431	2.419	167.9	6.2
Oct 05	00	33	43.4	−09	24	40	1.441	2.424	166.0	6.3
Oct 10	00	29	10.8	−09	48	15	1.458	2.429	162.2	6.4

NEAR-EARTH ASTEROID APPROACHES

A list of some of the most interesting numbered, named and provisionally designated close-asteroid approaches during 2015 is presented in the following table. It should be borne in mind that the visibility of close-approach asteroids is highly dependent on whether they are close to the solar glare and from which hemisphere the observer is based, but by their very nature they move rapidly across the sky. The table gives the closest separation between the Earth and the asteroid in AU (1 AU = 149.6 million km) and the date of that closest approach and also the constellation in which the object can be found when closest. The brightest magnitude achieved and the corresponding date and constellation for that condition are also given. The reader might wonder why the dates when the Near-Earth asteroid is closest and when it is brightest are not the same. Well, sometimes the dates do coincide, but if an object is not behind the Earth, it may have a poor phase (like the crescent Moon) with most of it in shadow, regardless of its proximity to the Earth. Some of the objects listed are as faint as magnitude 17 at their best and so present a real challenge, even to advanced amateur astronomers using CCDs. However, some are much brighter and the objects 2004 BL86, 1998 WT24, 2008 CM, Icarus, 1999 JD6, Sekhmet and 2002 FG7 are within visual telescopic range. Icarus is the most famous 2015 close encounter and was first discovered in June 1949. It tends to have closest approaches to us at intervals of nine, nineteen and twenty-eight years, but always in June. Icarus is just over one kilometre across and will come within 8.1 million kilometres of Earth on 16 June.

Scores of other tiny asteroids, with provisional designations, will come within advanced amateur CCD range during 2015, many of them undiscovered as this Yearbook goes to press.

In total there are some 200 known minor planets that will come within 0.2 AU (30 million kilometres) of Earth in 2015, some tiny and others not so tiny! Twelve of these are classed as Potentially Hazardous Asteroids (PHAs), meaning that they have an absolute magnitude of 22.0 or higher (corresponding to a size greater than 110–240 metres) and can pass within 0.05 AU (7.5 million kilometres) of the Earth. Many of the best, brightest and closest encounters for 2015 are listed in the table.

Some interesting numbered, named and provisional designation close-asteroid approaches during 2016. The peak brightness dates for 2005 YQ96 and (310442) 2000 CH59 are for 29 Dec 2014 and 2 Jan 2016 respectively. All other dates are in 2015.

Asteroid (number)/ designation/name	Closest (AU)	2015 date when closest	Constell. closest	Peak mag.	Peak date	Const. brightest
2005 YQ96	0.02652	Jan 2.91	Scorpius	17.3	29 Dec	Boötes
(85990) 1999 JV6	0.08331	Jan 5.47	Crater	16.2	18 Jan	Leo
(162004) 1991 VE	0.10433	Jan 17.15	Draco	16.0	12 Jan	Ursa Major

Asteroid (number)/ designation/name	Closest (AU)	2015 date when closest	Constell. closest	Peak mag.	Peak date	Const. brightest
2012 BZ1	0.03944	Jan 22.14	Auriga	17.9	20 Jan	Gemini
2004 BL86	0.00802	Jan 26.68	Cancer	9.2	27 Jan	Cancer
(90416) 2003 YK118	0.07823	Feb 27.81	Cetus	16.0	16 Feb	Taurus
(141527) 2002 FG7	0.04403	Mar 13.43	Chamel.	14.2	16 Mar	Centaurus
(2063) Bacchus	0.19520	Apr 7.80	Auriga	16.4	17 Apr	Lynx
(285331) 1999 FN53	0.06776	May 14.48	Taurus	17.3	1 May	Camelop.
(5381) Sekhmet	0.1613	May 17.36	Cor. Aust.	14.2	18 May	Scorpius
(1566) Icarus	0.05384	June 16.65	Canes Ven	13.5	19 June	Boötes
2010 LN14	0.04829	June 21.80	Cetus	17.3	18 June	Grus
2010 NY65	0.04399	June 25.90	Leo	17.9	30 June	Boötes
(164202) 2004 EW	0.07051	July 11.95	Caelum	17.9	2 July	Tucana
(242191) 2003 NZ6	0.08333	July 18.75	Canes Ven	16.7	12 July	Draco
2011 UW158	0.01644	July 19.61	Ursa Maj.	14.6	22 July	Ursa Min.
(85989) 1999 JD6	0.04842	July 25.20	Camelop.	13.8	22 July	Cepheus
(232691) 2004 AR1	0.07481	Aug 13.65	Eridanus	17.0	8 Aug	Fornax
(250458) 2004 BO41	0.14729	Aug 31.65	Auriga	16.5	9 Sept	Cepheus
(281375) 2008 JV19	0.04472	Sept 1.51	Indus	15.6	24 Aug	Pisc. Aust.
2012 CL19	0.04471	Oct 3.98	Mensa	17.5	28 Sept	Phoenix
2009 TK	0.04510	Oct 5.49	Aries	16.8	6 Oct	Aries
(86666) 2000 FL10	0.16888	Oct 10.27	Capricorn	14.9	5 Oct	Aquarius
2009 FD	0.0419	Oct 29.43	Aquila	18.0	2 Nov	Pegasus
1999 TT16	0.0385	Nov 10.10	Lyra	15.1	7 Nov	Lacerta
(138852) 2000 WN10	0.12589	Nov 11.47	Eridanus	17.3	17 Nov	Eridanus
2005 UL5	0.01527	Nov 20.18	Sextans	14.4	19 Nov	Hydra

Asteroid (number)/ designation/name	Closest (AU)	2015 date when closest	Constell. closest	Peak mag.	Peak date	Const. brightest
2009 WB105	0.03891	Nov 25.74	Taurus	17.0	26 Nov	Taurus
(33342) 1998 WT24	0.02799	Dec 11.55	Taurus	11.3	11 Dec	Taurus
1995 YR1	0.04346	Dec 23.78	Hydra	16.3	22 Dec	Sextans
(163899) 2003 SD220	0.07297	Dec 24.55	Virgo	14.8	14 Dec	Virgo
(310442) 2000 CH59	0.07073	Dec 26.73	Antlia	16.1	2 Jan	Puppis
(294739) 2008 CM	0.05846	Dec 29.01	Leo	12.6	31 Dec	Hydra

Accurate minor planet ephemerides can be computed for your location on Earth using the MPC ephemeris service at: www.minorplanetcenter.org/iau/MPEph/MPEph.html.

Figure 6. The minor planet 1998 WT24 makes a close approach to the Earth on 11 December this year. Fourteen years earlier this author captured it speeding through Auriga on 14 December 2001, using a 30-cm Schmidt-Cassegrain. The asteroid trails represent seven sixty-second exposures and the gap in the centre is due to cloud interference. The image is twelve arc minutes wide. Image: Martin Mobberley.

Meteors in 2015

JOHN MASON

Meteors (popularly known as 'shooting stars') may be seen on any clear moonless night, but on certain nights of the year their number increases noticeably. This occurs when the Earth chances to intersect a concentration of meteoric dust moving in an orbit around the Sun. If the dust is well spread out in space, the resulting shower of meteors may last for several days. The word 'shower' must not be misinterpreted – only on very rare occasions have the meteors been so numerous as to resemble snowflakes falling.

If the meteor tracks are marked on a star map and traced backwards, a number of them will be found to intersect in a point (or within a small area of the sky) which marks the radiant of the shower. This gives the direction from which the meteors have come.

Bright moonlight has an adverse effect on visual meteor observing, and for about five days to either side of full Moon, lunar glare swamps all but the brighter meteors, reducing, quite considerably, the total number of meteors seen. Of the major showers in 2015, the maxima of the Quadrantids, Eta Aquarids and Ursids will all suffer interference by moonlight. Quadrantid maximum coincides with a Moon only two days from full, which is a pity as the peak is near midnight, a good time for observers in the UK. Although the Eta Aquarids coincide with a Moon just past full, shower members may be observed in the eastern sky as the radiant rises before dawn, with the Moon descending in the west. The Ursids coincide with a waxing gibbous Moon, but visual observers may minimize the effects of moonlight by positioning themselves so the Moon is behind them and hidden behind a wall or other suitable obstruction. The complex of showers which peak in late July or very early August, e.g. the Delta Aquarids, Piscis Australids and Alpha Capricornids, are also unfavourably placed with respect to the Moon this year.

There are many excellent observing opportunities in 2015. The peak of the April Lyrids occurs just after New Moon, so the shower will be observable with no interference from moonlight. The peak of the Perseids, one of the year's most active showers, in August coincides with new Moon, so conditions are very favourable. With maximum expected at around 06h, the pre-dawn hours of 13 August are likely to yield the best observed rates, but activity should also be high on 11/12 August and 13/14 August. The autumn of 2015 also looks good for meteor observers with both the Orionids in October and the Leonids in November being observable in reasonably dark skies.

In December it is hoped that observers will take advantage of the very favourable conditions for the Geminids which are now the richest of the annual meteor showers. This year's maximum occurs just after new Moon so there will be no interference by moonlight at all this year. Unfortunately, the peak occurs during daylight hours from the UK, so the highest-observed rates are most likely in the pre-dawn hours of 14 December and during the following evening. Past observations show that bright Geminids become more numerous some hours after the rates have peaked, a consequence of particle-sorting in the meteoroid stream.

As always, observations away from the major shower maxima, and of year-round sporadic activity, are every bit as important as those obtained when high rates are anticipated.

The following table gives some of the more easily observed showers with their radiants; interference by moonlight is shown by the letter M:

Limiting Dates	Shower	Maximum	Radiant RA		Dec.	
			h	m	°	
1–6 Jan	Quadrantids	4 Jan, 00h	15	28	+50	M
19–25 Apr	Lyrids	22 Apr, 23h	18	08	+32	
24 Apr–20 May	Eta Aquarids	5–6 May	22	20	−01	M
17–26 June	Ophiuchids	10,20 June	17	20	−20	
July–Aug	Capricornids	8,15,26 July	20	44	−15	
			21	00	−15	
15 July–20 Aug	Delta Aquarids	29 July, 6 Aug	22	36	−17	M
			23	04	+02	
15 July–20 Aug	Piscis Australids	31 July	22	40	−30	M
15 July–20 Aug	Alpha Capricornids	2–3 Aug	20	36	−10	M
July–Aug	Iota Aquarids	6–7 Aug	22	10	−15	
			22	04	−06	

Limiting Dates	Shower	Maximum	Radiant RA		Dec.	
			h	m	°	
23 July–20 Aug	Perseids	13 Aug, 06h	3	13	+58	
16–30 Oct	Orionids	21-24 Oct	6	24	+15	
20 Oct–30 Nov	Taurids	5,12 Nov	3	44	+22	
			3	44	+14	
15–20 Nov	Leonids	18 Nov, 07h	10	08	+22	
Nov–Jan	Puppid-Velids	early Dec	9	00	−48	
7–16 Dec	Geminids	14 Dec, 13h	7	32	+33	
17–25 Dec	Ursids	22-23 Dec	14	28	+78	M

Some Events in 2016

ECLIPSES

There will be five eclipses, two of the Sun and three of the Moon.

9 March:	Total eclipse of the Sun – Indian Ocean, Indonesia, South Pacific and Micronesia.
23 March:	Penumbral eclipse of the Moon – Alaska, North America, Hawaii, Pacific Ocean, Australia, New Zealand, Australia and Eastern Asia.
18 August:	Penumbral eclipse of the Moon – North America, Pacific Ocean and Australia.
1 September:	Annular eclipse of the Sun – Atlantic Ocean, Gabon, Zaire, Tanzania, Mozambique, Madagascar and Indian Ocean.
16 September:	Penumbral eclipse of the Moon – Australia, Asia, Eastern Africa and Europe.

THE PLANETS

Mercury may be seen more easily from northern latitudes in the evenings about the time of greatest eastern elongation (18 April) and in the mornings about the time of greatest western elongation (28 September). In the Southern Hemisphere the corresponding most favourable dates are 16 August and 11 December (evenings) and 7 February and 5 June (mornings). There will be a transit of Mercury across the Sun on 9 May.

Venus is visible in the mornings from January until May. It passes through superior conjunction on 6 June and will then become visible in the evenings from July until the end of the year.

Mars is at opposition on 22 May in Scorpius.

Jupiter is at opposition on 8 March in Cancer.

Saturn is at opposition on 3 June in Libra.

Uranus is at opposition on 15 October in Pisces.

Neptune is at opposition on 2 September in Aquarius.

Pluto is at opposition on 7 July in Sagittarius.

Part Two

Article Section

Observing and Imaging the Sun

PETE LAWRENCE

INTRODUCTION

The Sun is unique in that it gives us an opportunity to observe a star in close-up. At a distance of 150 million kilometres, this 1.4 million-kilometre ball of predominantly hydrogen gas presents itself as a disc measuring approximately half a degree across in the sky. That's more or less the same apparent diameter as the Moon and thirty times the largest apparent planetary diameter possible, presented by Venus when close to inferior conjunction.

At this size, it's possible to see things on and around the Sun which also have tangible size and detail in their own right. As the Sun is a dynamic body, these features can be seen to alter shape and dimension over relatively short time periods, typically measured in days but sometimes altering in just hours, minutes and even in real time.

As we become more reliant on electronics, communications and satellite location services, so there is a greater focus on studying the effects of the Sun's output on our fragile world. These effects go under the collective header 'space weather' and there are now many detectors around Earth and in space that monitor and report on the various outputs from the Sun that may result in adverse space weather effects.

Some of these reports are of great use to amateur solar observers as they provide a heads-up warning to something interesting going on. Examples here include the GOES X-ray flux monitoring system (http://www.swpc.noaa.gov/rt_plots/xray_1m.html), and the excellent GONG (Global Oscillation Network Group – http://halpha.nso.edu/) solar activity resource.

Interest in solar observing is a growth area in astronomy, and for good reason. Unlike observing the stars at night, it's an obvious requirement that you can only observe the Sun during the day when the weather is reasonable (a rainy day doesn't tend to result in much sun!) and the temperature comfortable. That said, a clear sunny day in winter can often carry quite a chill!

Daytime astronomy also provides an abundance of light, which can be a double-edged sword. Lots of light avoids the headaches of night-time astronomy where a certain amount of careful planning is required in order to avoid leaving things out or behind when you pack up. However, if you plan to use something with a screen, such as a laptop or a digital camera, then you'll need to prepare yourself with some means to black out the Sun's bright light. This can range from an elegant 'tent' made from a frame and a photographer's blackout curtain, to an inelegant, but often highly efficient, cardboard box with a slit cut in it. Here, the laptop goes in the box and you view the screen through the slit.

Figure 1. The author's makeshift solar observing 'tent'. (All images courtesy of the author.)

The abundance of light from the Sun also leads to another issue that solar observers need to take very seriously. This is, of course, eye and equipment safety. I'll cover this in more detail in a moment, but it's worth bearing in mind that the Sun is the only normal astronomical object that can do you serious harm if you don't give it constant thought and respect. Being in the wrong place at the wrong time when a meteorite hits may also fall into this category, but this is extremely unlikely and not what would be classed as a normal event!

The growth in amateur interest in solar astronomy has led to the development of many new products designed to help observers view the Sun safely and in exotic wavelengths. The filters that are required for specialist wavelength views were once the preserve of the rich or well-funded university department, but demand has created a niche market that has driven the cost of such items into a more affordable price bracket.

The net effect is that it's now quite common to see groups of daytime astronomers equipped with specialist equipment such as hydrogen-alpha (H-alpha) filters, offering the general public their first ever view of our nearest star in this exciting wavelength.

SAFETY

It hopefully goes without saying that observing the Sun can be dangerous. Staring at the Sun without adequate protection can result in damage to the light receptors in the retina, leading to distorted vision or, in extreme cases, permanent blindness. Optical equipment is also at the mercy of the Sun, and pointing any optical instrument directly at its disc should be avoided unless adequate protection has been taken. Usually, adequate protection would constitute a certified solar safety filter or dedicated solar filter for specialized wavelength observation – e.g. an H-alpha or calcium-K filter.

The exception here is the technique of solar projection, which requires a telescope to be deliberately pointed towards the Sun with no filter in place. The technique requires the telescope to be set up with an eyepiece in place, and the resulting image is simply projected onto a screen of some sort. This can be as simple as a hand-held piece of white card.

The technique of solar projection is only suitable for refractors, typically smaller than, say, six inches in aperture. Reflectors and catadioptric telescopes, such as SCTs or Maksutovs, are not suitable for this method and using them may result in serious equipment damage. In addition, it is necessary to cap or remove any finders fitted to the telescope being pointed at the Sun.

The simplest method of aligning a filtered telescope to the Sun is to look at its shadow on the ground. When the optical tube's shadow area is minimized, the scope should be pointing in the general vicinity of the

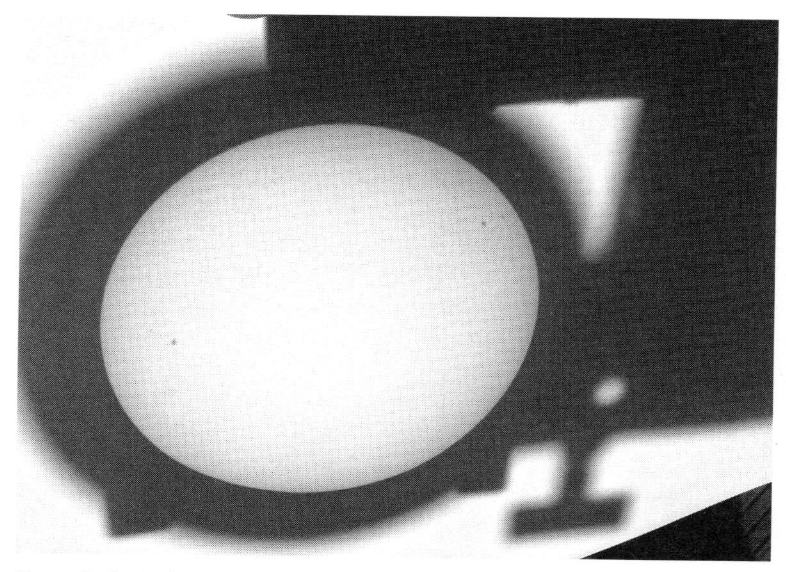

Figure 2. The Sun's image projected onto a piece of white card.

Sun's disc. Alternatively, there are now solar finders available which can be used in place of the telescope's normal optical finder.

Solar observing isn't dangerous as long as you observe a couple of important guidelines. If you're not sure about what you're doing, then the advice is, don't do it. Instead get advice from someone who does know. Common sense is imperative when observing the Sun. As obvious as this may sound, it's very easy for common sense to evaporate if you've been up all night star-gazing and then decide to set up for solar work. In such circumstances, it's wise to double- or triple-check everything you do, just to be sure.

THE WHITE LIGHT SUN

The most common method of observing the Sun is to view it in white light. The projection method described earlier, which is suitable for refractors of aperture of less than six inches (150mm), is one way of doing this. At one time, this was by far the most common method for safe solar observing, but it has been replaced by another technique which uses a certified solar safety filter.

There are various companies which offer such filters, but the one which has become almost synonymous with solar white-light observing is Astrosolar film from Baader Planetarium. This is a thin film material (not Mylar as it's often mistakenly described) with a reflective coating designed to dim the Sun's intensity by 99.999 per cent. The filter material is typically bought as an A4 sheet and it's up to you to create your own filter. This isn't as difficult (or dangerous) as it may initially sound and there are plenty of instructions available online and in publications to assist you (e.g. http://www.baader-planetarium.com/sofifolie/bauanleitung_e.htm on the Baader website).

The material is available in two strengths. Optical density (OD) 3.8 is suitable for imaging applications and not designed for visual use. The correct version for visual observing is OD5.0 film which further dims the Sun's light and is safe for extended sessions. OD5.0 Astrosolar film can also be used for imaging, so if you're unsure what to get, go for the more commonly available OD5.0 version for safety.

The basic premise for its use is that it needs to be made into a filter that can slip over the end of a telescope to entirely cover the telescope's

Figure 3. A homemade white-light solar filter in situ.

aperture. A common way to do this is to create a card cylinder that can be slipped off and on the front of the scope. The filter material is then cut to size and securely fixed over one end of the card ring. A second ring is typically created to finish the filter off, and to tidy up any loose ends of film stuck outside the inner ring.

It's imperative to check the filter's integrity before use. If there are any holes or rips in it, the filter should be discarded and a new one made. An A4 sheet of Astrosolar film typically costs less than £25 at the time of writing, which, in normal astronomical terms, isn't that expensive. If you cut the film carefully, it's possible to get several filters out of one sheet.

A big advantage to this type of filter is that it allows any type of telescope to be used for white-light solar observing. As long as the entire front aperture is covered, you're safe. If you have a telescope with an aperture larger than an A4 sheet, the solutions are to make an aperture mask or buy a larger roll of film. At the time of writing, a 1.0 x 0.5 m roll of Astrosolar film is available which costs around £80. If you know of several fellow observers who want to try white-light solar astronomy, it may be possible to come to an arrangement to spread the cost and create several filters out of one large sheet.

Figure 4. An offset aperture mask solar filter.

An aperture mask is basically an opaque mask that entirely covers a telescope's full aperture. A smaller circular hole is then cut in this mask and it's this hole that's completely covered with the safety film.

Look after a white-light solar filter and it'll give you years of use. But do remember to check its integrity before use!

In the interests of fairness, there are other types of solar safety filters available, such as those manufactured by Thousand Oaks (http://www.thousandoaksoptical.com/solar.html). In addition, there are also other ways to observe the Sun in white light too. An example here which has regained popularity in recent years is to use a Herschel Wedge. This is a device which, again, is only suitable for use with a refractor. It fits in the eyepiece end of the telescope and diverts virtually all of the incoming sunlight away, leaving a safe level to view visually.

WHITE-LIGHT FEATURES

The white-light Sun shows a multitude of interesting features. The Sun's disc represents the visible surface of the Sun's gaseous globe and

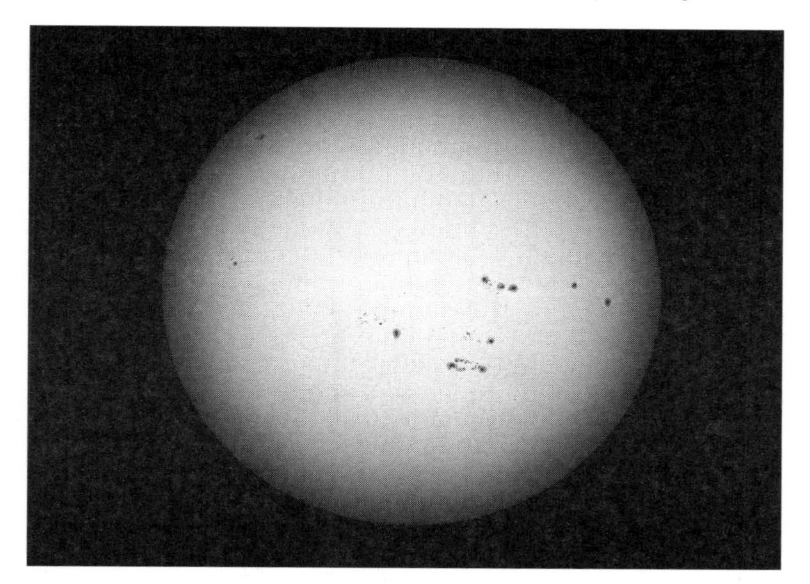

Figure 5. The Sun in white light with a good show of sunspots.

is called the photosphere (sphere of light). With good seeing, it's possible to see that the photosphere is not smooth but, rather, has a textured appearance, a bit like an eggshell or rice paper. Internal convective cells rise to the photosphere where we get to see them end on. It's the collective effect of these cells that gives rise to this texture. The fine pattern is called solar granulation and each granule typically measures 1,500 km across and lasts for between 8 to 20 minutes. It's estimated that around 4 million granules cover the Sun's globe at any one time.

Another interesting effect can be seen when looking at the entire solar disc or at least a significant part of it which includes the limb. As the limb is approached, so the photosphere appears darker, an effect called limb darkening. This occurs because towards the edge of the Sun, the depths of solar gas you are looking into are cooler than at the centre.

Concentrations of magnetic fields poking through the photosphere can thin the gas, making it appear more transparent. When this occurs, it's possible to see deeper inside the Sun and this greater depth means you're looking at hotter material which appears brighter. In white light, these hotter windows into the inner Sun are best seen when contrasted against the limb darkening. Here they appear as faculae which are seen on their own or clustered around sunspot regions.

Sunspots are the most dynamic and interesting aspect of the white-light Sun. Individual spots may occur in isolation or in groups collectively known as an active region. The magnetic fields in an active region can be relatively simple or highly complex. The Sun's magnetic field becomes concentrated and intense as a consequence of the Sun's internal make-up. This leads to concentrations of magnetic flux lines poking out through the photosphere. This can interfere with the convective flow of plasma from the hotter internal regions below the photosphere with the net effect that at the photosphere the plasma cools relative to its surroundings. The cooling causes the region to appear darker than the surrounding photosphere, giving rise to a sunspot.

A typical sunspot has a dark inner core known as the umbra, and a lighter outer region known as the penumbra. The spot may appear circular and symmetrical although, more often than not, a large collection of sunspots will have a shared and complex penumbra with equally complicated umbral shapes.

It's fascinating to look at how sunspot groups change over time as well as recording their movement across the Sun's disc as it rotates. The Sun undergoes differential rotation where the polar zones rotate more

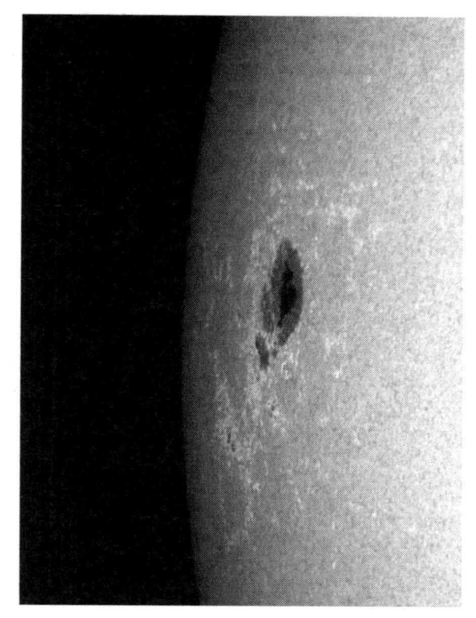

Figure 6. A large sunspot close to the Sun's limb showing an umbra, penumbra and bright faculae.

slowly than the equatorial regions. On average, it takes a sunspot about twenty-seven days to rotate back to the same position on the Sun's disc. Typically, a sunspot group will decay before this occurs but some large groups may survive for several rotations.

IMAGING THE WHITE-LIGHT SUN

Imaging the white-light Sun is relatively straightforward, thanks to the simplicity of construction of a white-light solar filter. As there is an abundance of light, even with a white-light filter fitted, it's possible to image the Sun's disc with a multitude of cameras ranging from camera phones, right through to DSLRs. Cooled astronomical CCD cameras can also be used, but for high-resolution work, a high frame-rate planetary camera is best as this can circumvent some of the issues associated with seeing effects. Solar granulation can be further enhanced by using a monochrome (greyscale) planetary camera fitted with a green or solar continuum filter. Good granulation results may also be obtained by using a white light filter in conjunction with a deep-sky OIII filter.

NARROW BAND FILTERS

While white-light features are definitely interesting to observe and image, there are also exotic and more dynamic phenomena occurring at specific wavelengths which are effectively swamped in white light. Think of a radio which can be tuned to all stations simultaneously. Although there may be plenty of interesting stations in their own right, the cacophony of noise resulting from hearing them all at once masks them from your ears.

As the Sun is predominantly hydrogen, one interesting wavelength is that known as hydrogen-alpha, the principle excitation wavelength of hydrogen in the visible part of the spectrum. Hydrogen-alpha, or H-alpha, is also commonly met in deep-sky astronomy. Huge clouds of hydrogen, excited by nearby hot, young stars, glow with the characteristic red of H-alpha.

In order to see the Sun in the light of H-alpha, it's necessary to use an H-alpha filter. These come in a variety of forms and where they were once prohibitively expensive, they are now well within the reach of amateur solar observers and imagers. Before continuing, it would be prudent to point out that an H-alpha solar filter is completely different to the deep-sky H-alpha filter used for imaging nebulae at night. The deep-sky version is not suitable for observing or imaging the Sun when used on its own.

The range of wavelengths that an H-alpha filter passes needs to be very narrow; typically, 1 angstrom or less (1 angstrom = one ten-billionth of a metre), centred on 6562.8 angstroms. This level of precision places great demands on the manufacture of such filters, and this is why they carry such a high price tag.

A budget H-alpha solar telescope with an aperture around 40 mm can typically be bought for less than £700. Larger dedicated solar telescopes or filter systems, which can convert a night-time telescope for H-alpha solar use, typically cost thousands of pounds.

A fairly recent addition to the market which seems to be causing a stir is the Daystar Quark. This is an H-alpha eyepiece system, designed

Figure 7. The Personal Solar Telescope, or PST, is a 40 mm, f/10 dedicated solar telescope.

to fit in the back of a refractor. Use the unit on an 80-mm or smaller refractor for non-prolonged periods and there's no requirement to have front-mounted protection at all. Use a larger-than-80-mm aperture refractor and you should cover the front aperture with a suitable energy rejection filter.

There are two Quarks currently available – one optimized for the chromospheric 'surface' and one optimized for prominences. Removing the front-filter aperture limitation for around £800 each, these eyepieces may herald yet another jump forward in affordable high-resolution, H-alpha solar viewing and imaging.

A typical H-alpha telescope uses a Fabry–Pérot etalon to remove the bulk of unwanted wavelengths. It does this by causing interference in a high-precision optical cavity. A plot of wavelengths passed through the etalon shows peaks where constructive interference occurs and troughs of almost zero intensity where destructive interference occurs. The narrowness of the peaks and intensity of the troughs defines the finesse of the filter. Obtaining narrower peaks and lower trough intensity requires high precision, resulting in greater cost.

The passed wavelengths plotted on a graph look like the teeth of a comb. Multiple peaks are not useful and only the H-alpha peak is

Figure 8. An etalon fitted to a night-time telescope via a custom adapter.

actually required. The unwanted peaks are removed by a second optical device known as a blocking filter. The blocking filter is centred on the H-alpha wavelength but has a wider, and therefore less technically demanding, bandpass; the term used to describe the range of wavelengths passed. The blocking filter's bandpass is wide enough to allow the wavelength range from the troughs either side of the main H-alpha peak to pass, but not so wide as to extend to the neighbouring peaks. Consequently, all that gets through is the H-alpha peak.

THE H-ALPHA VIEW

The Sun seen in the light of H-alpha is dramatically different to that seen through a white-light filter. Looking at the H-alpha Sun, you're no longer looking at the photosphere but rather a thin layer of hydrogen that blankets it, which is called the chromosphere. The name 'chromosphere' literally means 'sphere of colour' and this comes from the fact that the chromosphere can be seen visually at the time of a total solar eclipse. As the last vestiges of the Sun disappear behind the Moon's disc immediately before totality, or as the Sun is re-emerging immediately

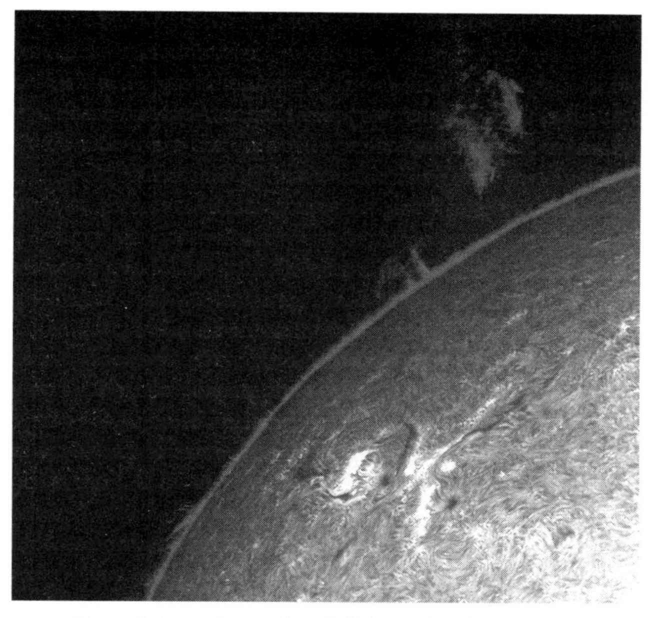

Figure 9. A prominence 'leaps' off the Sun's limb in H-alpha.

at the end of the total phase, a thin red-pink arc can be seen. This is the chromosphere and it is the intensity of its H-alpha colour that gives it its name.

The chromosphere is about 2,000 km thick. The region immediately above it represents an interface between the chromosphere and the Sun's atmosphere, or corona. This region extends out to around 10,000 km above the photosphere and is called the transition region. Here, jets of hydrogen plasma occur, typically measuring 500 km across and lasting up to 15 minutes. These spicules can be seen at the edge of the Sun with high-resolution equipment where they resemble a sort of solar 'fur'.

The chromosphere blankets the lower photosphere, hiding it from view. A common misconception is that the chromospheric 'surface', seen through an H-alpha set-up, shows photospheric granulation. In reality, on disc spicules and fibrils of plasma give a coarse, mottled texture on a much larger scale than photospheric granulation. This background chromospheric texture is known as dark mottling.

Longer dark clouds of hydrogen can often be seen snaking across the chromosphere. These are known as filaments and may appear connected with active regions or simply on their own, silhouetted against the chromosphere below. Quiescent filaments have formation associated with the remnants of decayed active regions and may persist for several solar rotations. A typical quiescent filament will change little over its lifetime. Over time, filaments may migrate towards their closest solar pole to form what's known as a polar crown.

Active region filaments (ARF), as their name suggests, are associated with active sunspot regions and these tend to be more dynamic. An ARF may change shape significantly over a relatively short space of time or erupt and disappear from view completely. The eruption may result in material being ejected away from the Sun in what's known as a coronal mass ejection or CME. A CME directed towards the Earth may have an effect on Earth's magnetic field, resulting in problems for communications satellites, large-scale power outages and, of course, the beautiful aurora.

Figure 10. A full solar disc in H-alpha exhibiting many different chromospheric and inner corona phenomena.

The principal mechanism for the release of a CME is a magnetic reconnection event where field lines effectively break and reconnect. The net result is a huge release of energy. Another by-product of a magnetic reconnection is the appearance of a flare. This is a dynamic brightening event which can change in brightness enormously over a short space of time ranging from minutes down to seconds in extreme cases. Flares tend to occur in regions of high magnetic flux, such as you'd get in and around an active region.

An active sunspot region looks completely different when viewed in H-alpha to how it looks through a white-light filter. The distinct and high-contrast appearance of a sunspot umbra and penumbra against the brighter photosphere is somewhat lost in H-alpha. The chromosphere blankets much of the region, camouflaging individual spots from view. The chromospheric blanket opens to reveal the main umbra below. Close to the limb this can produce a dramatic scene when the opening is seen somewhat sideways on and foreshortened.

Despite the loss of contrast between the sunspots and their surroundings, the view around an active region in H-alpha has a drama of its own. Here you'll find larger fibrils magnetically contorted by the strong fields that have caused the spot group to form. Hot and bright regions known as plages permeate active regions. Short timescale events such as Ellermann Bombs may also appear briefly. A typical Ellerman Bomb appears like a tiny pin-point star in or around an active region and may last for less than five minutes.

The dramatic H-alpha detail in and around an active region tends to change on timescales much shorter than those required to see changes in the white-light view. This makes them ideal for time-lapse imaging. A short cadence of just a few seconds between frames will typically show lots of small-scale movement and feature evolution.

H-alpha features occur from the chromosphere up into the solar corona where they may extend for many tens of thousands of kilometres. Filaments, for example, are not flat on the chromosphere but rather hang above it, like giant cool hydrogen clouds. As the Sun rotates, so these elevated features appear to move towards the Sun's limb. As they reach the limb, their elevated nature becomes obvious because they appear to protrude beyond the Sun's disc. At this point, the term filament is replaced with the term prominence.

Prominences are a highlight of the H-alpha Sun. They change on a daily basis, new versions appearing over the Sun's east limb, while old ones rotate out of view around the west limb. They can change in shape remarkably quickly and under exceptional circumstances can be seen to reshape or detach and leave the Sun's vicinity in real time.

A large, relatively quiet prominence can stick around for several days and it's fascinating to watch one slowly reveal itself over the eastern limb, similar to how an oncoming ship appears over a distant seaward horizon. At first you only get to see the towering heights of the feature but as time passes, the full, side-on drama of the prominence is revealed. As the Sun continues to rotate, so an eastern prominence may have enough density to survive the limb transition to become a dark

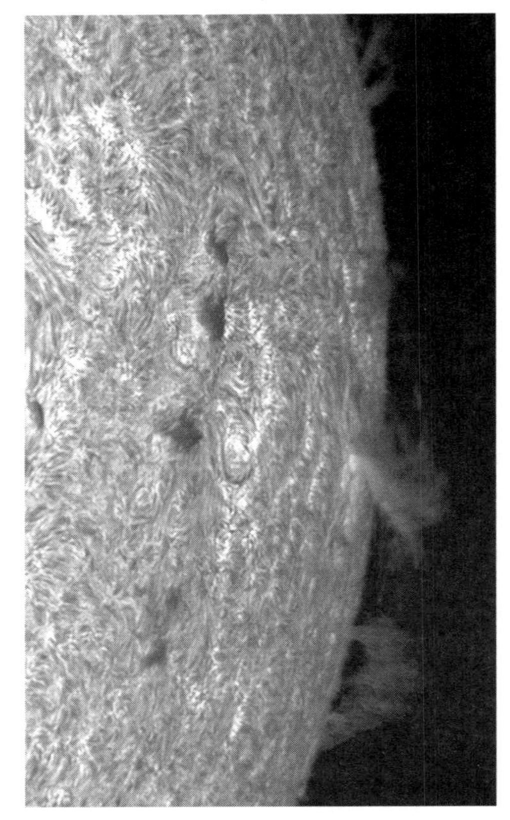

Figure 11. A 'filaprom'; part-filament and part-prominence.

filament. Many do not, simply becoming lost amongst the general magnetic mayhem of the chromosphere below.

A strong prominence may be seen bridging the transition line defined by the Sun's limb and appear as both part-prominence and part-filament. A term coined by amateurs that describes this hybrid feature is a 'filaprom'.

IMAGING THE H-ALPHA SUN

The best method to obtain an image of H-alpha solar features is to use a high frame-rate, monochrome (greyscale) planetary camera. Colour cameras struggle with the narrow red wavelength passed by the filter. DSLR cameras can be used, but the nature of the colour filters employed on what is essentially a monochrome sensor also creates difficulty.

As mentioned above, one popular and rewarding presentation method is to image H-alpha features with a high-resolution set-up to capture multiple frames which may then be used to create a time-lapse animation. Just a few tens of seconds between frame captures is normally enough to show considerable movement.

SUMMARY

What has been presented here is a short introduction to solar observing and imaging. Solar observing is a fascinating and dynamic part of astronomy and, as mentioned right at the start, provides us with a unique way of studying a star in close-up. As long as you observe the safety guidelines, the Sun can become a fascinating and daily companion. Its ever-changing range of features and dramatic, sometimes unexpected events make daily observing totally compelling.

The Astounding Horace Dall

MARTIN MOBBERLEY

Over the years, I have found that, as you get older, you become more appreciative of the elder statesmen of astronomy whom you met in your youth. At the time, I found, they never seemed quite so important. Yes, many of these characters were giants in the world of amateur astronomy, and some were certainly a bit daunting, but they were also from a different generation and so it was sometimes hard to see that you had much in common with men old enough to be your grandfather. However, decades later, and with these characters long since departed, you treasure the few memories you have of them, realizing that life is short and you will soon be as old as they were when you first met them! You realize that your astronomical youth was a unique and fleeting time when you could meet these fascinating characters, some fifty or sixty years older than yourself, in the brief period when your lifespans overlapped. As time passes you savour the encounters that you made with these astronomers from a bygone era and truly regret that others, whom you could have met, were not tracked down with a lot more effort.

I first met the legendary Horace Dall at a BAA Provincial meeting in Norwich on 15 September 1973, when he gave me some sage advice on testing my own home-made telescope mirror. Eleven years later, in February 1984, I spent a wonderful day with him, his Siamese cat and his wife Helena, at his magical Luton home, study, workshop and observatory. Despite being eighty-three years old, Horace was very sprightly. He was a small and slender man and this seemed to make him very agile. At that time he was still, after almost half a century, regarded as the UK's best lunar and planetary photographer and he was certainly the most knowledgeable authority on amateur astronomical telescopes. I imagined Horace would still be around in his nineties and so I was shocked, just two years later, when I heard that he had suffered a stroke, while trying to repair a friend's microscope, and had died in hospital. I took many photographs on the day I visited Horace Dall's home but, fortunately, two years earlier, one amateur astronomer had realized that a film of Horace just had to be made. In 1982, Robin Scagell used an early portable video recorder to capture Horace, his attic workshop and his observatory. The film is available at http://vimeo.com/28160854. Just watching Robin's marvellous video brings the memories of my own visit flooding back. However, Horace had been an active amateur astronomer for more than sixty years when it was recorded. So let us now start at the beginning and look back at his earliest days in astronomy and engineering.

THE YOUNG HORACE DALL

Horace Edward Stafford Dall was born on 5 January 1901 and died on 9 May 1986, thus living just long enough to witness the second return of comet Halley in his lifetime, although comets were definitely not his main area of interest. At the time of Horace's birth, his parents lived in Chelmsford. His father, John Dall, briefly worked as an instrument maker for the young Guglielmo Marconi. The young Horace Dall's early life was far from smooth, but a tough upbringing can often produce a highly unusual individual, sometimes embittered, but sometimes very resourceful too; fortunately, Horace fitted into the latter category. Horace's mother (born Millicent Stafford) died in 1904, with the result that he and his brother were looked after by many different relatives before his father finally found permanent employment with the George Kent company in Luton and married for the second time. It would seem that his father's engineering trade was a huge influence on the young Horace as, during the Christmas period of 1914/1915, he started work with the aeroplane manufacturer Hewlett & Blondeau, based at their Omnia Works factory at Leagrave, near Luton. At the time Horace was just thirteen years of age (he turned fourteen during his second week with the company) and it seems incredible that a thirteen-year-old could secure full-time employment as a draughtsman/technician!

However, Horace was no ordinary thirteen-year-old boy and, with the First World War in progress, the demand for planes for the war effort was enormous. The company had secured a huge government aircraft contract a few months earlier and they were desperate for skilled staff. Horace, even at thirteen, fitted their requirements. Over

Figure 1. The teenage Horace Dall. (Image courtesy of Peter Hodds.)

the next five years the company would produce 800 aircraft and employ up to 700 staff, one of whom was Horace. However, when the war ended, the orders tailed off and, despite trying to switch production to agricultural machinery, the company went bankrupt in 1920. But, by then, Horace had moved on; technical qualifications he had gained at night school allowed him to secure a better job, with the Luton firm where his father worked – George Kent Ltd. He was to work at that company from January 1918 until the end of 1965; a period of almost forty-eight years! At Kent's he specialized in developing fluid-flow meters, and he rapidly became a world authority on their design.

When I visited Horace in February 1984, he took some delight in projecting a 'live' image of a distant building, with his chimney pot Camera Obscura, onto the white table in his blacked-out attic. With a bit of tweaking he zoomed onto part of the building and told me that was where he had sat at his desk for almost half a century designing fluid-flow meters. Minutes later we were viewing a somewhat foreshortened cricket match taking place a mile away!

Horace seemed fascinated by all manner of instruments and so it is hardly surprising that he soon developed an interest in telescopes and microscopes. During his teens he had already purchased a one-inch

(25-mm) spyglass telescope and assembled a home-made 2.5-inch (60-mm) refractor from available lenses, as well as buying a second-hand microscope. He was also very fit indeed and walked or cycled incredibly long distances to visit areas of interest. His first view of a Camera Obscura was achieved when he cycled from Luton to Clacton!

THE FIRST BIG TELESCOPE

Even in the early 1900s, with e-mail and the Internet not even dreamed of, people with a scientific disposition found ways of sharing their enthusiasm. The weekly *English Mechanic* newspaper was awash with tips and ideas for anyone interested in making scientific equipment and, of course, the British Astronomical Association had existed since 1890, with the aim of helping all those who wanted to observe the night sky. In addition, if you were determined to unearth useful data, then libraries and universities could be approached. On one occasion, Horace cycled from Luton to Oxford just to read William Herschel's notes on mirror-making.

By his late teens, and now settled in full-time employment, Horace decided that he would really like to own a decent astronomical telescope to study the Moon and planets. However, reflectors of a decent aperture, and on a solid equatorial mounting, were very expensive and, despite his engineering background, he simply did not have the skill to make an entire telescope. The only solution was to buy a used Newtonian, preferably a Calver reflector, as there were many in circulation and they had a good reputation. Even so, such an instrument would be hard to afford. Fortunately, Horace was a member of a local scientific society and knew a fellow member of that society, a musician, who also wished to own a decent telescope. They decided to join forces and buy a telescope together and, eventually, in 1920, they purchased an 8.5-inch Calver Newtonian from the experienced Manchester amateur astronomer William Porthouse. The telescope mirror had been signed as being made by 'G. Calver' in 1877 and originally it had been owned by the observer Nathaniel Green. The equatorial mount was dated 1882. It is interesting to note that while Dall was now involved in the design and construction of instruments to measure water and sewage flow at George Kent Ltd., Porthouse was in charge of the administration of the local and general acts covering river conservancy

Figure 2. Horace Dall with his 8.5-inch Calver Newtonian at 186 Dunstable Road in Luton. Photograph taken around 1920. (Image courtesy of Peter Hodds.)

and sewage disposal in the Manchester area; this seems more than a coincidence to me.

The 8.5-inch Calver Newtonian that Dall jointly owned with his friend had a long focal ratio of around f/9 and, at the time of its purchase, Horace was nineteen and still living with his father and stepmother at 186 Dunstable Road in Luton. The instrument was valuable, the mirror needed regular re-silvering if exposed to damp air, and Horace had no observatory. Therefore, this unwieldy reflector had to be assembled for use when the weather looked good, and dismantled when it looked bad. Fortunately, at f/9, it probably held optical collimation very well. Maybe it is just me, but I think that the picture of Horace with his 8.5-inch f/9 Calver, shown in Figure 2, bears a striking resemblance to a famous picture of Clyde Tombaugh (1906–1997) seen posing with his home-made 9-inch f/8.8 Newtonian, some eight years later. Horace was aware that, at the time he acquired his reflector, the telescope's maker, George Calver (1834–1927), was still alive, and during 1923 he exchanged some letters with the eighty-nine-year-old craftsman, regarding the art of making telescope mirrors. Eventually

Horace would purchase his musician friend's share of the Calver telescope, making him the sole owner.

OPTICAL EXPERTISE FOR HIRE!

In 1925, having become fascinated by the techniques needed to make telescope optics, Horace tried his hand at grinding and polishing mirrors and lenses and instantly he became hooked. In that same year, he joined the British Astronomical Association and, despite the monthly meetings being held at the extremely inconvenient time of a Wednesday afternoon, at Sion College on the London Embankment, he did manage to attend some of them. One factor that made this possible was that the George Kent company realized what an exceptional employee Horace was, and so permitted him flexible working hours and holidays; this was virtually unheard of in that era unless you were a boardroom member.

It is true to say that Horace was more of a telescope builder than an observer in the 1920s and 1930s. He obviously liked solving problems in engineering and optics. He also realized that he could make money from his new hobby and, by the late 1920s, was advertising his services in a number of magazines, including *Scientific American*, and also via the BAA. It is quite possible that Horace could have started up his own telescope-making company, but he chose to stick with the day job and concentrate on providing an excellent one-to-one service for amateur astronomers who wanted something a bit different in the way of folded Cassegrain-style telescopes, portable telescopes, and high-performance eyepieces. Mass production for profit was not his style at all, but excellence, customer satisfaction, restoration work and innovation definitely was. In many ways Horace simply did not think like the typical astronomer of the time, who favoured either an unwieldy long focus refractor, or a long and cumbersome Newtonian with an eyepiece so high that a set of steps was needed to reach it. He disliked the way astronomical telescopes were not only cumbersome and user-unfriendly, but also turned the image upside down. He vastly favoured the Cassegrain style of telescope but felt it could be made more easily and with an erect image and a degree of portability. These beliefs would define his direction in telescope making for the rest of his life.

Some of the first travel telescopes Horace made for clients were what he called his 'hip-pocket Gregorians'. These were tiny telescopes no bigger than a modern two-inch barrel eyepiece and resembling a miniature modern-day Schmidt Cassegrain, but with an achromatic lens at the front. He supplied them along with a standard money purse for protection when they were in the owner's pocket! For example, one design consisted of a 1¼-inch (32-mm) achromatic doublet lens, a ¾-inch (19-mm) primary mirror and a 3/8-inch (10-mm) secondary mirror cemented onto the rear of the doublet lens. The instrument was less than 3 inches (76 mm) long and provided 20× magnification, with a field of view of a degree and a half (or three times the diameter of the full Moon).

Figure 3. An early Horace Dall 'Hip-pocket Gregorian' with travel purse. (Image courtesy of Peter Hodds.)

Rather than making a bigger Newtonian than his 8.5-inch Calver, Horace decided that he wanted to make himself a large telescope of a Cassegrain design. However, high performance Cassegrains were tricky for a beginner to construct, even an engineer as skilled as Horace, because a complex secondary mirror had to amplify the focal ratio. The traditional Cassegrain amplification was achieved by grinding and polishing a convex hyperbolic surface onto the secondary mirror, to work with the parabolic primary mirror. In 1928, just three years into his telescope-making hobby, Horace realized that if you simply made the secondary mirror surface spherical then the primary mirror surface could be elliptical and figured to match the secondary mirror; such a system was far easier to grind and polish. There was a penalty though, which was noticeable off-axis coma. However, if narrow field studies of the planets and lunar craters was your main interest, this was not a big issue. In later years, Dall did not seem to think that his optical solution was any more than totally obvious, to him at least. No big fanfare was made of his Dall Cassegrain design work, by him, or in the BAA, and, in fact, he first mentioned it to the editor of *Scientific American* magazine, Albert G. Ingalls. The design was eventually referred to in that publication as the Dall-Kirkham, as another telescope enthusiast, the Oregon amateur Allan Kirkham, first cracked the maths behind the Dall system in the June 1938 edition of *Scientific American.*

It is clear that Horace had a master plan in mind, which was to build himself a very large and user-friendly Cassegrain style of telescope, which would avoid the stepladder requirement of all large long-focus Newtonians of that era. By the end of 1930, he had constructed a prototype 14-inch (35-cm) f/21 Classical Cassegrain of his own, with a square tube made from tea chests. The primary mirror was f/5.0 and, because even he had yet to work out how to safely drill a central hole in such a large, thick and expensive piece of glass, the Cassegrain focus did not actually reach the primary surface; instead, a small prism, held just above the primary mirror, diverted the light cone sideways through the tube at roughly the same place where the declination axis occurred on his fork mount. This, of course, was perfect for user comfort. Horace described this instrument in the BAA Journal in 1931, but it was to turn out to be merely a halfway house in his quest for the ultimate Cassegrain.

Around the same time, Horace had started making his own Tolles eyepieces and at the February 1931 BAA meeting, one of his customers praised them highly. This was the former BAA President Captain Maurice Anderson Ainslie (1869–1951), a large and bearded naval instructor who was referred to by his nickname of 'The Skipper'. Ainslie gave an account of the design of the Tolles eyepiece at that London meeting, explaining that a member of the Association, Horace Dall of Luton, had sent him three eyepieces of the Tolles design 'and they are so good that I think members should know something about them'. Ainslie explained that even on a reflector 'as fast as f/7', the performance of Dall's Tolles eyepieces was extremely good and their only disadvantages were that the design prohibited the use of a micrometer, and that the eye had to be held very close to the lens. In conclusion, he told the BAA members that they should congratulate Mr Dall on his optical skill, and congratulate the Association that the eyepieces were the work of one of its own members. The Reverend T. E. R. Phillips added that he had tried one of these eyepieces on his 18-inch reflector with excellent results and he, too, was gratified to find a member of the Association making eyepieces this excellent. Walter Goodacre also confirmed the good opinions of these eyepieces, saying he had tried one on his 10-inch refractor.

HORACE THE GLOBETROTTER!

As well as Horace's goal of owning a huge lunar and planetary telescope he was increasingly fascinated by the concept of international travel and ultra-portable travel telescopes which he could take on his foreign holidays. In addition to his interest in engineering and astronomy, Horace wanted to see as much of our planet as possible, even in the years well before cheap air travel. Typically, from the age of thirty onwards, he would have a spring cycling holiday to a European country with plenty of mountains, like Switzerland, followed by a more ambitious foreign cycling holiday later in the year. When I say 'ambitious', well, maybe 'death-defying' would be an equally valid term! For example, at the start of the 1930s, Horace decided he would walk and cycle across the Atlas Mountain passes of Morocco and Algeria. During this trip he was arrested and then eventually rescued by the French Foreign Legion. The local authorities had simply not believed that an Englishman, wearing a suit and tie, would want to trek across a North African mountain range in temperatures up to 50 °C! He trekked across Lapland, too, at times sharing a tent with the locals.

If Horace could take a pocket telescope with him on holiday, then so much the better. So, at the 29 June 1932 BAA meeting at Sion College, he brought along the latest compact, fully portable 6-inch (152-mm) Cassegrain he had constructed and described it to the members present. The telescope weighed only 5½ lb (2.5 kg) and was amazingly compact by the standards of the 1930s. Principally, it was intended for terrestrial use as its baffle tube incorporated a triple element achromatic transfer lens which gave an erect view. However, Horace preferred an erect view for observing astronomical objects too. The primary mirror of the Cassegrain was a very fast f/3.2 (an almost unheard of f-ratio in the 1930s) and the final f-ratio was only 15, rather less than other Cassegrains of that era. After Horace finished describing his telescope to the members (and he loathed speaking in public), Captain Ainslie was, once again, full of praise for the young man and his instrument, stating that, at f/15, normal eyepieces could be used with the instrument. The two men became firm friends and Ainslie visited Horace many times, first at Horace's Dunstable Road address in Luton, before he married, and then later at his marital home of 166 Stockingstone Road.

In 1933, Horace decided to raise the bar even further regarding his world travel exploits and, amazingly, he decided that he would attempt to become the first person to cross central Iceland on a bicycle! Horace had already visited the country the previous year, which was when he had formulated the plan. As usual, his trusty three-speed Raleigh Roadster bicycle was taken along, although, as things turned out, he had to push it for most of the five days of the 150-mile Iceland trek, because much of the route was over formidable terrain. He was armed with a tiny amount of food, a small stove, a sleeping bag, a compass, shoe polish, a razor, a bicycle puncture-repair kit, a camera and a map,

Figure 4. A self-portrait by Horace Dall, taken on his 1933 Iceland trek.

but had decided that a tent was not needed! Some self-portraits he took of himself on the trip show him in his usual attire, namely a suit and tie and highly polished black shoes! Horace's first act on reaching the northern part of Iceland and relative civilization was to send a telegram to the friendly farmer who had rowed him and his bicycle across the Tungna river at the start of his trip. The farmer was convinced he was sending a madman to his certain death!

THE HOUSE ON THE HILL

With his job at George Kent and his income from telescope making, Horace had a very healthy income by the 1930s and so he decided he would hire an architect to build him a house on almost the highest point of Luton, on Stockingstone Road, one mile north of the town centre. The site Horace chose was midway along that road, as it climbs steeply and heads south, some 150 metres before the modern-day turning on to Felstead Way. The plot of land was to the east of the road, and even from the garden gate to the house there was a very steep climb just to get to the front door! In addition, Horace wanted his house to be taller than any neighbouring ones so that he could have a substantial workshop built into the attic and also so that his planned Camera Obscura could look down on the neighbouring land without obstruction. The view to the north and west of the house was very impressive and, even today, there are no houses on the west side of the road to spoil the view of trees and grassland. At the time of writing, the house at 166 Stockingstone Road still stands out as being taller and narrower than its neighbours. Now a property owner (rare for a young man in the 1930s) and with two sources of income, Horace married his fiancée Vivien Andrews in 1934.

It might be thought that the first project for Horace, in his new custom-built home, would be an observatory containing his planned huge Cassegrain telescope. However, this instrument was clearly second on the agenda. The Camera Obscura had to come first as this type of instrument had fascinated him since seeing the one at Clacton when he was a teenager and, well, the whole design of the house was based around the siting of such an instrument in the roof! Essentially, Dall's Camera Obscura was a 4¼-inch refractor (and tilting, swivelling mirror) that poked vertically through a hole in his house roof and could be retracted when not in use. He had acquired special Crown & Flint glass melts for the lens, from the company Chance Bros., which enabled the instrument to have unusually good colour correction, virtually to apochromatic standard. From outdoors the instrument just looked like a chimney pot with a hole in the side and what looked like (and was) an angled mirror sitting behind that hole. The mirror could be tilted down, to direct the light from nearby Luton streets and buildings into the instrument, or it could be tilted up to look at sunspots. Obviously the false chimney pot could rotate too, to provide 360-degree azimuth coverage. Inside Dall's rooftop attic there was one high-quality window in the side wall as well as a small window in the roof. These could be blacked out so that the image projected by the Camera Obscura became even more dramatic. Dall's attic workshop had a very high ceiling and the inside of the sloping roof was far too high to reach by hand, so long metal control rods were used to raise and retract the false chimney pot and control the elevation and azimuth of the mirror feeding the refractor lens. Beneath the chimney was a circular table several feet across, with a brilliant white surface, to display the whole field of view of the Camera Obscura. Even with such a reflective table surface much light is lost by scatter, so for high-power views Dall had made a sliding magnifier which could glide across the table surface, around the huge focal plane, and collect the light before it hit the table. The light was then directed into an eyepiece, where an observer could be positioned. Ingenious! The table could also move up and down for precise focussing.

Horace also set up a study on the first floor of the house, directly below the attic. While seated in that study, which overlooked his back garden and observatory, Horace, even in his eighties, could grab a pole, hook the end into a metal hoop in the ceiling, and pull down a previously invisible staircase, leading to his loft workshop. For a young visitor, it was like visiting Professor Dumbledore's magical office! When I visited Horace in 1984 he was eighty-three and I was twenty-six, yet I was very unsure of my footing on that staircase, even though Horace nimbly glided up it with his hands by his sides! His whole appearance and demeanour in that workshop reminded me of the first Doctor Who, played by William Hartnell, at home in his own personal TARDIS! Anyway, travelling back in time to 1935, with the Camera Obscura completed, Horace could devote his efforts to the second major project: that big Cassegrain, optimized for lunar and planetary photography; but first he agreed to build another telescope, for his BAA friend 'The Skipper'.

THE JACK-KNIFE TELESCOPE

At the same time as Horace was planning his biggest erect-image Cassegrain, he was asked by Captain Ainslie, already an owner of many Dall eyepieces, if he would help with another optical challenge. Ainslie was a telescope maker himself and enthused about Newtonian reflectors. However, he had been bequeathed an 8.5-inch (216-mm) Object Glass by the Hon. Lionel Guest and, not wishing to construct a long refractor tube, asked Horace if he could check the quality of the achromatic lens and design a folded system, one which he could comfortably use while standing up, like a Newtonian. Horace found that the lens, despite being made by Grubb (and passed as OK by the new Grubb Parsons company), was simply not up to his own personal standards.

Figure 5. The folded refractor Jack-Knife telescope, designed by, and with optics by, Horace Dall, was made for Captain Maurice Ainslie in the 1930s. This picture, probably taken by Dall in 1937, shows the large Captain Ainslie and, on the left, the smaller figure of Eric Perry. Perry constructed the mechanics for the telescope and also built Horace Dall's Observatory Dome around the same time.

So, he reworked all four lens surfaces for Ainslie. The two men decided that folding the 10-foot (3.05-metre) light path in half would work best and so Horace produced a high-quality 6-inch-diameter flat mirror to give a couple of 5-foot light paths from lens to mirror and mirror to eyepiece. At the top of the tube they decided that a high-quality star diagonal would be best for user comfort and also for giving a second reflection to avoid mirror-flipping the image. Horace had a friendly neighbour, who also lived in Stockingstone Road, who was highly skilled with woodwork. This neighbour was Eric W. Perry, who built the entire tube and mounting so that Horace only had to design the system and provide the optics. The telescope was fully functional by 1936 and became the main instrument in Ainslie's Bournemouth garden. It became quite famous in BAA circles of the 1930s and 1940s and was known by its nickname of 'The Jack-Knife telescope'.

OBSERVATORY RESEARCH AND DEVELOPMENT

Reading between the lines it is clear from Horace's published papers of the late 1930s that he had left no stone unturned in the design of his telescope and observatory. The BAA, of course, had established experts in telescope construction and maintenance. Dr W. H. Steavenson was, arguably, their most experienced equipment expert and Captain Ainslie was another. There were a number of other optical authority figures too, such as F. J. Hargreaves. Nevertheless, Horace liked to prove design concepts himself, by experimentation. He was keen to advance the science of telescope making in the quest to build himself the perfect lunar and planetary photography system of the 1930s.

So, Horace carried out a number of careful scientific trials on aspects of his new observatory's design. For example, he even carried out rigorous tests on the best paint to use for his proposed dome to prevent thermal problems! Two metal boxes and two cardboard boxes were mounted in his back garden, featuring ventilation similar to that experienced in an observatory. Inside each box he installed maximum and minimum thermometers and made a note of the weather conditions in the daytime and at night during the trials. One box of each type was painted with white paint and the other with aluminium paint. The trials told Horace that if an observatory was to be used for solar work, the white paint was far superior, but if it was to be used primarily for night-time lunar and planetary work, the aluminium paint was preferable, for a rapid cool down after a sunny day. The results of these tests were published on page 117 of the January 1938 BAA *Journal*. Horace decided that he would like a 3.6-metre-diameter dome, but, unusually, mounted atop a 3.8-metre-square building. He simply stated: 'There's more room inside a square building.' The dome would rotate on wheels extracted from a large number of roller skates and it also incorporated a bedspring and a bell-crank lever into the design. Although Horace set out the basic requirements for his observatory he once again used the services of his skilled neighbour, Eric W. Perry, to design and construct the building and Perry did an excellent job. (After Horace's death, the observatory building eventually ended up with Jim Hysom of AE Optics, but in 2012 (in a rather grim state) it was transferred to Norfolk telescope enthusiast Mark Stuckey, at Cromer, and has now been fully restored. Not only that, but in 2013 Mark fully restored Horace's 39-cm Dall-Kirkham Cassegrain too, and installed it back inside the original dome: telescope and dome reunited exactly as they had first looked seventy-five years earlier!)

Figure 6. Horace Dall inside his brand new dome in 1937.

Another matter which concerned Horace was the lack of contrast, at a telescope's diffraction limit, claimed by planetary observers when using telescopes with central obstructions. Typically, the planetary observers of that era either used highly expensive refractors (with no central obstruction) or long focus Newtonians, with a secondary mirror so tiny that it was barely large enough to collect the light from the primary mirror. With Horace's liking for Cassegrains this was a slight concern, as the proposed telescope had to have a secondary mirror of at least 20 per cent central obstruction (4 per cent in area terms) to work. So, another set of tests was devised by Horace. In this case he set up a test rig in his attic workshop using a tiny aperture to simulate a telescope. The aperture was stopped down to only 2.8 mm in diameter, but this meant that artificial planets, just seven feet (2.13 metres) from the aperture, were, to all practical purposes, at infinity. Despite the small aperture size Horace still worked out a way to simulate different central obstructions in the system using precision-cut thin brass discs and fine wire! Fourteen feet (4.27 metres) beyond the aperture, on the opposite side of the attic, Horace set up some photographic plates to record the experimental results at the focus of the stopped-down lens. So, Horace had erected an optical test rig, twenty-one feet in length, simply to determine the effect of varying central obstructions on his proposed Cassegrain. Incredibly, what he had built in his attic workshop was, effectively, a 1/140th-scale model of his future telescope, with the artificial planets at an equivalent distance of a fifth of a mile away. To cut to the chase, he concluded, from this elaborate experiment, that for obstructions above 20 per cent, a noticeable falling off in crispness results, but below 20 per cent, no serious loss of detail occurs. A big Cassegrain with a 20 per cent obstruction was, therefore, not going to have any major problems at the diffraction limit. Horace wrote his findings up in a paper for the February 1938 BAA *Journal*.

THE 394-MILLIMETRE CASSEGRAIN

Without a doubt Horace's 15.5-inch (394-mm) Cassegrain is the telescope he was most renowned for in the BAA from the late 1930s onwards. It was a huge aperture amateur telescope for that era. Yes, there were certainly British amateurs around with much larger telescopes, such as W. H. Steavenson and J. H. Reynolds, but Horace's telescope was user friendly, cooled quickly, and, most importantly, produced impressive photographic results. Beyond this though, it was the most extraordinary-looking telescope that most amateur astronomers had ever seen. In many references it is simply referred to as his 390-mm DK Cass.

Horace's prototype 14-inch f/21 Classical Cassegrain had taught him much about the desirable and undesirable characteristics of such a big telescope. As he was planning on building an observatory to house his Cassegrain the eventual telescope did not need any form of tube to protect it from the elements, as the dome would do that. Horace would have been well aware that Newtonian reflectors of that era were renowned for being plagued by tube currents, so he took the concept of an open tube one stage further, by not having any tube at all.

In addition, he would be using a thin mirror that would cool rapidly to the night air, but with a unique mirror cell to support the mirror perfectly at all angles and, unlike his 14-inch prototype, it would have a perforated centre. Horace preferred an erect image which showed the planets with north at the top; this was at odds with almost all of the other planetary observers of his era. So, and to give some variability of power, he designed a 'transfer lens' inside the Cassegrain baffle tube, to give an erect image as well as providing the option of using different lenses to give different focal lengths. Despite the aperture of the new telescope, Horace wanted to use the 15.5-inch Cassegrain for long-distance terrestrial daytime views from his hillside vantage point. His transfer lens baffle tube gave excellent freedom from sky flooding, so it worked well in daytime, and the original arrangement gave a variable system f/ratio between f/10.5 and f/26, simply by adjusting the mirror and transfer-lens spacing. With a special eyepiece of almost 3-inch focal length, he could even fit the whole Moon into the field at 60 x: an unheard-of achievement with such a big Cassegrain.

For the longest focal-length planetary work it should be borne in mind that the films of the 1930s and 1940s were not only fairly insensitive, but also very grainy. While film manufacturers would claim resolutions of 100 lines per millimetre by the 1960s, corresponding to a modern 10 micron CCD pixel diameter, this specification was misleading, as the tests were carried out on very high-contrast black and white line-test charts. In practice, at low planetary surface contrasts, the resolution was more like twenty lines per millimetre at the emulsion surface, or half a millimetre on a photographic print enlarged ten times. Horace was well aware of the effects of film grain and developing both black and white and colour films was another one of his skills. So, eventually, the 394-mm Dall-Kirkham Cassegrain in his observatory would allow, using a selection of transfer lenses and secondary mirrors, f-ratios of up to f/200! These ratios were achieved before the final focuser position which meant that there was an extraordinary range of tolerance within which the film plane was in focus. At f/200 Dall's camera could sit within a couple of centimetres of the focal plane and the image would still be perfectly focussed! In practice, Horace reserved f/ratios of 200 for brilliant Mars, preferring a maximum f/ratio of 120 for fainter targets like Jupiter and faint Saturn, but still with a lengthy corresponding exposure from half a second to several seconds.

Despite his engineering skills Horace did not make a new equatorial mounting for his 394-mm Dall-Kirkham Cassegrain. Perhaps surprisingly, he decided that the mounting for his old 8.5-inch Calver Newtonian would suffice. It would, however, need substantial modification. The old Calver had been mounted on a fork fixed to a short pedestal. The fork tines were only about one foot apart and so, clearly, were not wide enough to contain even a skeleton tube for a 15.5-inch

Figure 7. Horace Dall with his 39-cm Cassegrain in around 1960. (Image courtesy of C. Cook.)

Cassegrain. However, the new telescope, despite its size, would actually weigh less than the 8.5-inch Calver and far less than the 14-inch Classical Cassegrain prototype. Therefore, Horace decided to mount the 15.5-inch Dall-Kirkham Cassegrain's lightweight frame on top of the Calver fork. This was achieved by removing the old tube of the Calver Newtonian and running a solid shaft across the declination axis between the two fork trunnions, to increase rigidity. The extremely open tube of the Cassegrain simply consisted of three long metal poles, a bit like the framework of a wigwam. These attached to the base, the east and the west sides of the mirror cell. At the opposite end of the 'tube', these three poles converged towards the base of the imaginary tube's underside but were cleverly designed to just avoid obstructing the light path to the main mirror. When the telescope mirror was in use the primary mirror cover simply hinged down and clipped to the poles: ingenious! Where the three poles met, a single strut poked into the light path to hold the 20 per cent-obstruction spherical secondary mirror's holder. In another unique development Horace later added a tiny third mirror inside the unused central part of the secondary mirror. This third mirror was tilted so that a light from the side of the telescope tube could be switched on by the user from the mirror cell end. The light would shine onto the 'tertiary mirror' and illuminate the eyepiece field with a gentle glow; in this way the full extent of the field of view could be determined when working with narrow field eyepieces. Ingenious again! The wigwam-style tube was attached to the modified Calver fork in three places, namely at the declination axis shaft centre and at the axis fork trunnion points where the old Calver used to sit. As the reader might have realized, a telescope perched on the top of a fork mount is horribly out of balance and so Horace attached counterweights on the exterior of the Calver fork to balance the tube weight. He also made the wigwam structure more rigid by using additional thinner metal poles which passed under the main 'tube' and through the fork tine gaps. The resulting telescope could be pointed anywhere between horizontal and vertical, but not past the vertical – but then, as Horace was only interested in the Moon and planets, any objects higher than thirty degrees in declination, or transiting at more than seventy degrees in altitude, were of no interest to him at all. As a final refinement Horace improved the telescope mounting by adding gravity-compensating friction rollers at either end of the Calver polar axis and also a very low-wattage Synclock drive, to minimize heat being generated near to the telescope!

THE WAR YEARS AND BEYOND . . .

The completion of the new telescope and observatory in 1937 did not mark an end to Horace's quest for engineering innovation. If anything, pushing the boundaries of optical and engineering science was more of an interest to him than astronomy itself. He was, by now, renowned as an expert in measuring liquid flow, building telescopes and building microscopes too. By using optical oil to immerse his microscope samples, and using violet light, he could resolve details even finer than the wavelength of green light, sometimes experimenting with powers of up to 8000 x. He also devised a world-record-breaking method of scratching microscopic letters with a diamond at a size so small that he estimated the content of almost 300 Bibles could theoretically fit on to a square inch!

He was not finished with his quest to construct increasingly portable small-aperture telescopes either and so, just before the start of World War II, he completed a tiny Dall-style Cassegrain weighing only eight

Figure 8. Horace Dall with a variety of home-made Maksutovs in around 1960. Note the tiny pocket-sized version in Dall's hand! (Image courtesy of C. Cook.)

ounces (0.23 kg). This weight even included the instrument's small stand! The telescope had a 3¼-inch (83-mm) aperture and an unusually wide (42°) apparent-field cemented triplet eyepiece and was exhibited at the 26 April 1939 BAA Exhibition meeting, although there is no evidence that Horace was at that meeting in person. The telescope folded up so flat when not in use that it could be carried in a small pocket.

Of course, the Second World War disrupted the lives of all those living in the UK and Horace's expertise was called upon in various capacities. With German optical equipment unavailable, many old microscopes and telescopes were sent to Horace for repair and as an acknowledged expert on both liquid and gas flow he was consulted by the military as part of a team investigating V2 rocket parts that had survived impacts on British soil.

When the war finally ended, Horace devised one of his most important optical ideas, namely the so-called 'Null test' for parabolic telescope mirrors. He first described his method at the BAA meeting held at Burlington House, Piccadilly on 28 May 1947. On that Wednesday afternoon, Horace explained to the audience that astronomical mirrors of relatively short focus were not easy to test with sufficient accuracy using the Foucault test at the centre of curvature, because the zonal curvature radii change rapidly at the outer zones. In Horace's opinion the existing well-known methods of null testing at the final focus sacrificed the convenience of centre-of-curvature tests, as careful squaring-on was necessary and auxiliary flats were required. However, Horace had found that by counteracting the aberrations of the paraboloid near the centre of curvature with those of a simple plano-convex lens placed in the converging beam, a null-testing method became available without sacrificing any convenience, other than the need for alignment of the small lens. Horace suggested that a suitable lens was usually available from low-power Huygenian eyepieces. The pinhole light could be rendered approximately monochromatic using a red filter, and data could be plotted on a simple curve from which to derive the position in the converging beam at which to place the lens. The null test devised by Horace would be used by many amateur and professional mirror-makers in the decades to come.

Horace's fascination with mirror-making and the techniques of the man who had originally inspired him, George Calver, stayed with him throughout his life. In 1952, some twenty-five years after Calver's death, Horace visited what remained of his final optical workshop in the Suffolk village of Walpole, not far from the coastal town of Southwold. Eventually, Horace would write an article on Calver for the BAA *Journal*, which was published in the December 1975 edition. Up to the mid-1950s, Horace always walked or cycled to places of interest whenever he could. A ten-mile walk or a fifty-mile cycle ride was quite routine for him, and, with no children, he and Vivien had little need to purchase a car. He did eventually concede to purchasing a Vespa motor scooter though.

Throughout the 1940s and right through to the early 1980s Horace would remain a huge human resource of advice on telescope design. At the 29 December 1954 BAA meeting, at Burlington House in Piccadilly, a discussion was started by Dr W. H. Steavenson on telescope design and techniques. Horace, despite having the most open-tubed telescope in existence, had much to say about the design of telescope tubes for instruments that were not protected inside an observatory. He strongly advocated the use of square wooden tubes which, while not aesthetically pleasing, had 'large corner spaces which conduct thermal currents out of the path of rays to and from the mirror'.

THE TRIPLE JOVIAN SHADOW TRANSIT

On 21 April 1956, it was a clear night at Stockingstone Road in Luton, and Horace decided to try some photography of the giant planet Jupiter with his large 394-mm Cassegrain. Jupiter was two months past opposition and, at barely forty arc seconds across, well past its largest apparent diameter. However, it was at +15 degrees in declination and so was reaching a healthy fifty degrees in altitude in the early evening, above the thickest air mass of the Earth's turbulent atmosphere. As the planet transited the south meridian, around 1940 UT, the giant moon Ganymede was just ending its transit of the disc as Io started its own transit. In addition, the shadow of Callisto was very obvious near the centre of the Jovian disk. Horace decided to take some photographs as conditions were quite reasonable, but by 20h 50m UT, the shadow of Io itself also slid onto the disc. This was splendid as having two moon shadows on the Jovian disc was not that common and while Io itself was not a high enough contrast object for the film to record, a photograph capturing two shadows on the disc would be very nice indeed.

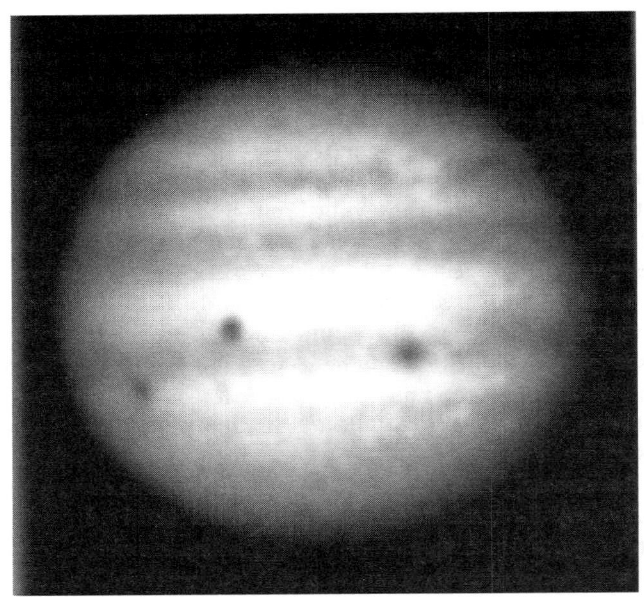

Figure 9. The rare Jupiter triple shadow transit of 21 April 1956, photographed by Horace Dall.

But the show was not over, because the Luton skies stayed clear, and by 21h 05m UT, the shadow of Ganymede was also on the disc. A remarkable three dark, Jovian satellite shadows were simultaneously transiting the planetary disc, along with Io itself. Horace was mesmerized at the sight and took his best shot at around 22h 15m UT, by which time Io had finished transiting. His photograph of the three shadows in transit became one of his most highly regarded.

Other BAA observers saw the rare spectacle, including Mr Bill Granger of Peterborough (and even pupils from Townsend Girls' School in St Albans), but Horace's photograph was the only one reported. Such triple-shadow events are exceedingly rare, occurring, on average, every six years or so, although two can occur in the same year. To have one occur on a clear night in the UK was unbelievably fortunate. The astronomer and mathematician Jean Meeus listed all such transit events, from 1901 to 2092, on page 280 of his book *Mathematical Astronomy Morsels*.

LUNAR AND PLANETARY PHOTOGRAPHY

During the 1940s and 1950s, Horace Dall's most photographed object must surely have been the Moon. At those times when it was high in the sky, namely the first-quarter phase in spring and the last-quarter phase in autumn, Horace would always be keen to take some high-resolution lunar photographs if the atmospheric 'seeing' was good. Planets were far less contrasty targets and the stark black and white lunar features were far easier to record with the grainy and insensitive films of the post-war era. Horace was always interested whenever a new film was released by Kodak or Ilford and he seemed to favour the Ilford brands for much of his lunar and planetary work. He was especially enthusiastic when, in September 1959, Ilford announced their medium-speed FP3 Series II film, and he immediately set about testing the new emulsion. Horace presented his findings, after several months of trials, to the BAA audience at the meeting of 27 January 1960. He told the members that from his tests he had found that the silver grains in the new film were no smaller than before; however, there was less clumping and more uniformity, hence slightly higher resolving power. As Horace explained, even a 10 per cent gain was very welcome. Horace had devised a theoretical factor of merit for potential lunar photography films based on the rated speed multiplied by the resolving power squared. This was logical, as if the film resolution was twice as good, then half the focal length could be used, so a film speed a quarter as fast should give the same optimum exposure time. Whenever he applied this formula the fastest (but grainiest) films always came out best, but Horace stressed that this did not tie up with practice when used on the Moon and planets, as they were not black and white test charts but real objects. In practice, he found that medium speed films, like the new Ilford FP3 (64ASA/ISO), worked best. Another prominent BAA lunar photographer of the 1950s and 1960s, Mr W. Rippengale, had also tested the new film and Horace confirmed that his results had been similar. Horace recommended f/ratios from 40 to 80 with the new film and exposures of a second or less, when used to record the lunar surface. (The reader may like to compare that with today's typical CCD image rates of 60 frames/second at f/30 and hundreds or thousands of frames captured per final image!)

Some five months after Horace described his tests with Ilford FP3 to

BAA members at Burlington House, he was back, on 29 June 1960, to tell them about his latest invention: a device to counteract the effects of atmospheric dispersion experienced when trying to photograph the Moon and planets at low altitudes. This time the venue was The Royal Institution in London. Horace explained that by de-centring one plano-concave lens of an achromatic doublet inside the transfer lens drawtube of his 15.5-inch Cassegrain, atmospheric dispersion could be eliminated. To prove his point he showed a photograph of Saturn he had taken when the planet was at an altitude of just twelve degrees. This was yet another remarkable Dall invention decades ahead of its time!

On the morning of 28 September 1964, at 01h 50m UT, Horace would secure one of his most memorable lunar photographs, of the giant lunar crater Clavius. Apart from being the largest well-preserved crater formation on the Earth-facing side of the Moon, it was of special interest to Horace because it was one of the few lunar craters to have been famously photographed by the 200-inch Mount Palomar telescope. Horace's 15.5-inch may only have been a thirteenth of that behemoth's aperture, but he was convinced that he could take a picture just as sharp. Using twenty-first-century CCD techniques both

Figure 10. The lunar crater Clavius, photographed by Horace Dall on 28 September 1964, at 01h 50m UT. (Image courtesy of the BAA archives.)

photographs can be beaten by a 100-mm aperture telescope in 2015; however, for its time, Horace's photograph was very special and one he always loved to show people visiting his study. The last quarter Moon was only at forty degrees altitude at the time, and almost four hours before it would transit at its highest point, but the seeing was extremely stable, which made all the difference. Horace's photograph of Clavius would appear in many journals and books during the coming years.

UV VISION AND HYDROGEN-ALPHA

From a personal perspective, 1964 had been a very sad year for Horace as his wife of thirty years, Vivien, had become ill and, tragically, died. After three decades of marriage it was a mental wrench and with no children or grandchildren there was little that Horace could do but immerse himself in his hobby and obey the old adage of 'life must go on'. Fortunately, he was never short of things to occupy his mind but now he had a health issue of his own to cope with. During the early 1960s, Horace had noted some deterioration in his vision which he swiftly diagnosed as being due to cataracts. He had, by this time, spent many holidays in very sunny climates which may have been a factor. Of course, for a man whose life revolved around testing optics, such a condition had to be fixed and so Horace was booked in for a cataract operation at his local hospital. Horace was so curious about optics that he even devised an optical arrangement to enable him to study his own eye in detail and, when he was about to be wheeled into the operating theatre, he presented the eye surgeon with detailed sketches of where and how the cataracts had grown! In early 1965, with the two cloudy lenses completely removed, Horace was keen to take full advantage of his new eyesight. As well as his general vision being clearer he appreciated that the removal of the lenses, which had evolved to protect the human retina from ultra-violet rays, should give him far greater vision at the UV end of the spectrum. This proved to be the case and Horace determined that whereas he could only see down to a wavelength of 4,000 angstroms prior to the operation, he could now reach 3,300 angstroms. He even made himself a set of 'Wood's Glass' spectacles specifically ground to match the new UV focus position of his eyes and filtered to just let violet light through! Horace soon found that, telescopically, he could now see the cloud features in the Venusian atmosphere that were invisible prior to his operation.

From late 1965 onwards, Horace was retired from his day job at George Kent, although he still offered his services as a consultant to that company. He also inspired and advised Jim Hysom, the master optician of the new company Astronomical Equipment (AE), which sprang up in Luton and Harpenden during the 1960s. With Horace Dall's influence it can be no surprise that during the 1970s and 1980s the Luton and Bedford areas became a centre for British-made amateur telescope companies and, in the 1970s, AE even offered a special Dall-Kirkham-Dall closed-tube telescope for lunar and planetary specialists. With far more spare time in the day Horace started thinking about solar photography, especially photography of solar prominences in H-alpha light. In fact, originally Horace had been inspired by an article in the July 1962 edition of *Sky & Telescope* in which a Mr Klaus of Switzerland had seen and photographed solar prominences with his own telescope. Working with the BAA Solar Section Director Bill Baxter, Horace had evaluated the commercial narrow-band H-alpha filters that were available at the time. These filters were not the sub-angstrom types available today, which enable surface features to be seen; the 1960s were a world away from the affordable H-alpha telescopes of the twenty-first century! In 1963, Horace had experimented with 30 angstrom bandpass H-Alpha filters, but after Bill Baxter had visited the US, he brought back a couple of 4 angstrom units, one for Horace and one for himself. During 1966, Horace constructed a 3-inch aperture 'Prom scope' in a tube some five feet long, which he mounted alongside the 15.5-inch Cassegrain in his main observatory. The telescope was all Horace's own work, except for the pricey filter, which was tuned by tilting it inside the tube. Sophisticated energy rejection filters were not available in the 1960s and so the standard method was to place the blazing Sun, at the focus, behind a circular (conical in cross section) metal field stop, so that the solar disc hid behind the field stop and the expensive filter was not subjected to the direct sunlight and heat. Thus, only the deep red H-alpha prominences, arching above the surface on the limb, were available for study. Of course, some nifty telescope drive-adjustment work was needed to keep the Sun perfectly behind the field stop, so preserving the lifetime of the filter. In the coming years Horace submitted some splendid solar-prominence photographs to the BAA solar section. The Sun was not his only interest in this period either, as Horace was still fascinated by gadgets and he even constructed a barometer so sensitive that it could detect the change in air pressure in his study, from the floor to the top of the bookshelf! He even wired up an early image intensifier to dabble with imaging faint night-time objects for a time.

HORACE IN THE 1970S AND 1980S

There can be few Englishmen who meet their wife-to-be in a café in the Patagonian rain forest, but Horace met his second wife, Helena Amy Thurley, in such circumstances, during 1968! Aged sixty-seven years, Horace had decided to travel the entire length of South America and, remarkably, bumped into another senior citizen, Helena, undertaking a similar expedition. Shortly after *Apollo 13* survived its near-disastrous Moon mission, they were married, in 1970, at Caxton Hall in Westminster, with Patrick Moore acting as Horace's best man! In addition, Horace had suddenly inherited three adult step-children and many step-grandchildren, a totally new experience for him.

Such was Horace's prominence within the BAA that, as he approached his seventieth birthday, friends insisted that there should be a special party. It was held, eleven days after his birthday, at the Dorchester Hotel in London, on 16 January 1971. A total of 150 guests turned up, many of them BAA members. Patrick Moore was quoted as saying at the party that 'No one has ever gone to Horace for help and come away empty-handed'. It may surprise the reader that Horace never became the BAA president, despite being a vice-president and one of the association's most popular and respected members for more than half a century. However, Horace was a shy and modest man who had no interest whatsoever in titles or perceived 'status'; he was simply happy helping others, working hard and pushing technical boundaries. When the comet discoverer Jack Bennett came to England in May 1971, he visited Horace to discuss, amongst other things, comet-sweeping telescopes. Horace told me that the equally remarkable George Alcock had travelled from Peterborough to Luton to meet Jack at 166 Stockingstone Road. Both comet discoverers had apparently drooled over a compact low-power 7-inch Maksutov telescope, with a two-degree field, stored in Horace's attic; he had designed and built the instrument specifically for low-power activities, such as comet hunting.

Many people would perhaps take life a bit more carefully from seventy onwards, but not Horace Dall, or his equally adventurous new wife. He travelled, with Helena, on the famous *Monte Umbe* total solar eclipse cruise in June 1973 and was filmed balancing his camera on his nose for *The Sky at Night* as totality approached, because he considered this was as good a method of coping with the swaying of the ship as any other. He might have been retired and on a cruise ship, but when he returned home, there were plenty of telescope restoration and repair jobs awaiting him. Amateur astronomers, just like those of the twenty-first century, had quickly learned that while anyone will sell you a telescope, complete with lots of hype about it having the best optics ever made, there are only a handful of people who really know what they are talking about and can really analyse and improve an optical system. Horace now had a reputation as the quiet but knowledgeable man from Luton: a man of action, not words, who understood optics better than anyone else. Thus, when Orwell Astronomical Society, at Ipswich, wanted the huge object glass of their historic 10-inch Tomline refractor cleaned and checked, they sent it to Horace in May 1973, just before his eclipse trip. Horace kept precisely to the schedule and supplied a detailed account of what he had found, and altered, to all of his clients. He not only tested the historic Tomline lens, but also cleaned it and refigured it and studied the notes of W. H. Steavenson, who had last examined the lens in 1936. He also cleaned and measured the giant refractor's six eyepieces. The optics were returned, as promised, in August 1973 and Orwell A. S. chose Horace for two further cleaning operations in the years to come.

In May 1974, one year after their eclipse cruise trip, Horace and Helena both travelled to the Mongolian People's Republic to visit the National Observatory situated some 20 kilometres from Ulan Bator. At that time it was extremely difficult for any private citizen to enter Mongolia, but eventually, via the Sri Lankan BAA member Herschel Gunawardena, and a Polish astronomer, Dr Ziolkowski, Horace obtained the necessary entry permits and travelled to the region on the Trans-Siberian railway. The Mongolian Observatory Director, Dr Baasanzhav, showed Horace all of the instruments, including a 135-mm Zeiss zenith telescope, a 200-mm Coronograph and a Maksutov. The fact that no one spoke English did not seem to matter as Horace knew more than anyone else about telescopes. It was strictly forbidden for foreigners to travel unescorted in Mongolia during the 1970s, but

Horace and Helena did so for four days! It was also strictly forbidden to take photographs out of the country but Horace managed to smuggle many of his holiday films out, even though a few were seized by the 'jobsworth' customs officials.

Throughout his seventies Horace continued to be the undisputed king of UK planetary photography, still using the 394-mm Cassegrain which he had built during the 1930s. Decades ahead of his time, he was even stacking the sharpest 35-mm film frames of planets in the enlarger to reduce film grain and even experimenting with photographic unsharp masking. On 28 January 1976, aged seventy-five, he took what was by far the best UK Saturn photograph of that era. In those days even cleanly resolving the Cassini division was something special.

Figure 11. Saturn photographed by Horace Dall on 28 January 1976 with his 39-cm Dall-Kirkham. This was a three-second exposure at f/120 on Kodak Plus-X film.

Following the *Apollo* lunar landings, laser reflectors were left on the lunar surface and professional US observatories were bouncing lasers off the Moon to record its distance. Horace decided to have a go himself and, via contacts in the US, he acquired all the equipment needed, including a dangerously powerful 1970s gas laser and an image intensifier. Horace was highly amused when his huge gas laser arrived at Luton with the customs declaration describing this 'weapon' as a toy! Remarkably, despite mere photons arriving back at his telescope, Horace succeeded in bouncing his laser pulses off the *Apollo* reflectors!

Various other projects kept Horace active well into his late seventies, and many younger amateurs (such as myself in 1984) were occasionally invited to his home for the day, which was a real thrill. Telescopes and microscopes were still routinely repaired and the world was still explored. In 1980, Horace collaborated with Jim Hysom and Colin Ronan on the investigation of a rare Dollond-Wollaston telescope discovered in the BAA Instruments collection by the curator Steve Anderson. Six years later, it would be Jim Hysom who would write a couple of splendid obituaries for Dall in the BAA *Journal* and in *Sky & Telescope*.

However, despite his apparent indestructibility, in 1981, Horace's trip across the whole of America from the southern end of South America to northern Canada, was, by his own admission, an expedition too far. His health was never as strong after that gruelling expedition and he told me about a serious throat disorder which had subsequently developed when I visited his home a few years later. Nevertheless, his big Cassegrain, almost fifty years into its life, was still in fully working order even in the 1980s.

Horace Dall's final contribution to the BAA appears to have been in the February 1986 *Journal* where he added some extra comments on visual spectroscopes, following a paper on the subject written by Ron

Figure 12. Horace Dall's 39-cm Cassegrain in February 1984; photographed by the author.

Figure 13. Horace Dall, aged eighty-three, outside his observatory in February 1984, photographed by the author.

Figure 14. Horace Dall's telescope inside his dome, both lovingly restored by Mark Stuckey in 2013. (Photograph courtesy of Mark Stuckey.)

Livesey. Three months later he suffered a stroke while repairing a friend's microscope, and died shortly afterwards in hospital. At their centenary meeting in October 1990, the BAA awarded a new medal, bearing Horace Dall's name, to the telescope builder and asteroid discoverer, Brian Manning; a man who was, in many ways, following in Dall's pioneering footsteps.

During the early twenty-first century, there has been a resurgence of popularity of the Dall-Kirkham design, with standard, corrected and optimized Dall-Kirkhams now being manufactured by Takahashi, Planewave and Orion Optics. Also, thanks to the hard work of Mark Stuckey, Horace's telescope and dome are now back in the same state that they were in when first completed in 1938. So, the great man's name lives on and, surely, we will never, ever, see the like of Horace Dall again!

Ripples from the Start of Time?

STEPHEN WEBB

INTRODUCTION

Which discoveries in physical science could have the greatest impact on our understanding of the universe and humanity's place in it? If I restrict myself to discoveries that could plausibly be made with current experiments, the following would be my 'top three': first, extraterrestrial intelligence (I need hardly spell out the ramifications of detecting extraterrestrial intelligence); second, extra spatial dimensions (the discovery would greatly expand our understanding of the physical world); third, primordial B-mode polarization of the cosmic microwave background (new vistas of research would open for cosmologists, astronomers and physicists).

The implications of finding extraterrestrial intelligence or extra dimensions might boggle the mind, but we can all at least readily appreciate what is involved in the search. It is less easy to comprehend the meaning of that third discovery. The main aims of this article are to describe what is meant by the phrase 'primordial B-mode polarization of the cosmic microwave background' and to explain why the discovery of primordial polarization would be so important. In order to meet those aims, however, we first need a framework in which we can discuss the concept: this means we need to look at some of the consequences of general relativity, contemplate why the cosmic microwave background implies the existence of a hot Big Bang, and examine the way in which ideas from relativity can fix a major problem with the hot Big Bang model. With that behind us, we can then make sense of B-mode polarization.

There is another important difference between the discoveries on my list: two of them might never be made. Humanity might be alone, after all, and it would hardly surprise anyone if it turns out that the universe contains precisely three spatial dimensions. The discovery of primordial B-mode polarization, however, might already have taken place. On 17 March 2014, a team of astronomers working on the BICEP2 experiment presented evidence of a slight twist in the light from the cosmic microwave background radiation – a twist that might have had its origins in the beginning of the universe. A further aim of this article, then, is to briefly describe the BICEP2 experiment and explain what it is the team claims to have found.

A word of caution: at the time of writing, the BICEP2 result is unconfirmed. The implications of the result are so profound that different experiments must replicate the findings before we can accept that a groundbreaking discovery has been made. This is how science works. Readers might remember how, in 2011, the OPERA experiment claimed to have observed neutrinos travelling faster than light. This would have been a discovery with *tremendous* impact. OPERA had indeed made a discovery – but it turned out to be of a loose fibre-optic cable in a timing circuit rather than faster-than-light neutrinos. Several critics claim that the BICEP2 result will also be explained in mundane terms. Nevertheless, the BICEP2 team consists of experienced observers who spent a long time analysing their results. Furthermore, the signal they see is so strong that – if it is real – other experiments will soon see it. By the time you read this it is likely that the BICEP2 conclusion will have been independently confirmed or else discredited. Even if the conclusion *is* withdrawn, the BICEP2 experiment has been a heroic effort; its tale remains worth telling. For the purposes of this article, however, I shall assume that the conclusions of the BICEP2 team have indeed been confirmed and that a new door has opened for science.

SPACE CAN RIPPLE, SPACE CAN STRETCH

The story starts a century or so ago, when Einstein came up with the theory of general relativity. For years Einstein had struggled with an inconsistency between his special theory of relativity and the common understanding of gravity. According to special relativity, no signal can travel faster than light, but the gravitational force between two objects was assumed to propagate instantaneously. Einstein eventually reconciled the two viewpoints through his general theory of relativity. In general relativity, space–time is a dynamic participant in physics; it is a fabric, capable of being deformed, twisted, stretched. Indeed, the deformation of space–time causes the phenomenon we call gravity. The presence of mass (or, in general, any type of energy) causes space to curve; and mass, moving freely through deformed space–time, takes a path that follows the curvature. We think of the Sun as exerting a force that causes Earth to move in an elliptical orbit. What's really happening is that our planet moves freely through a region of space curved by the Sun's mass.

Einstein soon pointed out one consequence of his ideas: a deformation in the fabric of space can propagate as a wave: *space can ripple*. For example, if two dense, compact objects are in orbit around each another, then they radiate gravitational waves, in the same way that oscillating electric charges radiate electromagnetic waves. Gravitational waves have been seen indirectly by Russell Hulse and Joseph Taylor who, over a period of many years, made accurate observations of one of the pulsars in a binary pulsar system. Hulse and Taylor found that the pulsar orbit is decaying exactly as predicted by general relativity,

Figure 1. A computer simulation of the gravitational waves emitted by two orbiting black holes. The 'shells' represent ripples in the fabric of space itself. Orbiting black holes produce gravitational waves with a tiny amplitude. When the holes finally coalesce, however, they emit gravitational radiation that the current generation of gravitational wave detectors could directly detect. (Image courtesy of Henze, NASA.)

radiating energy in the form of gravitational waves. Their discovery, which showed that deformations of space can indeed propagate as waves, earned them the Nobel prize in 1993.

Einstein looked for other consequences of his theory by applying it to the universe as a whole. He was dismayed to learn that general relativity predicts a dynamic universe: space expands or contracts. His dismay arose because he subscribed to the then-common belief that the universe was static, looking the same now as it did in the distant past and as it would in the distant future. Einstein therefore tried to 'fudge' his equations by adding a term meant to balance the dynamic tendency. This was a shame, because by doing so he missed out on what would have been the most impressive prediction in all of science: he could have predicted the expansion of the universe.

In 1929, Edwin Hubble discovered that distant galaxies are moving away from us, and the more distant the galaxy, the faster the speed of recession. The easiest way to explain this observation is in terms of an expanding universe. Think of small dots marked on a sheet of rubber; the dots represent galaxies and the rubber represents space. Stretch the rubber and the dots move apart – and the bigger the separation between the dots, the quicker they recede from each other. Although Einstein missed out on the prediction, his equations of general relativity contain within them a description of this expansionary behaviour. *Space can stretch.*

THE BIG BANG AND THE MICROWAVE BACKGROUND

If the universe is getting bigger, then it must have been smaller, and presumably hotter, in the past. The notion of a 'Big Bang', a dense, hot state from which the universe began, is thus quite natural given Hubble's discovery and Einstein's equations. Support for the notion came in 1964, when Arno Penzias and Robert Wilson observed the cosmic microwave background (CMB) – faint radiation that permeates the entire sky. The radiation was consistent with a black body at a temperature just under 3K. It was an observation for which they were awarded the Nobel prize in 1978.

Why is a black-body microwave background an indication of a hot Big Bang? Well, imagine what the universe must have been like in the past. If we had been living a billion years ago, we would have observed much the same as we observe today: stars and galaxies, dust clouds and gas clouds, all arranged in vast filamentary structures. Furthermore, the background radiation would have been only slightly hotter, at a characteristic temperature of about 2.9 K. Even at the time our Solar System was forming, some 4.5 billion years ago, the galaxies would look much the same as today and the background radiation, though hotter, would still be less than 4 K.

As we travel further back in time, however, the visible universe continues to reduce in size, the average density of the universe continues to increase and the background radiation continues to heat up. About 13 billion years ago there would have been no stars or galaxies, just a hot gas of primarily hydrogen atoms. At even earlier times the density of the hydrogen and the temperature of the background would be even greater. At some point, about 13.79 billion years ago, the characteristic temperature of the background was about 3,000 K – a temperature at which photons possess enough energy to ionize hydrogen atoms. Under these conditions an atom of neutral hydrogen, which consists of a negatively charged electron orbiting a positively charged proton, could not survive for long; collisions with photons would quickly knock the electron from the proton's grip. At even earlier times, then, there existed a hot, dense, interacting mix of electrons, protons and photons – a black body.

According to the Big Bang scenario, therefore, the universe was opaque for the first 379,000 years of its existence: photons could not move far before scattering off free electrons and, to a much lesser extent, free protons (see Figure 2). The stretching of space stretched the wavelength of radiation and the universe cooled. Once the temperature had dropped sufficiently, neutral hydrogen atoms could form and the universe became transparent. This event is known as the 'era of recombination'. After this time, photons were no longer blocked by free electrons and could therefore continue serenely on their way. Although the photon wavelength continued to increase as the universe expanded, the radiation kept the form it possessed at the end of recombination. Since the radiation had a black-body form at recombination, it should have a black-body form today. The difference between now and then is that space has stretched a thousandfold: the radiation's characteristic temperature has thus dropped from about 3,000 K at the era of recombination to about 3 K today.

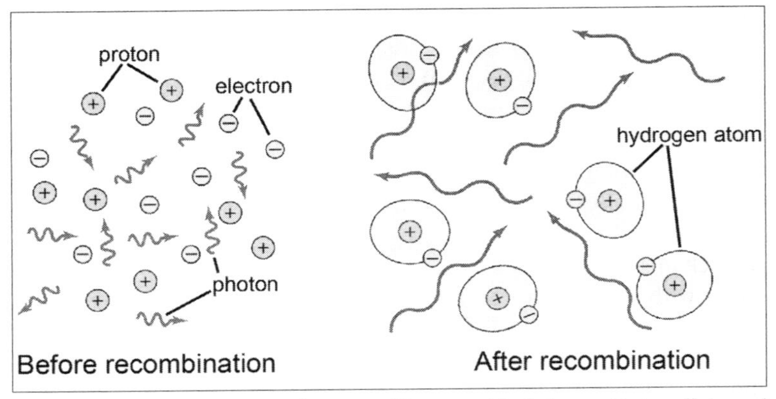

Figure 2. Before recombination, photons could not travel far before scattering off charged electrons (and, to a lesser extent, off charged protons). After recombination, photons could travel for a long distance before being scattered. Thus we cannot use photons to see what happened before recombination any more than we can see inside a cloud; the cosmic microwave background is rather like the surface of a cloud – it is the furthest we can penetrate using light.

Only one theory has been able to explain the observations of black-body microwave background radiation. The existence of the CMB is hard evidence of an expansionary Big Bang cosmology.

OBSERVING THE CMB

The existence of the CMB has a profound implication for astronomy because it means we can never see further back than 379,000 years after the Big Bang. Photons from earlier times are forever hidden by the fog of plasma that existed back then. In order to learn more about the early universe we must squeeze every last bit of information from the CMB. Unfortunately, microwave astronomers face a significant challenge because Earth's atmosphere absorbs most of the interesting wavelengths. A few observing 'windows' are available for ground-based telescopes, but an in-depth, all-sky study of the CMB requires a dedicated space-based mission.

Thirty years or so after Penzias and Wilson made their discovery, NASA's COBE satellite demonstrated that the CMB is an almost-perfect black body with a characteristic temperature of 2.725 K. By doing so, it cemented the Big Bang picture as the only convincing model of the early stages in the evolution of the universe. This was work for which John Mather and George Smoot, principal investigators on COBE, were awarded the Nobel prize in 2006.

The astronomical community proposed follow-up missions to COBE, and NASA's Wilkinson Microwave Anistropy Probe (WMAP) was launched in 2001 (observing the CMB for nine years) while ESA's Planck mission was launched in 2009 (observing the CMB for over two years with exquisite accuracy). Data from these missions were used to create a heat map of the entire sky, and tiny variations were found in the 2.725 K temperature of the CMB. In some places the sky is slightly hotter, in other places slightly cooler. In other words, at the time when the universe became transparent, the plasma was not a perfect black body. It was *almost* perfect – the universe was remarkably uniform – but some regions were slightly denser than average (and thus slightly hotter) and other regions were slightly less dense than average (and thus slightly colder). We shall consider what gave rise to these 'waves of density' later, but for now it is enough to note that those tiny irregularities, which existed when the universe was just 379,000 years old, are what gave rise to the structure we see today: clusters of galaxies arranged in filaments, between which are voids. Ultimately, these tiny irregularities gave rise to us.

The significance of the WMAP and Planck observations lies in the fact that cosmologists have developed various different models of the Big Bang, and each model generates a different heat-map pattern in the CMB. The models can now be tested against reality. For example, thanks to Planck we now know that 26.8 per cent of the mass-energy content of the universe must be in the form of dark matter, some mysterious substance that is as yet unknown to science. Various lines of evidence point to the existence of dark matter, but it is observations of the CMB that can pin down precisely how much dark matter the universe contains.

Satellites might provide the only way of obtaining multi-wavelength, all-sky maps of the CMB, but such missions are years in the planning and expensive to implement. Therefore the groundbreaking results of the COBE, WMAP and Planck satellites have been supplemented by a range of ground-based telescopes observing regions of the sky at specific wavelengths. Astronomers wishing to observe the CMB must place their telescopes at high altitude and in as dry an atmosphere as possible, since atmospheric water vapour absorbs most microwave wavelengths. The South Pole is a good site for CMB observatories, as is the Atacama region of Chile; high-altitude balloons offer another option. A plethora of CMB experiments are planned or currently in operation.

By mapping this oldest light in the universe, then, scientists have constructed a stunningly successful standard model of cosmology based on the hot Big Bang. However, there is a fundamental problem with the Big Bang picture – a problem much discussed soon after Penzias and Wilson first detected the CMB.

A PROBLEM WITH THE BIG BANG

The isotropy of the CMB – the fact that it is uniform in all directions – cries out for an explanation. The obvious and natural explanation is that all regions of the sky interacted in the past and were driven to thermal equilibrium. We employ this sort of explanation every day. For example, if we see a cup of tea at room temperature we naturally assume that the tea has been interacting with its environment for a while, and that heat has flowed from the initially hot tea into the surroundings until thermal equilibrium has been established – simple. However, this perfectly natural sort of explanation cannot hold for the microwave background.

Imagine pointing a microwave telescope due west in order to observe the CMB: the photons trapped in your telescope will have been travelling for more than 13.7 billion years. Now point the telescope due east and perform the same observation: once again, the photons you detect will have been travelling for more than 13.7 billion years. Those two regions of the sky are too widely separated for any signal to have passed between them; there simply has not been enough time for them to have established contact. In more formal terms, those two parts of the universe were not in causal contact when the universe became transparent. Indeed, calculations show that regions of the universe separated by more than about two degrees (just a few times the width of the full Moon as seen from Earth) were causally separated when the universe became transparent.

Unless one is willing to believe that energy and information can

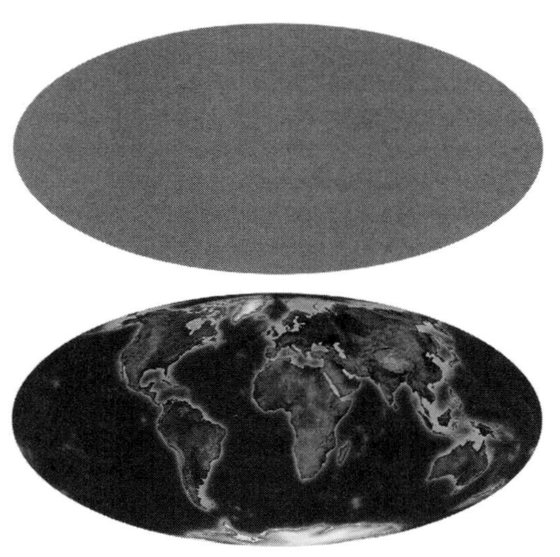

Figure 3. The lower part of the figure shows a Mollweide projection of the surface of the Earth; lots of structure is apparent. The upper part of the figure shows a Mollweide projection of the surface of the CMB; it is featureless to 1 part in 100,000. In the traditional Big Bang model, parts of the sky separated by more than a couple of degrees were too far apart to have exchanged energy or information at the time the CMB was formed. In the traditional Big Bang model, therefore, we expect to see a CMB that looks more like the lower figure. The Big Bang model cannot explain why we observe a CMB with an identical temperature across all parts of the sky. (Image courtesy of NASA.)

travel hundreds of times faster than the speed of light, this finding generates a fatal difficulty for the Big Bang model. Different regions of the sky cannot have had a shared history and therefore cannot have come into thermal equilibrium. And yet the CMB is smooth to 1 part in 100,000. How can we possibly explain this observation?

INFLATION TO THE RESCUE

In the Newtonian picture an object possesses only one property that can generate a gravitational interaction: its mass. In Einstein's theory there is a much wider variety of gravitational sources. An object's energy density acts as a source of gravity, and all types of energy contribute: mass (through $E = mc^2$), thermal energy, electromagnetic radiation – *every* contribution to the energy density must be taken into account when calculating the strength of a gravitational interaction. Perhaps surprisingly, an object's pressure is a further source of gravity. In most cases pressure makes a negligibly small contribution to gravitational effects, but there are situations where its contribution plays a key role. For example, the outward-pushing pressure in the central region of a high-mass star acts as a counterbalance to the inward pull of the star's gravity. Eventually, when the star exhausts its supply of thermonuclear fuel, it will contract and the pressure will increase so as to prevent further gravitational collapse. However, because pressure acts as a source of gravity, it can actually *hasten* the collapse of a dense object! Pressure thus plays an important, and rather subtle, role in the formation of stellar-mass black holes.

Unlike energy, pressure can be negative or positive. The notion of negative pressure might seem counterintuitive, but you will be familiar with an everyday example of the phenomenon: stretch a rubber band and it pulls in. So if an outward push is positive pressure then an inward pull, or tension, is negative pressure. The equations of general relativity state that a positive pressure generates a gravitational attraction; this accounts for the extra contribution to gravitational collapse from the central region of a compact star. A negative pressure generates what is essentially a gravitational *repulsion*. Negative pressure causes space to stretch.

It appears that there is negative pressure associated with space itself. This is the easiest way of explaining the discovery, made in 1999, that the expansion of the universe is accelerating. If the energy density of the universe were the only source of gravity, then the expansion would inevitably slow. Pressure must also be taken into account, however, and the repulsion from the negative pressure of space now more than counterbalances the attraction from matter and dark matter. The discovery of the accelerating expansion, a discovery that demonstrates the negative pressure associated with space, led to the award of the 2011 Nobel prize to Saul Perlmutter, Brian Schmidt and Adam Riess.

Two decades before cosmologists discovered that the expansion of the universe is accelerating, a young American theoretician called Alan Guth was studying a rather arcane problem in particle physics (the precise details of which aren't relevant here). His research led him to study what would happen if the very early universe contained a so-called fundamental 'scalar field' – a physical quantity defined by a number at every point in space. At the time of Guth's work, the notion of fundamental scalar fields was pure speculation. However, physicists have recently proved the existence of one fundamental scalar field in nature: in 2012, experiments at the Large Hadron Collider produced the Higgs boson, a particle manifestation of a scalar field that pervades the entire universe. This discovery led to the award of the 2013 Nobel prize to Peter Higgs and François Englert. Although physicists still do not understand the nature of the scalar field studied by Guth, or indeed whether it even exists, the discovery of the Higgs boson means that proposing the notion of scalar fields is not itself outlandish. (One of the many exciting implications of a confirmed BICEP2 result is that we could learn more about that field.) In the absence of any other knowledge, we can at least give the scalar field a name: it is usually called the inflaton – for reasons that will soon become clear.

Suppose that the inflaton field, to which we usually assign the symbol ϕ, has the same value at each point in space and that the value can vary with time. Further suppose that the field has a potential energy density represented by $V(\phi)$. In physics, the state of minimum energy is called the vacuum. Guth's idea was that, momentarily, the universe was trapped in a 'false' vacuum state: $V(\phi)$ took some non-minimum value and it was unable to quickly lower its value and reach the minimum. It was thus in an *effective* vacuum state – the inflaton field ϕ could not quickly lower its energy density and being unable to lower the state of energy density is, after all, the definition of vacuum. It was in a 'false' vacuum state because there was a state of lower energy density – the 'true' vacuum – that eventually it could reach. Guth came up with a complicated mechanism to explain how the universe could get trapped in a false vacuum state, but a much simpler mechanism might hold the truth. Suppose the potential energy density of the inflaton field has a simple shape, similar to the one shown in Figure 4. A small quantum fluctuation could have caused a tiny patch of space–time to jump 'up the hill'. As long as the potential is varying slowly at that point then the universe is in a 'false' vacuum. The space–time region will 'roll' slowly down the hill to reach the 'true' vacuum – but while it is rolling slowly, rather like a ball rolling on a flat hilltop, something quite extraordinary occurs.

The key point to note is that the false vacuum has negative pressure.

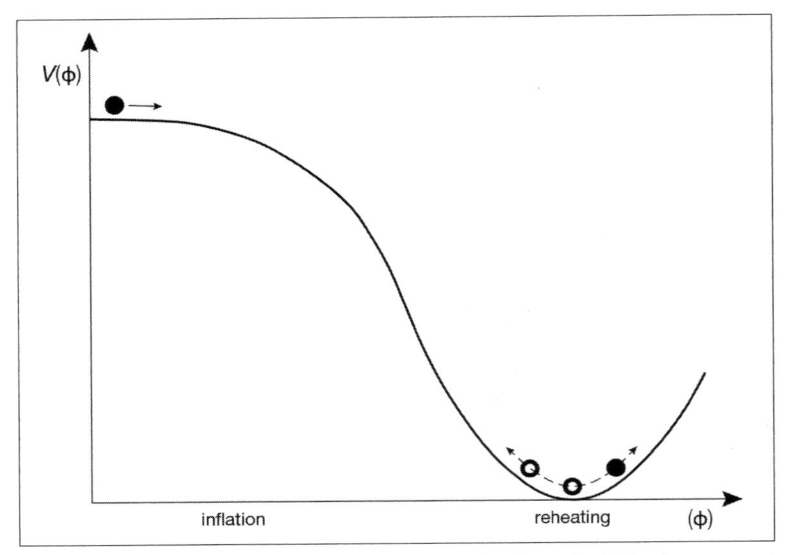

Figure 4. A small region of space might have found itself with a scalar field ϕ that possessed a non-zero energy density. If it took time for the scalar field to reach the energy minimum, perhaps because it could only slowly 'roll down the hill', then inflation would occur. So while the scalar field is on the plateau, the volume of space undergoes repeated doublings: in the almost vanishingly small time it takes the field to reach its minimum, a patch of space much smaller than a proton becomes as large as a grapefruit. While the field oscillates around its minimum, it dumps the tremendous amount of energy it contains into the creation of particles. This 'reheating' of the universe is what most of us mean when we talk about the Big Bang.

One can understand why this is the case by imagining a small cylinder of false vacuum surrounded by true vacuum. If the cylinder has a piston at one end, and the piston is rapidly pulled outward, then the volume of false vacuum inside the cylinder increases. If the cylinder were filled with a substance such as air, which has positive pressure and thus would push on the piston, then the energy density inside the cylinder would decrease. However, the energy density of the false vacuum cannot be lowered quickly – temporarily it is the state of lowest energy – and so the energy density remains constant. The total energy, therefore, increases. Since energy is conserved, the extra energy must have been supplied by whatever it was that pulled the piston: pulling against the false vacuum would feel like pulling against a rubber band. There is a tension there: the false vacuum has negative pressure. If one works through the mathematics of general relativity, one finds that the negative pressure of the false vacuum contributes three times as much to the gravitational field as the positive energy density: there is a large net repulsion.

Suppose, then, that a quantum fluctuation causes a region of space–time many billions of times smaller than a proton to be in a state of false vacuum at a time 10^{-36} seconds or so after the initial creation event. Until the inflaton field rolls down to the state of true vacuum the negative pressure causes the fabric of space to stretch exponentially. The precise numbers involved depend on the details of the model under consideration, but typically one can imagine the universe doubling in size every 10^{-34} seconds. After just 10^{-32} seconds there would have been 100 doublings. This might not sound impressive, but recall the story of the man who invented chess. His emperor was so pleased to have a new game to play that he let the man name his own reward. The inventor asked for one grain of rice for the first square on the chessboard, two grains of rice for the second square, four grains of rice for the third square, with the doubling to continue on each of the sixty-four squares. The emperor, who did not immediately appreciate the power of repeated doublings, agreed to the request. He was less pleased when he realized that the chessboard would contain a heap of rice larger than Mount Everest. The emperor had the inventor killed!

The idea, then, is that the universe started out incredibly small and then, because of the presence of a scalar field, it rapidly got large – a phenomenon Guth called 'inflation', which is, of course, why the scalar field has the name inflaton. In a sense, during inflation, space stretched much faster than the speed of light, but this is allowed since the expansion transmitted no information. Travel *through* space is limited by the speed of light; the stretching *of* space has no such limit. (As an aside, it should be noted that Alexei Starobinsky published a model of inflation slightly before Guth. However, Starobinsky was in the USSR and the political situation at that time meant that his work was not widely known. Furthermore, the Starobinsky and Guth models suffered key problems, which were fixed by Andrei Linde and, independently, by Andreas Albrecht and Paul Steinhardt. We have seen how several discoveries that are relevant to this problem have been rewarded by a Nobel prize. The idea of inflation itself, if it is proven true, is certainly worthy of a Nobel prize, but the prize committee might find it difficult to assign the credit!)

A vast amount of potential energy was locked in the inflaton field, but when the inflaton reached the bottom of the potential hill and started to oscillate around the minimum point, all that energy was transferred into the production of a hot soup of particles. It is *this* event, the reheating of the universe, that we usually refer to when we talk about the Big Bang. The origin of the universe happened *before* the traditional Big Bang.

For Guth, the realization that cosmological inflation could arise naturally from basic ideas in particle physics was a 'eureka' moment. Inflation not only solved the particular problem he was working on, but it also explained why the universe can look the same on opposite

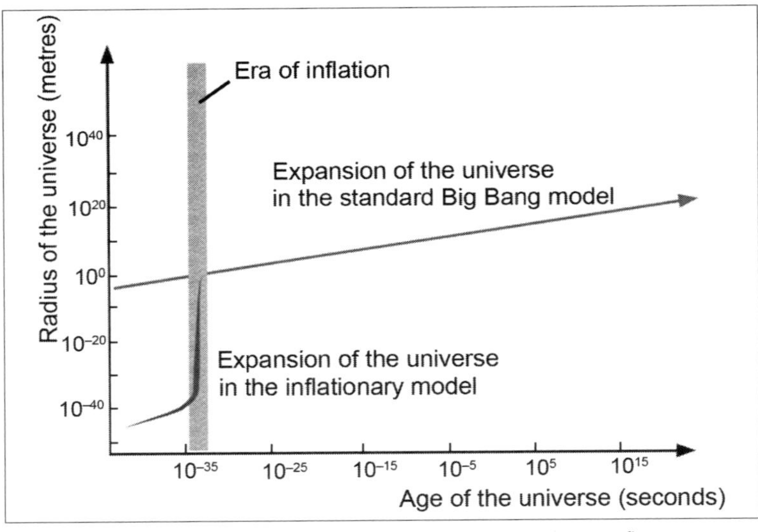

Figure 5. In the inflationary model, the universe inflates from a subatomic fluctuation to an object the size of a grapefruit in about 10^{-32} seconds. After inflation ends, the universe expands as in the standard Big Bang model. If we try to extrapolate the standard Big Bang expansion back to early times, we encounter the problem discussed earlier in the article: the early universe would have been too big for thermal equilibrium to have been established and so the observed isotropy of the CMB is inexplicable. If the universe started out small, then thermal equilibrium could have been established. Note that the numbers in the figure are indicative only; different models of inflation can give rise to slightly different values.

sides of the skies: before inflation happened everything that we now see was once extremely close; each part of space was near enough to every other part of space for energy and information to have been exchanged and thermal equilibrium to have been established. Inflation then ripped neighbouring points apart, beyond the reach of light. Different regions of the sky *were* in causal contact after all! Inflation seemed to cure various other ailments in traditional Big Bang theory too. Little wonder that this cure-all for cosmology, a panacea that drew its ingredients and motivation from a quite different area of physics, quickly became an established part of the standard cosmological model.

In 2013, the Planck team released a first analysis of its data and it provided support for the inflationary scenario. Theoreticians had shown how quantum fluctuations in the inflaton field would have led to a particular spectrum of matter density perturbations – the 'waves of density' mentioned earlier – when the inflaton decayed into particles at the end of inflation. Regions that were denser than average grew still denser thanks to their gravitational pull, and would eventually form the galaxy clusters we see today. The results of the theoreticians' calculations matched Planck's observations quite beautifully. It was a big hint that inflation occurred – but how would it ever be possible to get more than a hint? How would it ever be possible to investigate the universe when it was only 10^{-36} seconds old?

POLARIZATION

COBE, WMAP and Planck measured the temperature of the microwave background with ever-increasing precision. Could we get meaningful information by measuring some other property of the CMB besides temperature? Well, one could attempt to investigate the *polarization* of the background.

Light is a transverse electromagnetic wave: electric and magnetic fields oscillate perpendicular to each other and transversely to the direction of propagation. The polarization of an electromagnetic wave simply describes the orientation of the electric field. Light from the Sun is unpolarized: the orientation of the electric field changes rapidly and randomly. This is hardly surprising, since there is no reason why one orientation of the electric field in sunlight should be preferred over any other orientation. However, molecules of oxygen and nitrogen in Earth's atmosphere scatter sunlight. This scattering not only causes the sky to appear blue, it can also partially polarize the light: in other words, after scattering, the electric field oscillates in a slightly preferred direction. The extent of the polarization depends upon the scattering angle. The sky's polarization pattern depends on the position of the Sun, and thus it changes throughout the day and throughout the year. Insects use polarization patterns as a tool for navigation.

Just as scattered sunlight can be polarized, the microwave background might exhibit polarization if photons were scattered when the universe became transparent. And although a map of the CMB polarization pattern would be of little use as a navigational tool, we could use it to learn something about the conditions that prevailed during the era of recombination. So although WMAP and Planck were not equipped specifically to measure CMB polarization, their data were analysed with that in mind, and a variety of ground-based experiments have been deployed with measurements of CMB polarization as their sole focus.

What would a CMB polarization pattern look like? Well, researchers find it useful to talk about two distinct components, or modes, to a polarization pattern. For the CMB, one mode would consist of a set of arrows pointing towards or away from the source of polarization. Cosmologists call this an 'E-mode' because this pattern resembles the electric field lines emanating from or ending on a static charge, as shown in Figure 6. (The figure also shows that the E-mode can have a radial alignment.) The other mode would consist of a set of arrows angled around the source of polarization. Cosmologists call this a 'B-mode' because this pattern resembles the magnetic field, or B-field, surrounding a magnet. As can be seen from Figure 6, the difference between the two modes is one of 'handedness': the E-mode looks the same when reflected in a mirror while the B-mode looks different. The important point to note from all this is that quite different physical sources give rise to the two different modes.

We know that the early universe contained hot overdense regions (which held many free electrons) and cold underdense regions (which had fewer electrons) and that there would have been a flow of material from hotter to colder regions. Calculations show that for a short period of time, towards the end of the era of recombination when there were still free electrons from which photons could scatter, the conditions

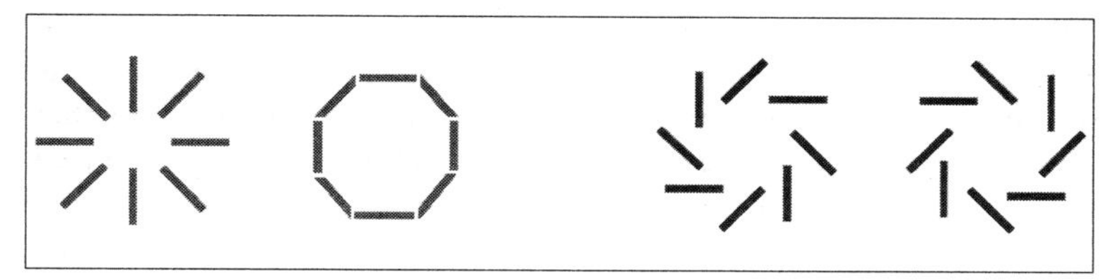

Figure 6. Left: the two mirror-symmetric E-mode patterns. Right: the two B-mode patterns. Density perturbations can only generate an E-mode pattern; gravitational effects can generate both E-mode and B-mode patterns.

would be right for E-mode polarization to occur. In other words, the temperature anisotropies in the CMB, which we know indicate the presence of regions that would grow into the large-scale structure we see today, should be slightly polarized. The decomposition of the polarization pattern from these sources would create a pure E-mode signal, which would have a peak at an angular scale of about 0.2°. The signal would be small – calculations suggest that it would be about ten times smaller than the temperature anisotropies – but in principle it could be detected. In 2002, the polarization of the CMB was indeed detected: John Kovac and colleagues working with an instrument called the Degree Angular Scale Interferometer (DASI), based at the Amundsen-Scott South Pole station in Antarctica, found convincing evidence for E-mode polarization. WMAP, Planck and other experiments have subsequently detected E-modes.

Is there another source of CMB polarization? Well, gravitational effects can polarize light. Gravitational lensing of the CMB could add a slight polarization effect by distorting the passage of light waves, but this would be a 'line of sight' effect rather than a phenomenon that could tell us about conditions in the early universe. However, any gravitational waves that were distorting space at the end of the recombination era would affect the propagation of light at that point; the alternate squeezing and stretching of space polarizes light, and the polarization from gravitational waves would arise at the same time as the polarization arising from scattering. We now encounter a good news/bad news situation. The bad news is that CMB polarization from gravitational effects is even smaller than that from density fluctuations: that makes their detection extremely challenging – although recent advances in detector technology have made it a challenge worth attempting. The good news is that gravitational effects polarize the CMB in both an E-mode and a B-mode: since density fluctuations cannot generate a B-mode, a 'swirling' B-mode polarization pattern constitutes a clear signal of something other than density perturbations.

In 2013, scientists using the South Pole Telescope (which, like DASI, is situated at the Amundsen-Scott station in Antarctica) observed a B-mode signal in the CMB. This was an important observation if only because it demonstrated that success in the B-mode search was technically possible. The signal observed by the South Pole Telescope occurred at an angular scale of less than one degree, which means it was caused by the gravitational lensing of a pre-existing E-mode signal generated by the usual density fluctuations. In other words, on its 13.7-billion-year journey towards us, light with an E-mode pattern encountered intervening galaxy clusters and then gravitational lensing subsequently generated a B-mode pattern. However, if astronomers could detect a B-mode pattern on larger angular scales, then this would be good evidence of gravitational waves propagating at the era of recombination. *This* is what the BICEP2 team claims to have seen.

We are almost in a position to discuss the BICEP2 result, but we still have one more question to consider: why should we expect to see large-scale gravitational waves propagating at the era of recombination? Is there any source of such waves?

GRAVITATIONAL WAVES FROM INFLATION

According to inflation, the universe began as an incredibly tiny patch of space–time. In the inflationary scenario, therefore, any description of the very early universe must inevitably take quantum effects into account, because at small distance scales the world is inescapably quantum in nature. One prediction is that quantum fluctuations in the inflaton field would have given rise to 'density waves' when the energy of the inflaton converted into matter. As we have seen, these 'density waves' left their imprint as anisotropies in the CMB – and in the large-scale structure of the present-day universe. That prediction seems to have been borne out. Another prediction is that the geometry of space itself would have experienced quantum fluctuations. Distortions of space propagate as gravitational waves, to which the early universe would be completely transparent. (As we have seen, this is not the case for electromagnetic waves.) Inflation would have stretched gravitational waves from the earliest times so that, by the era of recombination, they could be large. It follows that the quantum fluctuations which gave rise to primordial gravitational waves could produce a faint B-mode polarization pattern in the CMB. Furthermore, calculations show that the signal is expected to peak at angular scales of about 2.5° – much larger than the E-mode peak.

We have at last reached an understanding of the phrase 'primordial B-mode polarization of the cosmic microwave background'. Quantum fluctuations of space gave rise to primordial gravitational waves, which inflation made large; photons passing through these waves during the era of recombination would have been given a slight twist; and thus a faint B-mode polarization pattern would have been imprinted upon the CMB.

BICEP2

The BICEP2 instrument was developed solely to search for the B-mode signal. It consists of a small telescope with a wide field of view. Liquid helium cools the telescope to just 4 K while its 256 polarization detectors allow it to study the CMB with precision. BICEP2 follows BICEP1, which ran between 2006 and 2008, but possesses ten times the efficiency of its predecessor. The instrument is a next-door neighbour of the South Pole Telescope: the altitude, the cold and the dry, stable atmospheric conditions at the Amundsen-Scott South Pole station are perfect for such observations. Between January 2010 and December 2012, John Kovac and his colleagues used BICEP2 to study the so-called 'Southern Hole' region of the Antarctic sky, a region that is relatively free from dust and has little foreground emission.

Cosmologists find it useful to compare the size of the primordial gravitational-wave perturbations to the size of density perturbations, and they call this ratio r. The BICEP2 team expected that even if perturbations from gravitational waves did indeed exist, the value of r would be so small that not even the improved sensitivity of BICEP2 would detect it. Instead, they found $r = 0.2$ – a remarkably large signal – at the predicted location of the peak.

Figure 7. The Sun sets on the BICEP2 telescope (foreground) and the South Pole Telescope (background). Both telescopes are at the Amundsen-Scott South Pole Station, which is located at the geographic South Pole. The station is on a high Antarctic plateau, almost 3 km above sea level. It is an ideal place to conduct microwave astronomy. (Image courtesy of Steffen Richter, Harvard University.)

Figure 8 shows a map of the B-mode polarization pattern as discovered by BICEP2. The map shows subtle twists of polarization, with a claimed origin in quantum fluctuations in the geometry of space when the universe was just a trillionth of a trillionth of a trillionth of a second old.

SO WHY IS THIS IMPORTANT?

The BICEP2 result, if it is confirmed, will have an impact in many ways. First, the result is very strong evidence that the early universe really did undergo a period of inflation. It is one thing to develop an idea that fixes various puzzling aspects of the Big Bang; it is quite another thing to obtain experimental data about the idea. With confirmation of in-

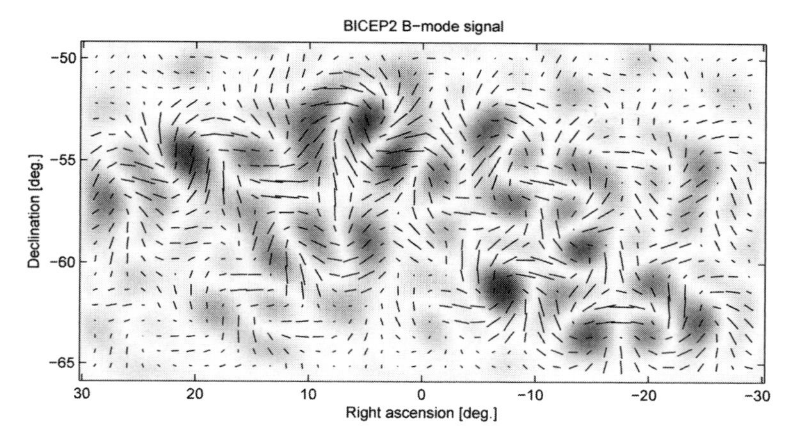

Figure 8. A signal from the dawn of time itself? A line in this diagram represents the direction of the electric field at that point in the CMB, after the E-mode signal has been subtracted. The length of the line represents the strength of the polarization (in other words, how much the average polarization deviates from zero) at that point. The shading in the original diagram is coloured red or blue; red shading highlighted areas of clockwise polarization and blue shading highlighted areas of anticlockwise polarization. This information is, unfortunately, lost in a black-and-white reproduction. (Diagram courtesy of BICEP2 Collaboration.)

flation we have an understanding of why the universe started out with a hot Big Bang, why it is flat, why it is so smooth and why it is so large.

Second, the universe is almost 13.8 billion years old. If the BICEP2 results are confirmed, then science will have progressed to the stage where we can investigate events that happened when the universe was just 10^{-36} seconds old. It is an astonishing thought.

Third, a quite natural outcome of most models of inflation is the notion that our universe is part of a much larger structure. Some would go even further and argue that we must take the idea of the 'multiverse' seriously.

Fourth, the result is very strong indirect evidence for the existence of gravitational waves. The observations of Hulse and Taylor had already provided indirect evidence of gravitational waves, but the BICEP2 observations originate from a completely different source. The result will encourage those astronomers who are hoping to *directly* detect gravitational waves with facilities such as LIGO and VIRGO. In the coming years we can expect gravitational-wave telescopes to open up a new window on astronomy.

Fifth, models of inflation suggest a relationship between the intensity of the B-mode signal and the energy scale of the cosmos at the time of inflation. If the BICEP2 measurement of *r* is correct, the energy scale at inflation was almost as large as it could possibly have been – only a few hundred times less than the Planck scale, the point at which quantum gravity effects become strong. The energy scale implied by the BICEP2 result is far beyond the energy that accelerators such as the Large Hadron Collider can reach but, intriguingly, it is the scale of interest for grand unification – an attempt to unify all fundamental non-gravitational interactions – which got Guth interested in inflation in the first place. For the first time, physicists might have evidence for phenomena that take place at energies ten trillion times greater than occur in the Large Hadron Collider. They will hope to learn how physics works at this utterly extreme scale.

There are many further implications of the BICEP2 result, but I shall end with just one more. As emphasized in the sections above, quantum effects produced gravitational waves during the period of inflation. The detection of primordial gravitational waves demonstrates that gravity is quantized. It does not tell us what the correct theory of quantum gravity might be – but there must be one.

Studies of CMB primordial B-mode polarization thus have the potential to revolutionize science. But is BICEP2 too good to be true?

WILL THE BICEP2 RESULT TURN TO DUST?

It is the essence of science that any extraordinary claim must be subject to thorough checks before being accepted. Soon after the B-mode polarization claim was made, several astronomers expressed concern that the BICEP2 team had underestimated the effects of dust in our Galaxy. No one doubted that BICEP2 had detected a B-mode polarization pattern, but the difficulty was that emission from nearby dust can generate such a pattern. Dust could therefore account for some or all of any observed signal. Might the BICEP2 signal originate from a foreground rather than a cosmological effect? In order to confirm a cosmological origin, measurements would have been taken at a variety of frequencies – but BICEP2 only observed at the single frequency of 150 GHz.

In September 2014, as this Yearbook went to press, scientists released a preliminary analysis of dust-emission measurements taken by the Planck satellite. Unlike BICEP2, Planck was not optimized for polarization observations; however, it made measurements across the whole sky and over a range of frequencies. The Planck data showed that BICEP2 had been observing in a part of the sky that contains much more galactic dust than previously thought. Indeed, the Planck dust maps demonstrate that the entire BICEP2 signal is consistent with foreground contamination. So has Planck killed the BICEP2 claim of having detected evidence for inflation? Not quite; not yet!

The BICEP2 and Planck experiments used different signal processing and observation techniques and so, at the time of writing, it is impossible for Planck to say how much of the BICEP2 B-mode signal might have been due to dust. A final determination of whether BICEP2 saw only dust, or a mix of dust and gravitational waves from inflation, will require collaboration between the Planck and BICEP2 teams. The two teams are now working jointly on a detailed comparison of their results and, by the time you read this, their analysis will have appeared. We should then understand the origin of the BICEP2 B-mode signal.

Regardless of whether the initial BICEP2 claim is confirmed, this episode demonstrates the excitement and vigour of present-day cosmology: scientists can now make exquisitely precise measurements of the cosmos in a variety of ways, and debate their interpretations in an open fashion. If the signal seen by BICEP2 turns out to have been due to dust rather than primordial gravitational waves then that would be disappointing, but one would still have to concede that the experimental effort itself has been heroic. If even part of the BICEP2 signal is shown to have been due to inflation – well, whole new scientific vistas will open up.

The Solar Eclipse of 20 March 2015

JOHN MASON

The first total solar eclipse to be visible anywhere on Earth since that of 3 November 2013 (which was a hybrid or annular/total eclipse), and the first to be visible in northern Europe since that of 11 August 1999, will see both regular eclipse chasers and first-timers heading for the North Atlantic and the Arctic on 20 March 2015.

A total eclipse of the Sun is undoubtedly one of Nature's greatest spectacles. On average such events occur about seventy times per century; there was no total solar eclipse in 2014. All eclipses of the Sun are interesting, but for sheer grandeur, total eclipses are unrivalled. Only then can the solar chromosphere, the prominences and the corona be seen with the naked eye.

Total eclipses occur when the Sun, the Moon and the Earth are exactly lined up, so that the Moon's shadow reaches the surface of the Earth. However, the main cone of shadow (the umbra) is only just long enough to do this, and totality can be witnessed from only a very restricted area of the Earth's surface as the Moon's shadow sweeps along a narrow corridor known as the path of totality. To either side of this track only a partial eclipse will be seen. Moreover, totality is always brief. From any one site it can never last for longer than seven minutes and thirty-one seconds, and the record in recent times appears to be held by the 1955 totality as seen from the Philippine Islands, which lasted for seven minutes and eight seconds.

On 20 March 2015, the Moon's umbra will race across the North Atlantic Ocean at supersonic speed,[1] narrowly skirting the south-eastern corner of Iceland and making landfall at only two places – the Faroe Islands and the Svalbard archipelago (Figure 1). Eclipse chasers will need to make their way to one of these two locations, or they can observe from on board a ship, or perhaps from an aircraft flying along the path of totality. The greatest duration (where the total eclipse lasts the longest along the entire track) occurs at 09h 45m 1s UT on 20 March, from a location north of the Faroe Islands at lat. 64.2831°N,

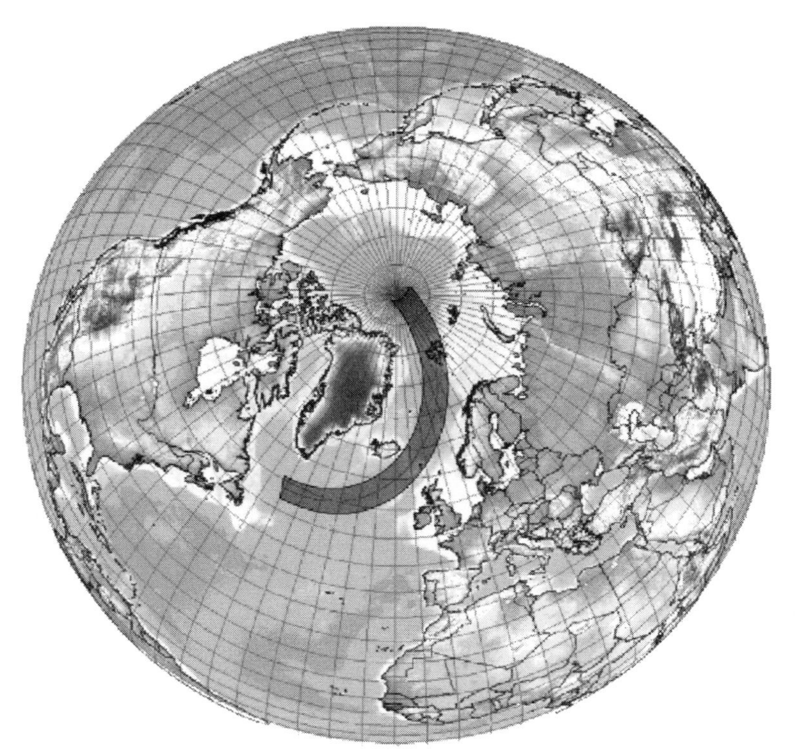

Figure 1. The path of totality for the solar eclipse of 20 March 2015 begins at sunrise at 09h 13m UT (Universal Time) in the North Atlantic, south of Greenland, and curves around between south-east Iceland and outlying north-western parts of the British Isles, making landfall at only two places – the Faroe Islands and the Svalbard archipelago – before ending at sunset at 10h 18m UT in the Arctic Ocean, close to the North Pole. (Map courtesy of Jay Anderson, University of Manitoba and http://home.cc.umanitoba.ca/~jander/tot2015/tot15intro.htm.)

long. 6.8932°W, where the duration of totality will be 2 minutes 46.9 seconds and the altitude of the Sun at mid-eclipse will be 18°.5. Of course, you could be located several hundred kilometres from this point of greatest duration and still witness a total eclipse lasting within a fraction of a second of the greatest duration, as long as you positioned yourself close to the centre line of the path of totality.

Good weather is the key to successful total-eclipse viewing, since it is better to see a shorter eclipse in a clear sky than a longer eclipse hidden behind cloud! One should always try to choose an observing location along the eclipse track that provides the best chance of a cloud-free sky during the eclipse. It is important to consult weather forecasts carefully for locations along the track when planning to travel to view an eclipse. In addition, if one has some flexibility of movement on eclipse day, it will be useful to carefully monitor the local weather forecasts in the days immediately leading up to the eclipse and even on an hour-by-hour basis to maximize one's chances of finding clear sky.

For a detailed overview of the weather prospects along the 20 March 2015 eclipse track, consult the University of Manitoba website at: http://home.cc.umanitoba.ca/~jander/tot2015/tot15intro.htm.

THE PARTIAL ECLIPSE

With the track of totality passing between Iceland and the Outer Hebrides on 20 March 2015, nowhere in the British Isles will witness totality itself,[2] but a very significant partial eclipse will be seen right across the region, ranging from a maximum 85 per cent obscuration in the south-east of England to over 97 per cent in the far north and north-west of Scotland; the whole event lasting well over two hours. Indeed, an obvious partial eclipse will be visible from every country in Europe and the partial phase will also be seen from places as widely spread as Newfoundland, North Africa and north-western Asia (Figure 2).

Given a maximum eclipse obscuration of 97 per cent visible from the Outer Hebrides and the Shetland Islands, this will be the largest partial eclipse to be visible from anywhere in the British Isles since 11 August 1999. (In the annular eclipse seen from Scotland in May 2003, only 94 per cent of the Sun's area was eclipsed.) Note that the term eclipse obscuration is the fraction (or percentage) of the Sun's area eclipsed, whereas eclipse magnitude – another term often quoted in the literature – is the fraction of the Sun's diameter eclipsed. There won't be another partial eclipse of this magnitude until 12 August 2026, so observers in the British Isles should make the most of the March 2015 event.

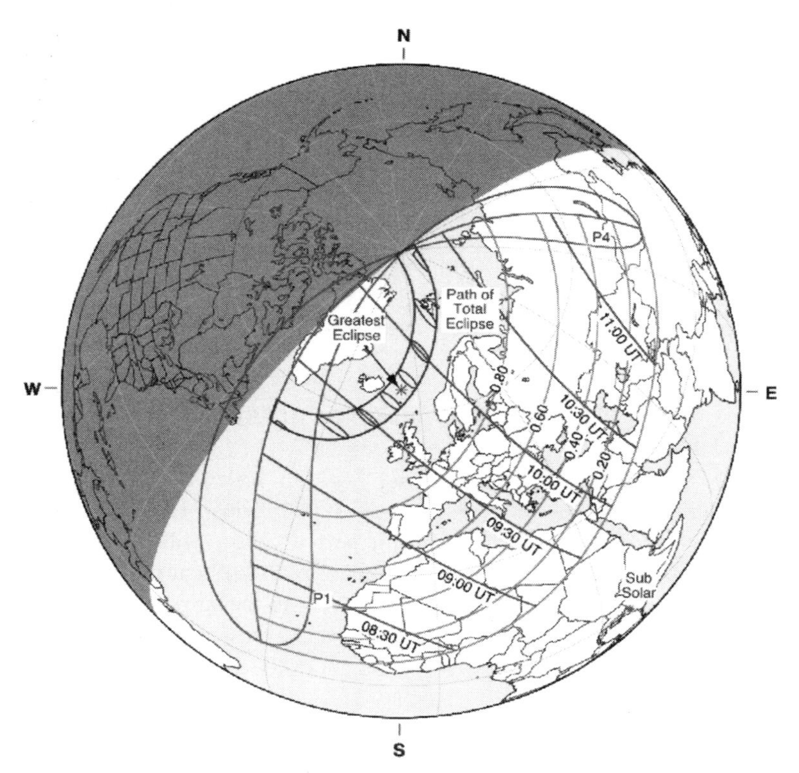

Figure 2. Orthographic map for the 20 March 2015 solar eclipse showing the path of totality, point of greatest eclipse and the extent and visibility of the partial phase with timings in Universal Time (UT). (Diagram courtesy of F. Espenak, NASA Goddard Space Flight Center and eclipse.gsfc.nasa.gov.)

Here are the local circumstances of the partial eclipse from various locations within the British Isles on 20 March 2015:

The Partial Eclipse from Locations in the British Isles on 20 March 2015

	Max. Obsc. %	Max. Eclipse UT	Begins UT	Ends UT
Aberdeen	94.0	09:37:53	08:32:54	10:46:06
Aberystwyth	88.8	09:28:37	08:23:37	10:37:41
Belfast	93.1	09:30:38	08:26:17	10:38:48
Birmingham	87.4	09:30:44	08:25:09	10:40:15
Cardiff	86.7	09:27:54	08:22:36	10:37:21
Dublin	91.6	09:28:16	08:23:54	10:36:37
Edinburgh	93.2	09:35:06	08:30:12	10:43:28
Galway	93.2	09:25:55	08:22:19	10:33:36
Glasgow	93.7	09:34:04	08:29:25	10:42:15
Leeds	89.1	09:33:09	08:27:35	10:42:25
Lerwick	96.9	09:43:09	08:38:33	10:50:30
Liverpool	89.5	09:31:14	08:26:00	10:40:18
London	84.5	09:30:52	08:24:46	10:40:56
Newcastle-upon-Tyne	90.8	09:34:58	08:29:33	10:43:55
Norwich	85.2	09:34:03	08:27:41	10:44:07
Plymouth	85.6	09:25:11	08:20:07	10:34:36
Southampton	84.5	09:28:40	08:22:53	10:38:36

Undoubtedly, there will be enormous public interest in viewing such a significant partial eclipse from locations in northern Europe that are outside of the narrow path of totality. But looking at the Sun is dangerous and can result in serious eye damage or blindness. The danger is not because of the eclipse – it is dangerous to look at the Sun at any time. You ***must*** always protect your eyes during the partial eclipse. However, an eclipse of the Sun ***can*** be observed safely by following the Dos and Don'ts of the Solar Eclipse Safety Code (see the fact box at the end of this chapter), but ***do*** supervise children closely at all times.

Normally, by the time the Sun is nearly half covered, anyone standing near a tree or bush will be able to see tiny crescent-shaped images on the ground around them. Gaps in the foliage act as 'pinhole cameras' and focus the images of the crescent Sun. (Unfortunately, both the Faroes and Svalbard are characterized by a complete lack of

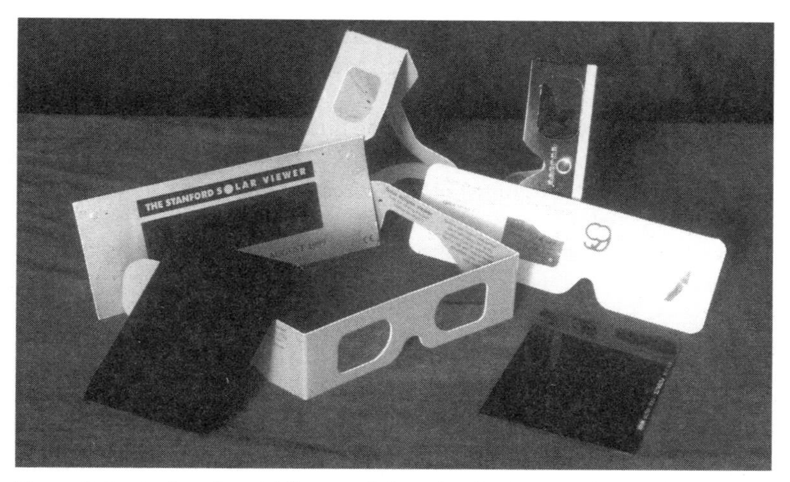

Figure 3. A selection of special filters made for safe solar viewing. Aluminized mylar filters or black polymer filters should be identifiable as suitable for direct viewing of the Sun, bearing the CE mark and a statement that they conform to European Community Directive 89/686/EEC. Welder's glass must be rated at No. 14 or higher. Always read and follow the manufacturer's instructions carefully and check filters thoroughly for any damage before use. (Image courtesy of the author.)

trees and very few woody plants, so this is not an effect that will be easily observed from these locations on 20 March 2015!)

It is also possible to make your own pinhole projector to observe the partial eclipse in safety. Make a small (4-mm diameter) hole in the middle of a large piece of card and use it to 'project' an image of the Sun onto another (preferably white) card screen positioned 1–2 metres away. ***Don't*** look through the hole – look ***only*** at the projected image on the (white) card screen.

TIMELINE OF A TOTAL SOLAR ECLIPSE

A total solar eclipse begins at first contact when a tiny 'notch' appears on the Sun's limb. Although the exact moment can be predicted with great accuracy, it takes a few seconds for the notch to become noticeable.

Even if you plan to do no more than 'look' at the eclipse, make sure that you have adequate dark filters, which are absolutely safe for solar viewing (Figure 3) and remember to take the greatest care at all times. Welder's glass, rated at number 14 or higher, is suitable for observing the partial phases of the eclipse, but it may be difficult to obtain. Another good material for safe solar filters is aluminized mylar. Mylar is a very tough plastic film, and solar filters are made by coating it with a thin layer of aluminium. The partial phases of the eclipse may be observed safely through a pair of mylar 'spectacles' or a 'solar eclipse viewer', but do make sure that any filters you use carry the 'CE' mark, and before each use check them very carefully for any damage.

As the partial phase progresses, the Moon gradually passes on to the face of the Sun. For a surprisingly long time there is no perceptible diminution in light (or fall in temperature), but when the Sun is more than half covered, these effects usually start to become evident. If there are any sunspots (which we won't really know until the day of the eclipse), compare them with the darkness of the Moon; the lunar disk will be seen to be much the blacker.

As totality approaches, the whole scene changes with amazing rapidity. Normally, the temperature falls, the sky darkens, birds may start to roost, some types of flowers close, and the shadow of the Moon may be seen rushing across the landscape towards the observer. Again, much of this is unlikely to be witnessed in the Arctic in March 2015, and one doubts whether eclipse watchers in Svalbard will venture close enough to see if the polar bears exhibit any unusual behaviour as totality approaches!

Shadow bands, which are purely atmospheric phenomena, may appear briefly shortly before (and sometimes shortly after) totality, but only if the conditions are absolutely right. Just before the last sliver of the solar photosphere disappears, we see the effect termed the 'Diamond Ring', a brilliant point which lasts for an all-too-brief period (Figure 4). It is usually better seen at the end of totality.

The Moon's limb is not smooth and moments before totality, the sunlight comes to us through the lunar valleys on the limb, and the result is a series of bright points of light known as Baily's Beads.

At second contact, the brilliant disk of the Sun is completely hidden and totality has begun. The corona and prominences flash into view; the sky is dark, and almost at once the brightest stars and planets can be seen. There is a strange, somewhat eerie calm. The corona (Figure 5) is

Figure 4. The last bead of sunlight shines down a deep valley on the limb of the Moon moments before second contact in the total solar eclipse of 11 July 2010, which the author observed from the Island of Hao in the Pacific Ocean. (Image courtesy of the author.)

Figure 5. The pearly-white solar corona is visible all around the dark disk of the Moon at mid-totality on 11 July 2010. The magnetic poles of the Sun are at the 2 o'clock and 8 o'clock positions in this image. A number of fine streamers were visible within the corona, but a very thin layer of cloud reduced the clarity of these features. (Image courtesy of the author.)

not always of the same form. Near sunspot minimum it sends out long streamers, while near maximum it is more symmetrical – though of course no hard-and-fast rules can be laid down. By March 2015, the Sun will be a year or so past sunspot maximum (although the current cycle exhibited a double peak in sunspot numbers), so it will be hard to forecast the form of the corona during totality. Neither can we predict what prominences will be visible, if any, so as far as March 2015 is concerned we shall have to wait and see.

Although no two total eclipses are alike, the corona normally gives out about as much light as the full Moon. This means that direct viewing, even with a telescope, is safe, but ***only*** while the Sun is totally eclipsed. Pause to look round the sky during totality; Venus (magnitude –4.0) will be a spectacular object in Aries to the east of the eclipsed Sun (and somewhat higher in the sky). From Svalbard, Jupiter (magnitude –2.4) may also be seen low towards the north-north-east in Cancer. Mars (magnitude +1.3) is rather faint in Pisces, but may be glimpsed under very transparent skies about two-thirds of the way from the eclipsed Sun to Venus.

Totality ends as suddenly as it began, at third contact. Baily's beads should reappear briefly (Figures 6a and 6b), and then – the Diamond Ring. This is one of the most glorious moments of the entire eclipse, and everyone will be on the alert, because it lasts for so short a time (Figure 7). In only a few seconds the photosphere starts to reappear; the corona fades from view, and the Diamond Ring is lost. This is certainly the most dangerous moment for the careless observer. The slightest sliver of the main photosphere is as dangerous as the uneclipsed Sun itself – and so if you have been viewing directly, make sure you take your eye away from the 'danger-zone' in time. Remember, too, that an SLR camera acts in the same way as a telescope or binocular lens.

Gradually the Moon moves off the face of the Sun. The sky quickly brightens and everything seems to 'wake up'. Sometimes you may see the effects of the receding shadow of the Moon in the form of a curved dark, sometimes purplish, patch covering part of the sky.

At fourth contact, the Moon finally moves off the Sun, and everything is back to normal; the eclipse is well and truly over. In fact, it often happens that few observers wait to see fourth contact; they are too busy packing up their equipment and comparing notes.

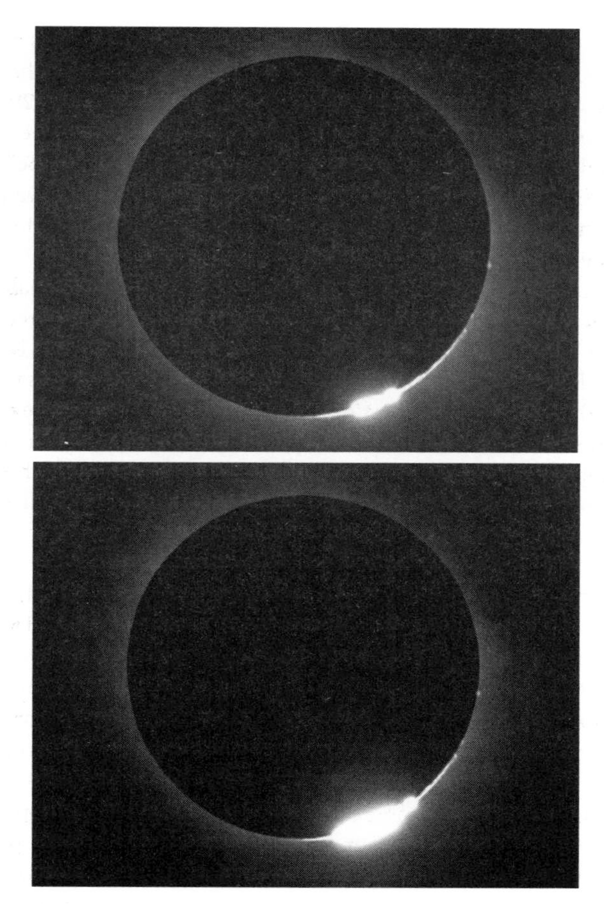

Figure 6a (top) and 6b (bottom). Two views of third contact at the end of totality witnessed from the Sahara Desert in Libya on 29 March 2006. These images, taken just three seconds apart, show Baily's beads merging to produce a single brilliant point of light moments before the full 'Diamond Ring' appeared. (Images courtesy of Martin Mobberley.)

TOTALITY FROM THE FAROES

On the face of it, the Faroe Islands would seem a good bet for eclipse chasers due to their close proximity to the point of greatest eclipse, with the Sun being at an altitude of almost 20° at mid-eclipse. In spite

Figure 7. Even thin drifting cloud cannot hide the glory of the 'Diamond Ring' at the end of totality on 11 July 2010, as the Sun dramatically bursts out from behind the dark disk of the Moon. (Image courtesy of the author.)

of their high northerly latitude (62°N), the Faroes are bathed in a branch of the Gulf Stream and have a moderate climate, with frequent rainfall – over 260 days per year on average. They also lie in the path of North Atlantic low-pressure systems and stormy conditions are possible at any time. According to eclipse weather guru Jay Anderson of the University of Manitoba, Canada: 'The cloud-cover statistics (for the Faroes) paint a grey picture with a mean cloudiness of 75 per cent and a chance of possible sunshine of only 24 per cent at Tórshavn in March. For a mobile eclipse seeker, these statistics can be improved by a judicious choice of observing site. The first consideration should be given to selecting a site in the lee of a line of hills so that the airflow is compelled to descend, helping to dissipate the low-level clouds, but the site selection must be made using the forecast wind direction in the days ahead of the eclipse; a site on the lee side of an obliging line of hills will help increase the odds of a view of the Sun.' If fog and very low cloud are the main problem, then the best approach will be to select a site on the top of the hills and hope that the Sun breaks through.

Local circumstances in the Faroes are as follows:

	1st contact	2nd contact	3rd contact	4th contact	Alt.	Dur.
Tórshavn	08:38:51	09:40:54	09:42:54	10:47:40	20°	2m 00s

Times are UT. The duration is corrected for lunar limb profile.

The key to success in the Faroes would appear to be mobility and a willingness to choose a suitable observing site at short notice based on the prevailing weather conditions and wind direction. The Faroes are very hilly, with many bays and fjords, but travel is comparatively easy, due to a network of tunnels and bridges that link the islands and cut beneath the peaks. This ease of movement will be of great benefit if it turns out that some last-minute manoeuvring and a change of observing site are needed to avoid troublesome clouds during the eclipse. There is a good chance of at least some openings in the cloud cover unless a large-scale weather system is affecting the islands. Quick thinking and an element of luck will be needed!

TOTALITY IN SVALBARD

The other landfall of the 20 March 2015 total eclipse is in the islands of the Svalbard archipelago, most notably Spitzbergen, the largest island of the group (Figure 8). At this latitude (78°N), polar easterly winds prevail, but passing low pressure regions can bring a southerly flow with thick cloud and steady precipitation. As with the Faroes, the Svalbard island terrain is steep and the valleys and fjords control wind direction, cloudiness and precipitation. The degree of surrounding ice cover is also of great importance here because sea-ice blocks the moisture from the ocean surface and the weather is colder and clearer – good news for eclipse watchers.

Most travellers will probably head for Longyearbyen, the main town on Spitzbergen (with a small population of around 2,000 inhabitants) and the area's primary port, administrative centre and the only airport on the islands. The location is stunningly beautiful; large glaciers cover much of the island and icebergs float around the town. On 20 March, the Sun will rise to an altitude of 11° at totality from Longyearbyen, but because of the 1,000-metre-high mountains to the south and southwest, the eclipsed Sun will be hidden from most of the town, and

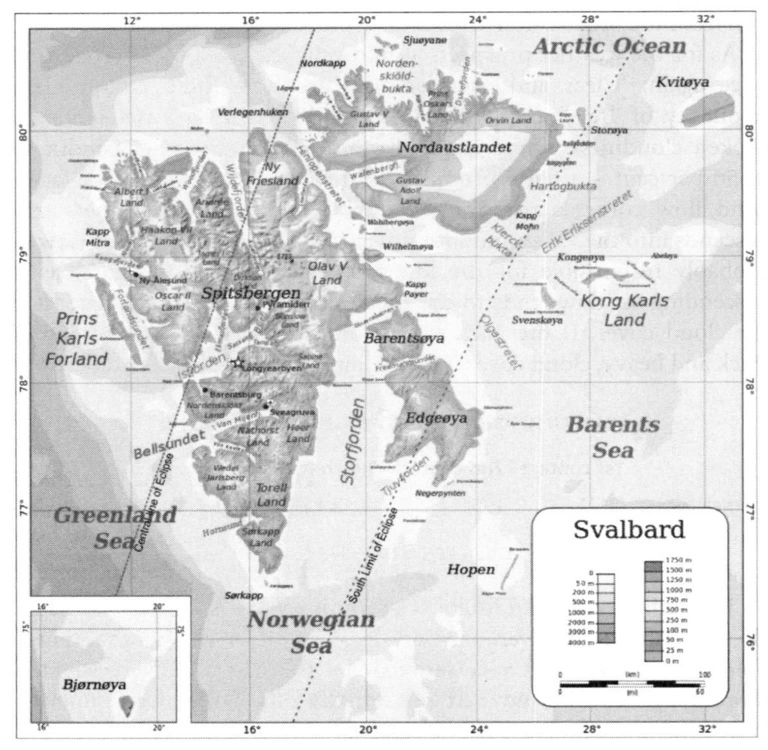

Figure 8. The path of totality over the islands of the Svalbard archipelago on 20 March 2015. The centre line of the track passes right across Spitzbergen, the largest island of the group. All of the main islands are within the track, although the south-eastern part of Edgeøya lies beyond the south limit of eclipse and Kvitøya and the islands of Kong Karls Land are completely outside of the track. (Map courtesy of Oona Räisänen, Wikimedia Commons, Jay Anderson, University of Manitoba and http://home.cc.umanitoba.ca/~jander/tot2015/tot15intro.htm.)

viewing sites must be chosen very carefully to ensure that the Sun is visible during totality. Longyearbyen lies at the mouth of the 30-km-long Advendalen valley, and the main road from the town runs south-east along the valley, providing access to a flat plain that is good for eclipse-viewing. Steep hills to the south side of the road will hide the Sun from many places on the road itself, but valleys between the hills may enable the eclipse to be seen. However, there is an obstruction-free view of the

Sun from the broad 3-km-wide area to the north of the road.

As for the weather prospects at mid-eclipse, to quote Jay Anderson once again: 'Clear and overcast skies are rare; there is a modest frequency of "few" and "scattered" clouds and a high frequency of broken cloudiness. On average, this amounts to a mean cloudiness of 56 per cent – a fairly promising statistic . . . and the above-valley wind flow comes across the mountains behind Longyearbyen and descends into the valley and into Adventfjorden. This prevailing flow is probably responsible for the low frequency of overcast skies, as a descending air flow tends to dry the atmospheric column and reduce the cloud cover. If the pack ice surrounding western Spitsbergen is thick and heavy, cloud cover is lighter and temperatures are colder.'

Local circumstances in Svalbard are as follows:

	1st contact	2nd contact	3rd contact	4th contact	Alt.	Dur.
Longyearbyen	09:11:53	10:10:43	10:13:11	11:12:21	11°	2m 28s

Times are UT. The duration is corrected for lunar limb profile.

Throughout Svalbard, there is a significant threat from polar bears all year round. However, they are legally protected, and shooting a polar bear is regarded very seriously by the police and investigated thoroughly. You are advised not to leave the settlement limits of Longyearbyen, which are clearly marked with signs bearing the picture of a polar bear, without the advice and assistance of local guides.

Given the general storminess of the North Atlantic in March, it is likely that some cruise ships carrying eclipse watchers will also head for the usually rather calmer waters near Svalbard, particularly around the southern shores of the islands which are often ice-free at that time of year.

All eclipses of the Sun are interesting, but for sheer grandeur total eclipses are unrivalled. Only then can the solar chromosphere, the prominences and the corona be seen with the naked eye – and the total eclipse of March 2015 will be the last such event to be visible in Europe until 12 August 2026.

And if the skies are clear as darkness falls on eclipse day, even with the long hours of twilight at these latitudes in late March, there is also the chance of viewing the northern lights from Svalbard. What a fabulous opportunity to view two of Nature's greatest spectacles on the same day! Let us hope that the weather co-operates.

SOLAR ECLIPSE SAFETY CODE

A total eclipse of the Sun is one of Nature's greatest spectacles. Most of us will see only one total eclipse in our lifetime.

But looking at the Sun is dangerous and can result in serious eye damage or blindness. The danger is NOT because of the eclipse – it is dangerous to look at the Sun at any time.

The Sun is the brightest object in the sky. In addition to the visible light, it sends out huge amounts of invisible infrared and ultraviolet rays which can harm your sight. To view the Sun safely, these rays must be blocked out by special filters.

The instructions in this safety code are intended for those viewing an eclipse of the Sun with the unaided human eye, e.g. WITHOUT any optical aid. They do NOT cover the use of special solar filters in conjunction with telescopes or binoculars.

A TOTAL ECLIPSE is when the Moon completely covers the brilliant disk of the Sun. A PARTIAL ECLIPSE is when any part of the Sun's brilliant disk, however small, can be seen. You MUST always protect your eyes during the partial eclipse.

An eclipse CAN be observed safely with the unaided human eye by following the DOs and DON'Ts of the SOLAR ECLIPSE SAFETY CODE, but DO supervise children closely at all times.

WHILE THE SUN IS IN PARTIAL ECLIPSE:

DON'T ever look at the Sun without proper eye protection.

DON'T view the Sun through sunglasses of any type (single or multiple pairs), or filters made of black and white or colour photographic film, or any combination of photographic filters, crossed polarizers or gelatin filters, CDs, CD-ROMs, or smoked glass. These are NOT safe.

DO view the Sun ONLY through special filters made for safe solar viewing, e.g. aluminized mylar filters, or black polymer filters, identified as suitable for direct viewing of the Sun, bearing the CE mark AND a statement that it conforms to European Community Directive 89/686/EEC, or use a welder's glass rated at no. 14 or higher. Always read and follow the manufacturer's instructions carefully.

DO check filters thoroughly for any damage BEFORE use. DON'T use them if they are scuffed, scratched or there are pinholes in them.

DO place the special filter firmly over your eyes BEFORE looking up at the Sun, and DON'T remove it until AFTER looking away.

DO use the special filters to view the eclipse while any part of the Sun's brilliant disk, however small, is still visible. This will apply at all times if you are not within the path of the total eclipse and are viewing only a partial eclipse.

DON'T stare through the special filter for more than three minutes at a time. Intermittent use of the filter for the duration of the partial eclipse is the best way to view the event.

Even with the special filter placed firmly over your eyes, DON'T ever look at the Sun through any optical instrument, e.g. telescope, binoculars or camera. Such devices concentrate the Sun's harmful radiation and will cause severe eye damage in a fraction of a second. Filters identified as suitable for direct viewing of the Sun are NOT safe for use in conjunction with any optical instrument. If you are not certain that a filter is approved and safe, or you have any other doubts, DON'T USE IT.

And Also:

DO observe the Sun INDIRECTLY by 'pinhole projection'. Make a small (4-mm diameter) hole in the middle of a large piece of card and use it to 'project' an image of the Sun onto another (preferably white) card screen positioned 1–2 metres away. DON'T look through the hole – look ONLY at the projected image on the (white) card screen.

→

While the Sun Is In Total Eclipse:

IF you are within the narrow path of totality, during a total solar eclipse, the Moon will for a short time (usually only a few minutes) COMPLETELY cover the Sun's brilliant disk.

AT THAT TIME ONLY:

DO view the totally eclipsed Sun directly without any filter. ONLY at this time can the faint and beautiful corona – the Sun's pearly white outer atmosphere – be appreciated.

BUT:

DO be alert to the reappearance of the Sun's brilliant disk at the end of the total eclipse. As soon as the first light of the Sun has reappeared (the spectacular 'diamond ring'), you MUST look away immediately and use the special filters once more.

REMEMBER that even within the path of the total eclipse, there will be a partial eclipse for about one-and-a-quarter hours before and after the brief total phase. You MUST always use the special filters during the partial eclipse.

Notes

1 The speed of the Moon's shadow across the surface of the Earth varies considerably along the path of totality. On 20 March 2015, the shadow speed is in excess of 5,000 kilometres per hour at either end of the track, but drops to just over 3,200 kilometres per hour at minimum, which is where the duration of totality is near maximum.

2 The last total solar eclipse to be visible from the British Isles was on 11 August 1999, and the next to be visible from mainland Britain will not occur until 23 September 2090, but the track of the total eclipse of 3 September 2081 will pass over the Channel Islands.

Lalande 36613

RICHARD BAUM

Lalande 36613 is a star with a curious history, described as 'a strange interlude' by the astronomical historian Willy Ley (1906–1969). Of the ninth magnitude and located in the constellation Sagittarius, the star was listed by the French astronomer Joseph Jérôme Lefrançois de Lalande (1732–1807) in his voluminous *Histoire Céleste Française* (1801) and used as a comparison star or reference point to determine the position of nearby objects. Physically, however, it is unassuming and, but for its centrality to a bizarre episode in the mid-nineteenth century, would be no more noteworthy than any other catalogue star or countless others unremarked. Yet for a time in the period 1850–52, it was at the centre of a sequence of events that set astronomical commentators chattering with exciting news; news that set the stage for another chapter in our exploration of the physical world, and gave further impetus to our insatiable quest to extend the frontiers of knowledge.

AN ASTEROID DISCOVERED

The affair began on the night of 12 April 1849. Comparing Hour XII of Steinheil's star map with the heavens, Annibale de Gasparis (1819–1892), assistant astronomer at the Specola di Capodimonte, Naples, chanced upon an uncharted object of the tenth magnitude. Unfavourable conditions then prevented an exact determination of its position. Better conditions prevailed on the 14th and 17th and comparisons were made with other stars, especially No. 23098 of Lalande's *Histoire Céleste*. This revealed that the object had retrograde motion and was heading towards the Equator.

On 11 May, Fabri Scarpellini (1808–1873), secretary of the Correspondenza Scientifica at Rome, announced the discovery and Heinrich Christian Schumacher (1780–1864) widely disseminated the news through the medium of his *Astronomische Nachrichten*, the publication which acted as the clearing house for astronomical discoveries. Ernesto Capocci (1798–1864), director of the Specola di Capodimonte, proposed the name Igea Borbonica after the goddess of Health, daughter of Aesculapius; for political reasons Borbonica was dropped and the object, named Hygiea (this is the modern spelling, formerly Hygeia or Hygea), took its place as the tenth asteroid to be found (it is now recognized as the fourth-largest asteroid in the solar system).

Just over a year after its discovery, award-winning James Ferguson (1797–1867), a Scottish-born astronomer on the staff of the National Observatory, Washington, DC (later renamed the US Naval Observatory), distinguished as the first American to find an asteroid (31 Euphrosyne on 1 September 1854), began a routine series of observations of Hygiea with the 24.4-cm Merz and Mahler equatorial refractor installed in 1844. During the summer and autumn months of 1850, he tracked Hygiea as it cruised through the rich star fields of Sagittarius, measuring its position by reference to nearby catalogue stars with a filar micrometer.

Figure 1. James Ferguson (1797–1867). First American astronomer to discover an asteroid. His observations of the asteroid Hygiea in 1850 led to a search for a supposed planet beyond Neptune. (Image courtesy of US Naval Observatory, Washington, DC.)

Then considered the astronomer's real arm of precision, a filar micrometer is an elegant device mounted in the focal plane of an equatorial refractor and employed to measure small angular distances on the celestial sphere. Its design remains largely the same as the original instrument, which was first applied to an astronomical telescope in the late 1630s by the English amateur astronomer William Gascoigne (c. 1620–1644), a Royalist killed at the Battle of Marston Moor on 2 July 1644, during the first English Civil War of 1642–6. Basically, the instrument consists of a flat, rectangular box across which two fine wires are stretched. These are fixed and at right angles to each other. Another wire parallel to one of the former is drawn taut across a frame free to slide within this box. The movement of the frame is controlled by a finely pitched screw connected to a graduated drum, from which the observer can read off the total of revolutions applied to the screw. Important to the case in hand, the Washington micrometer had a movable metal plate bearing three east-west wires, denoted 1, 2 and 3. This served to measure the angular distance (declination difference) between two objects, one of known position, the other not. To determine the position of the latter, the observer shifted the metal plate to obtain a bisection of the comparison star by one of the three wires. The other object was then bisected by a fourth but adjustable wire also in the east-west direction. The revolutions of the screw thus provided a measure of the angular distance traversed by the wires when the angular value of one revolution was known. Hence from the settings of the plate and the movable wire the declination difference could be calculated, and the position of the unknown object derived by computation.

In this way Ferguson followed Hygiea from 18 May to 14 November 1850, fixing its position on sixty-three nights, and describing it as 'very minute (12.13 mag.) . . . well seen' on 21 November. A report of the observations was drawn up and published in *The Astronomical Journal* of 18 January 1851 as 'Observations of Hygea made with the Filar-Micrometer of the Washington Equatorial'.

A STAR IS MISSING

And so it came to the attention of the rising young British astronomer John Russell Hind (1823–1895), foreign secretary of the Royal Astronomical Society. He had been an assistant in the Magnetic and

Meteorological Department of the Royal Greenwich Observatory, but in October 1844, he was head-hunted on the recommendation of the Astronomer Royal, George Biddell Airy (1801–1892), by the wealthy London wine merchant George Bishop (1785–1861), who hired him as supervisor of the small private observatory he had set up at his residence, South Villa, Inner Circle of Regent's Park, London, in 1836.

Figure 2. John Russell Hind (1823–1895). His misconception about a missing star induced a search for a possible trans-Neptunian planet in the period 1851–2. Hind discovered ten asteroids and in 1853 was appointed Superintendent of the Nautical Almanac. (From Ormsby MacKnight Mitchel's *The Orbs of Heaven*, London, 1857.)

Hind had a vested interested in Ferguson's report. He was familiar with the region traversed by Hygiea in 1850. 'For the past four years,' he noted, 'I have constantly had this very region of the heavens under close examination, having in August 1847 missed a small star in about RA 19h 1m 45s, NPD (North Polar Distance) 111° 56′ (1800).' But his interest was more focussed. Under the direction of the previous supervisor, the well-known Revd William Rutter Dawes (1799–1868), the priority had been binary stars. With Hind it switched to planet-hunting. At first his interest focussed on the rumoured body beyond Uranus (Neptune, discovered in 1846), but after the discovery of Astraea, the fifth asteroid, in 1845, he switched to the asteroid group. By now he

Figure 3. Private observatory of George Bishop at South Villa, London. Located in longitude 37s. 10 West, latitude 51° 31′ 29″.8. It housed a 17.78-cm Dollond equatorial refractor. Following Bishop's death, the observatory was dismantled in 1863 and moved to his son's residence at Meadowbank, Twickenham, and later donated to the Royal Observatory, Naples in 1877. (Frontispiece of George Bishop's *Astronomical Observations Taken at the Observatory South Villa, Inner Circle, Regent's Park, London During the Years 1839–1851* (London, 1852).)

had three discoveries to his credit, as indeed did de Gasparis. Together they accounted for six out of the eight such discoveries made since 1845.

Initially, Hind used the star maps prepared by the Berlin Academy of Sciences, but found them unsuitable as they only extended to –2° in declination and had a limiting magnitude of 10. Accordingly, he had instigated his own star-mapping project, the purpose being to produce a map of the zodiacal stars that covered three degrees both sides of the ecliptic to a limiting magnitude of eleven. Charts were issued as work progressed, building to twenty-four charts, one for each hour of right ascension. In the period concerned he had been 'filling in all stars for Hour XIX of our charts', the region navigated by Hygiea in October 1850.

It is important to note that many of the stars Ferguson used to fix the path of Hygiea were uncatalogued. So first he had to micrometrically determine their coordinates by reference to nearby catalogue stars and identify them with a temporary local prefix such as a lowercase letter or numeral. It was thus natural in pre-photographic times and incomplete stellar cartography for observers to compare results in matters of this nature.

Now in the course of his comparison, Hind noted that on the night of 21 October 1850, Ferguson had referred Hygiea to a star of magnitude 9.10 listed as 'k'. From three observations with the West Equatorial based on star 1719 of the *Greenwich Twelve-Year Catalogue*, he gave the adopted mean place of 'k' for 1850.0 as RA 19h 17m 40s.60 and Dec. –20° 45′ 05". 68.

Hind was puzzled. He failed to recollect 'any suspicious body in the vicinity'. Stranger still, 'k' could not be reconciled with any star on his chart of the region, nor yet could it be found when he came to search for it in the heavens. Had Ferguson made a mistake, either in his observations or reductions? How could he, an experienced observer, overlook a star of the ninth magnitude? Had he, in fact, substituted one star for another? Hind well knew the problem of identification in thickly populated star fields such as the planet had traversed in the summer of 1850. Ferguson himself had remarked: 'From this date, 26 August, till the termination of the series, there were many small stars about the path of the planet, making it necessary to observe more than one star. Occasionally the wrong star was observed, and the observation lost.' This happened on 20 September 1850, and he remarked: '. . . the wrong star observed. The single comparison given was accidental.' Again, on 19 October, he noted: 'There are many stars here and it is necessary to have the place closely computed, need to reexamine.' There were distractions too: 'Much noise, steam boats and singing.'

Even so, Hind reasoned, this line of thought had little to commend it to serious thought. In all ways Ferguson had shown himself to be a competent and reliable observer. Other than errors due to personal equation, his record seemed wholly trustworthy. The fact that the star could not now be found did not necessarily imply error. On the contrary, two other explanations were evident. Either there was an unknown variable star in that position or else, and here Hind's preference became obvious, Ferguson had chanced upon a new asteroid. In a letter to William Cranch Bond (1789–1859), director of the Harvard College Astronomical Observatory, he not only drew attention to the missing star (hereafter designated *k) but also speculated about its possible planetary character. It was all very circumstantial but Bond was intrigued enough to relay the information to Lt Matthew Fontaine Maury, Ferguson's chief at the National Observatory, Washington, DC.

*K IS ABSENT

Maury's immediate reaction involved Ferguson in a laborious session with the large refractor. Instructed 'to examine the proper part of the heavens for the star in question', he diligently checked the suspect zone on the night of 29 August 1851, only to report the star absent.

This circumstance induced Maury to audit the relevant entries in the observing journals for October 1850. Ferguson had first seen the star on 16 October of that year, when he was looking for Hygiea, and had compared it with Greenwich star 1719. It had again been sighted on the 19th, with two unknown stars; observed on the 21st, with Hygiea, though with regard to the latter Ferguson had remarked: 'Doubtful if planet, and very badly seen'; and finally, on the 22nd, it had been compared with Lalande's ninth-magnitude star 36878. Throughout, the Washington astronomer had designated the object as a star of 9.10 magnitude.

Significantly, these observations indicated the object had motion in right ascension, and was stationary at some time between 16 and 22 October 1850. Suspicion undoubtedly attached to the object, and Maury concurred with Hind's hypothesis. Accordingly, on 3 September 1851, he informed the Honorary Secretary of the US Navy, William A Graham: 'The star of comparison with Hygea on the night of 21 October, 1850, has disappeared. It is not now to be found where it then was. Hence I infer that it is an unknown planet.' In conclusion, on the assumption the observatory had missed a rare opportunity, and perhaps with thoughts of the Neptune affair in mind, he ended on a regretful note, that the observations, 'had there been sufficient force at the Observatory for their immediate reduction, would then have revealed to us the character of this star'. This was an oblique reference to the effect the war with Mexico (1846–8) had on staffing levels, especially at junior level.

A GROWING SUSPICION

During the summer of 1851, Hind and de Gasparis each logged their fourth discovery of an asteroid; Irene was found by the former on 19 May, and Eunomia by the latter on 29 July. In August, Hind discussed the case of the truant star with a visiting American astronomer, Benjamin Apthorp Gould (1824–1896), founder and editor of *The Astronomical Journal*, and mentioned his suspicion that it would not be explained 'on any other supposition than by assuming it to have been an unknown planet'. By which he meant an asteroid, as he had intimated to Bond. Maury certainly believed this to be so and after the sweep of 29 August, had encouraged Ferguson to continue the hunt on this premise.

In the meantime, Hind had reconsidered the matter. Having read Maury's 3 September letter to Graham printed in *The Astronomical Journal* of 22 October 1851 along with the observed places of *k on 16 and 22 October 1850, he realized the motion of *k was inconsistent with his original hypothesis. Instead it suggested something rather more exciting. Something he averred '. . . I conceive it highly desirable . . . should be in the hands of astronomers.'

Although Maury had refrained from publicizing the Washington search, Hind, who was actively searching himself, was aware of its progress from Gould: 'You tell me,' he wrote to Gould on 12 November, 'that the Washington observers are engaged in watching the small stars about the place of the one missing, but their silence up to the present date is discouraging.' He now seized the opportunity to acquaint Gould with the dramatic import of his latest findings. Recalling their conversation in the August of that year, he noted that Maury had examined the original observations and that further evidence in favour of the planetary nature of the object had been adduced:

> . . . the positions resulting from the differential observations of 16 and 22 October showing an increase in right ascension which is quite beyond any probable error of observation on the part of a skilful observer, such as Mr FERGUSON has proved himself to be. Now a mere glance at the places of the suspicious object, and of the sun at the time, will be sufficient to convince us, that, if there be a planet in the case, it could not have belonged to the prolific group between

> Mars and Jupiter, that its mean distance in fact must be greater than that of any known planet . . . If we assume (as I think we may safely do, on the further hypothesis of circular motion) that Mr Ferguson's star would be *stationary* within a day or two before or after 16 October, we shall find that a planet in this position must have a distance of more than 137 [AUs], and a period of above 1600 years.

By comparison, Neptune orbits the sun at a distance of around 30 AUs in a period of 164.79 years.

Startling in his forecast, Hind intimated what was rapidly becoming an article of faith to many astronomers. The concept of a small and finite solar system had been condemned in 1781 with the discovery of Uranus, and any remaining doubt banished by Giuseppe Piazzi (1746–1826) on 1 January 1801, when he found a strange object of the seventh magnitude amongst the stars of Taurus, and so initiated knowledge of the asteroids. Authentication of the trans-Uranian hypothesis with the discovery of Neptune in September 1846 induced further speculation. Scarcely a week had passed before the primary instigator of that discovery, the eminent geometer of Paris U. J. J. Le Verrier (1811–1877), confided to a colleague in Geneva that in his opinion Neptune did not mark the frontier of our planetary system. The fact the gravitational field of the sun extended far beyond made it certain other planets might yet be found in those remote regions. Thus in the case of *k, Hind transmuted conjecture into fact, and projected a view already current.

Still he was beset by suppressed doubt as he further remarked to Gould: 'I am at a loss to imagine how a slow-moving planet of 9.10 magnitude can have escaped me. Such a planet would be easily recovered, and I am convinced I must have seen it during the past summer, if it retained the same degree of brilliancy assigned by FERGUSON (9.10 magnitude). So far, then, as my own search has extended, I feel able to state confidently that there is no planet hereabouts of the 9.10 magnitude.' He added, 'The *second* observation on 21 October is not given in your journal; perhaps Lieutenant MAURY may favour us with the *original comparisons* by which the star's places are determined.' Almost as an afterthought, Hind mused, 'just suppose 'that its light is subject to variation so as to cause it to descend to the eleventh class, then I should think it necessary to institute a further examination before pronouncing an opinion upon the subject.'

THE MYSTERY DEEPENS

Hind's letter appeared in *The Astronomical Journal* of 1 January 1852, but Gould had already apprised Maury of its content. Consequently, the search for *k had been re-orientated 'upon the supposition that it is not an asteroid, but an exterior planet'.

A year had now elapsed since the fugitive was last sighted. Ferguson doubtless faced the prospect of its recovery with some disquiet, justifiably so in the circumstances. Photographic search methods were non-existent, whilst available stellar cartography gave scant coverage of the region concerned; the great *Bonner Dürchmusterung* was still eleven years in the future. If indeed a planet was abroad, then clearly its motion was carrying it deeper into the dense and largely unexplored star fields of Sagittarius, increasing the difficulties already in the way of its recovery.

Still the search began, optimistically enough with a general sweep of the suspect region. Spurred on by the thought of discovery, Ferguson spent four tedious months at the ocular of the large refractor, checking and re-checking a bewildering array of faint and largely unknown stars in the hope of detecting one that moved, betraying its planetary character, but to no avail – *k had vanished. Realizing the futility of continuing the search, Maury decided upon its abandonment on 11 December 1852. In the meantime Ferguson had examined all stars in the suspect area down to the eleventh visual magnitude, as recommended by Hind.

Writing to Gould on 5 January 1852, Maury noted:

> Since my last communication of the 3rd September, 1851, relative to the missing star of October, 1850, I have had a thorough examination made of the part of the heavens in which it was seen, but without its recovery . . . The search had been conducted by Mr Ferguson, with the large Equatorial, upon the supposition that the missing star was not an asteroid, but an exterior planet.

Ferguson's observations of October 1850, from which the apparent places of *k on the 16, 21 and 22 (corrected for refraction) had been deduced, were given. Accurate positions for the stars of comparison on 19 October were not available, so the place of the truant on that date was omitted.

And there the matter rested; a line was drawn under the ghostly affair and, as with all such things, *k slipped from currency and passed through the Ivory Gate, adding yet another mystery to the growing litany of astronomical myths. From the beginning many authorities had questioned the circumstances of the claim, inclining to believe they were underpinned by error. The mystery remained unsolved for almost three decades: and of the principals involved only Hind and Gould, by then astronomers of repute, were alive to learn the truth.

THE STORY OF THE WIRES

The solution unexpectedly turned up at the US Naval Observatory, Washington, DC, one morning in December 1878, a year of planet fever. It took the form of a letter addressed to the Superintendent, Rear Admiral John Rodgers. Dated 2 December, it came from Professor Christian Heinrich Friedrich Peters (1813–1890), the Danish-German director of the Litchfield observatory, Hamilton College, Clinton, New York.

Astronomer, mathematician and ex-soldier of fortune, Peters had actively supported Sicilian revolutionaries, but fled the country aboard an English ship bound for Malta when the revolution failed. He returned briefly to Sicily to help fortify Catania but was wounded when his command was driven from the town. Advisedly he shipped out for Marseille in a small fishing boat, and after a short interval made his way to Turkey, where he became scientific advisor to Reshid Pasha, the grand vizier of Sultan Abdul-Mejid II. There was talk of him heading a scientific expedition to Syria and Palestine but the outbreak of war in the Crimea in 1854 made it obvious there was little to gain by remaining in the Middle East and with a letter of recommendation from Alexander von Humboldt (1769–1859) in his pocket, at the suggestion of the American ambassador to Turkey, Peters set sail for the United States and spent the rest of his life there.

In 1858, he accepted the directorship of the Litchfield observatory at Hamilton College, then a small college for men in Clinton, New York, where he remained as professor of astronomy. He joined the college just as its observatory was completed and, soon after, embarked on an ambitious work, no less than to map stars 30° each side of the ecliptic down to magnitude 14, in much the same way as Hind had done and

Figure 4. Christian Heinrich Peters (1813–1890). Astronomer and former soldier of fortune who dispelled the misconception surrounding the supposed exterior planet of 1850–52. Photograph c.1880. (Image courtesy of Hamilton and Kirkland Colleges, Clinton, New York.)

with the same purpose of facilitating the discovery of asteroids. Unfortunately, however, the work remained incomplete, but Peters was one of the more successful asteroid hunters; at his death on 18 July 1890, he had forty-eight finds to his credit.

New, missing and hypothetical planets were popular topics for discussion in 1878. Twelve new asteroids had been found – five of them by Peters. After a lapse of several years, the intramercurial planet Vulcan was again headline news. During the total solar eclipse of 28 July that year, the Canadian astronomer James Craig Watson (1838–1880) and the American comet hunter Lewis Swift (1820–1913) each claimed to have sighted two small unknown objects close to the sun which they considered to be planets inside the orbit of Mercury. In reporting on these and other eclipse studies in a *communiqué* from Washington, DC on 8 August, English solar physicist J. Norman Lockyer (1836–1920) casually drew attention to other developments of a speculative nature by remarking, 'Prof. Watson, of Ann Arbor . . . broke off work on a planet beyond Neptune'. The reference was timely. As noted earlier, U. J. J. Le Verrier had envisioned such a possibility in 1846, barely a week after Neptune was discovered. Others, too, had taken up the challenge.

Figure 5. Litchfield Observatory, Hamilton College, 1872, of which Professor Peters was director. (Image courtesy of Hamilton and Kirkland Colleges, Clinton, New York.)

On thirty clear, moonless nights between 3 November 1877 and 5 March 1878, Amherst College astronomer David Peck Todd (1855–1939) had conducted a limited search with the 66.04-cm refractor at the US Naval Observatory for a body he believed orbited the sun at a distance of 52 AU in a period of 375 years. Across the Atlantic, Professor George Forbes (1849–1936) of Edinburgh was already advocating the possibility based on a suggestion by the French astronomer Camille Flammarion (1842–1925). It was thus merely a matter of time before someone remembered Ferguson's mysterious object of 1850. The circumstances of the affair along with the places of *k, framed within the uncertain argument of J. R. Hind, were recalled by the scientific weekly *Nature* on 31 October 1878.

Peters had rejected the Swift-Watson allegations of intramercurial planets, and, after reading various accounts of their observations, had decided upon a complete analysis of the available data. Indeed, he was highly sceptical of the intramercurial planet hypothesis, which, he remarked, 'may be accounted for by various other causes, without resorting to a mass inside, and moving nearly in the plane of Mercury'. Thus he was convinced the field reports of both Swift and Watson related to anything but inner planets. It was about this time he came upon the article about *k in the 31 October edition of *Nature*. His reaction was predictable. 'An anonymous writer in "Nature" stirs it up again, reasoning from the mentioned observations in favour of an exterior planet, using almost the identical, but very uncertain arguments of Mr Hind.'

Peters followed up his doubts and looked up the published correspondence, and checked the official volume of observations for 1850, verifying that Ferguson had indeed observed Hygeia on the night of 21 October 1850. After some mathematical detective work based on the description of the Washington micrometer used in the observations, an explanation began to take shape. But was it correct? So far he had worked only from published data. Would the original observing record support his findings? It was this research that prompted the letter to Washington, DC of 2 December, for, obviously, his conclusions could only be tested by comparing them with the original material at the Naval Observatory.

Having prefaced his purpose with a short history of the episode, he went on to say, 'In order, now, that nobody thereby might be induced to spend months and years upon a renewed search, I hasten to bring to your knowledge the errors I have detected in the Washington Observations for 1850 (on pages 320 and 321), and which alone have given rise to the misconception.'

The hub of his argument, he contested, resided in the published description of the filar micrometer used by Ferguson. The movable plate that served to measure the differences in declination carried three wires, denoted 1, 2 and 3, all of which were orientated in the east-west direction, with wire number 1 being nearest to the head of the micrometer screw.

The object in question, *k, was compared twice with a star lettered (a) on 16 October 1850; five times with another marked (c) on 19 October; three times with asteroid Hygiea on the 21st; and once on the 22nd with a star marked (e). Now it happened that, for all the eleven instances when *k was the object of pointing, the wire was wrongly logged as number 1, whereas, in fact, it should have been number 2. This only rectified the anomalies in declination, but those of right ascension were automatically reduced since the tilt of the north-south wires was now correctly described. Converting the daily means of the screw values into arc, and transcribing the differences in right ascension from the printed record, Peters tabulated his results for the comparisons with *k on 16, 19 and 22 October, stars (a), (c) and (e) respectively. Having applied these corrections, from the observations on the three nights specified, he determined the mean place of *k for 1850.0 as right ascension 19h 17m 41s.09 and declination –20° 52' 49".5. This, as he then disclosed, coincided with the position of a known star of no great significance – Lalande 36613! Ignominiously, *k had been downgraded from an exterior planet to the status of an ordinary catalogue star registered by Lalande, three times by Argelander in his Zones, and twice by Lamont.

After correcting a minor error in the second set of measures of 19 October 1850 and confirming by computation that Ferguson did indeed observe Hygiea on the night of 21 October, despite his doubts to the contrary, Peters concluded: 'In order to stop any further perpetuation of the credence, that a trans-Neptunian planet is revealed by the Washington Observations discussed, it would be very desirable to give publicity to the corrections indicated.'

Official response was swift. Rear Admiral John Rodgers instructed Asaph Hall (1829–1907), renowned for his discovery of the satellites of Mars in 1877, to immediately examine the log books of the equatorial. Writing in *Nature* of 27 March 1879, under the heading 'The Trans-Neptunian Planet', Hall noted Ferguson had indeed made a mistake, but went on to say; 'It is due, however, to Mr Ferguson to say that his record is full and complete, and that his changes in the reductions were honestly made. The record is in pencil and no figures were erased or rubbed out. They are crossed out, and the assumed figure is put by the side of the original one, while at the bottom of the page is a note with pen and ink, and in Mr Ferguson's handwriting, stating the changes that were made. Prof. Peters' ingenious discovery of the truth was made without knowledge of the observing-book.'

With the truth disclosed and confirmed, on 6 December 1878, Rodgers promptly forwarded Peters' critique to the *Astronomische Nachrichten*, and so, in due course, the explanation entered the public domain. But the world had not heard the last of the affair. For in 1931, the well-known German astronomer Ernst Zinner (1886–1970) published a list of suspected variable stars – entry 1689 of which turned out to be our old friend Lalande 36613. Obviously, Zinner had come upon Hind's communication to *The Astronomical Journal* and taken note of his alternative idea that there might be an unknown variable star in the case; being of a thorough disposition, he included it amongst his other specimens, totally unaware that the matter had already been explained.

It was a curious end to a remarkable incident; of a planet found by accident but lost in embarrassment; a bizarre compound of indiscretions set in a period when the discovery of planets was still relatively novel. But if Ferguson made a mistake and Hind was too quick to hypothecate, their actions exemplified the spirit of the time. Yet they also underscore something else – that astronomers, no less than others, are quite capable of giving 'to the airy nothing, a local habitation and a name'.

New Horizons to Pluto and Beyond

DAVID M. HARLAND

INTRODUCTION

This article reviews what we know of Pluto and describes the New Horizons mission that is scheduled to visit Pluto in 2015 and go on to investigate some of the objects in the Kuiper Belt.

PLUTO

Pluto was discovered in 1930 at the Lowell Observatory in Flagstaff, Arizona, during a photographic search for such an object by Clyde Tombaugh. It was a mere speck at 15th magnitude, and its slow progress across a series of plates indicated that it was in a rather elliptical orbit with an eccentricity of 0.25 that was inclined at 17 degrees to the ecliptic and had a period of 248 years. Given its distance from the Sun, at the suggestion of Venetia Burney, an eleven-year-old schoolgirl in Oxford, England, it was named after the god of the stygian underworld.

At the time of its discovery, Pluto was inbound for perihelion later in the century, at which time it would be within the orbit of Neptune. However, owing to the angles of their orbits, the paths of these bodies do not intersect. In fact, computer calculations show that Pluto's orbit is chaotic and its inclination, eccentricity and the orientation of its semi-major axis vary within certain bounds in a random manner.

Pluto was clearly very small, because it was only a star-like speck of light in the 100-inch reflector of the Mount Wilson Observatory, at that time the largest telescope in the world. Only with the introduction of the 200-inch on Mount Palomar in 1949 was its disk able to be resolved, enabling Gerard Kuiper to estimate its diameter at about 5,800 km. Photometric variations indicated a periodicity of 6.4 days.

Figure 1. The original blink-comparator plates from Clyde Tombaugh's discovery of Pluto. (Image courtesy of Lowell Observatory Archives.)

The first insight into the nature of Pluto's surface came in 1976, when infrared photometry established the presence of ices. Later, methane was identified.

In 1978, James Christie of the US Naval Observatory was analysing a series of images of Pluto taken in order to better define its orbit. The passage of that light through the Earth's atmosphere distorted the images, making them fuzzy. In trying to identify the centre of each blob, Christie noticed a pattern which suggested the presence of a satellite. This was named Charon, after the ferryman who sailed across the River Styx to Hades, the underworld ruled by Pluto. The centres of the bodies were about 20,000 km apart, and the orbital period of 6.4 days explained the photometric variation. Charon was resolved for the first time in 1983 using the technique of speckle interferometry.

The plane of Charon's orbit around Pluto is tilted about 120 degrees to the plane of their orbit around the Sun. At the time of Pluto's discovery, the system was oriented almost pole-on from the point of view of a terrestrial observer, but between 1985 and 1990 it was edge-on and, as Charon passed in front of Pluto and then behind it, the occultations enabled the sizes of the bodies to be refined. At 2,300 km across, Pluto was evidently considerably smaller than our Moon and had only about 1 per cent of the Earth's mass. Charon was estimated at 1,190 km across with 5–10 per cent the mass of its primary. An analysis of the albedo and spectroscopic variations during the long series of occultations provided the first crude map of Pluto. Methane ice was confirmed to be present on Pluto, but Charon had water ice. The two bodies are evidently tidally locked, each maintaining the same face towards the other. It was speculated that the stresses imparted in achieving this synchronicity may have generated sufficient heat to drive an episode of cryovolcanism. If so, then their surfaces could be similar to the large Neptunian satellite, Triton, 2,700 km across, which the Voyager 2 flyby of 1989 revealed to have an active icy surface that produces nitrogen geysers.

Given the eccentricity of Pluto's heliocentric orbit, the amount of energy in sunlight varies. As Pluto approached its September 1989 perihelion at about 30 AU, the chemical frost sublimated to create an extremely rarefied envelope. Nevertheless, observations by the Kuiper Airborne Observatory of the manner in which the light from a star was refracted during an occultation in 1988 indicated the presence of layers of haze extending far above the surface. As this envelope will freeze onto the surface as Pluto moves towards aphelion at about 50 AU, the process is cyclic. Furthermore, the seasons are extreme because the spin axis of Pluto, like that of Uranus, lies close to the plane of its solar orbit, with the result that for much of the time one or other of its poles continuously faces the Sun while the other is in darkness.

The Hubble Space Telescope was able to exploit its position above the turmoil of the terrestrial atmosphere to obtain sharper images of Pluto, and although the object spanned only a few pixels, computer processing was able to produce maps by direct viewing rather than from occultations. The results showed variation in albedo across the mottled surface, with dark (0.50) and bright (0.66) patches. Observations starting in 1994 and running for a decade showed the north polar region brightening and the south polar region darkening, no doubt as part of the seasonal cycle. Further spectroscopic observations revealed Pluto's surface to be predominantly (98 per cent) nitrogen ice, plus traces of methane and carbon monoxide, with the methane ice being on the hemisphere that faces Charon and the carbon monoxide ice on the opposite hemisphere.

When the density of Pluto was refined to 1.8–2.0 gm/cc, this suggested roughly 50–70 per cent rock and 30–50 per cent ice by mass (water ice is just under 1 gm/cc). Owing to the heat from the radioisotopes in the rock, the interior is very likely differentiated into a rocky core and an icy mantle. The mantle of water ice is evidently overlain by

Figure 2. Pluto and Charon as revealed by the Faint Object Camera of the Hubble Space Telescope on 21 February 1994. (Image courtesy of Dr R. Albrecht, ESA/ESO Space Telescope European Coordinating Facility; NASA.)

a crust of nitrogen and carbon monoxide ices, and it is these which participate in the seasonal cycle. The pressure at Pluto's surface has been observed to vary between 6.5 and 24 microbars (sea-level on Earth is 1 bar).

MISSION CONCEPTS

The first thought of sending a spacecraft to fly by Pluto arose in 1965 when Gary Flandro at the Jet Propulsion Laboratory, which manages NASA's deep-space missions, was investigating how a succession of gravity assists could enable a spacecraft to conduct a Grand Tour of the Outer Solar System. The requisite alignment of the planets occurred only once in 170 years, and, by an incredible stroke of luck, the launch window, lasting several years, would occur in the 1970s. One option was to launch in the summer of 1977 and perform flybys of Jupiter, Saturn and Pluto, with Pluto being visited in 1986. In comparison to a 'ballistic' trajectory, the gravity assists would shorten the flight time to Pluto from forty-five years to nine. However, since the spacecraft's heliocentric speed would be increased by the gravity assists, it would fairly race past Pluto. In the event, the trajectory chosen for the Voyager 2 mission did not include visiting Pluto.

A study for a specific Pluto mission began in 1989, just sixty days after Voyager 2 flew past Neptune. Because deep-space missions were becoming ever more complex and expensive, it was felt that a costly reconnaissance flight would be rejected, therefore the study called 'Pluto 350' envisaged a spacecraft with a total mass of 350 kg, which was half that of Voyager. The minimized payload would be a camera, an ultraviolet spectrometer, a plasma analyser, and a radio-science package. It would characterize the surfaces of Pluto and Charon, study the structure of the haze layers, and determine whether the perceived similarities between Triton and Pluto were real. It was soon noted that the nature of the system meant that if a mission were not launched soon, the science it could carry out would be diminished. In particular, whereas at equinox in the late 1980s almost all of the surfaces of the bodies were illuminated, by 2015 a large percentage of their southern hemispheres would be in darkness. Furthermore, since the planet would chill after perihelion, if the mission were to arrive after about 2020 the atmosphere would have collapsed onto the surface. When the proposal was criticized for its low mass and high-perceived risk, NASA studied a costly Mariner Mark II flyby mission with a spacecraft based on the architecture developed for the Cassini mission. On approaching Pluto, the spacecraft would release a small probe on a trajectory timed to arrive either half a rotation before or half a rotation after its mother ship to view the area that would be in darkness for the main event. However, in a time of shrinking budgets for scientific missions, Pluto 350 was recommended over the supposedly less risky Mariner Mark II proposal.

At this point, JPL carried out an investigation of the smallest possible spacecraft capable of mapping the surfaces of Pluto and Charon at a resolution of 1 km, making temperature maps of their visible hemispheres, and profiling the atmosphere. With a mass of only about 150 kg, the Pluto Fast Flyby mission raised the prospect of truly revolutionizing spacecraft design. In fact, it was so small that a Titan IV launcher would be able to place it on a direct solar-system escape trajectory to Pluto in a mere seven years. In the original concept, a pair of spacecraft would be launched on trajectories to provide radio-occultations of both Pluto and Charon and timed to reach Pluto half a revolution apart in order to inspect as much as possible of the surfaces of the two bodies employing a camera, an infrared spectrometer and an ultraviolet spectrometer. By 1994, it seemed likely that Pluto Fast Flyby would be launched in the first decade of the new millennium. However, due to the need for two Titan IV launchers, the overall cost of the mission would exceed the billion-dollar mark. Daniel Goldin, the NASA Administrator, stressed that the project would never be approved unless a means was found to reduce the launch costs. There were three options. One option was to stick with the Titan IV but send only a single spacecraft, accepting the loss of redundancy and a reduction in surface coverage. The second option was to use a smaller launch vehicle and use gravity assists. This would draw out the mission. The remaining option was to invite international collaboration, in particular by using a cheaper Russian launch vehicle in return for releasing small probes to directly sample Pluto's atmosphere. But the Russian economy was in dire straits and by 1995 this idea had fallen by the wayside.

The Pluto Fast Flyby was replaced by Pluto Express, which maintained the planned payload mass but halved the overall mass to 75 kg. If this mission could be launched in the first years of the new millennium then it should be able to reach Pluto early in the 2010s.

In 1992, the recent introduction of sensitive electronic detectors at the world's largest astronomical observatories enabled David Jewitt and Jane Luu to make the first discovery of an object in the Kuiper Belt. The existence of a reservoir of bodies near the plane of the ecliptic and beyond the orbit of Neptune was postulated by Kenneth Edgeworth in 1949 and independently by Gerard Kuiper in 1951. It is believed to be the main source of short-period comets. The object Jewitt and Luu found, (15760) 1992QB1, is estimated to be one-tenth the size of Pluto. By 1995, Kuiper Belt objects were being discovered at a rate of dozens per year. It was proposed that if Pluto Express had sufficient fuel after its Pluto flyby, it could go on to investigate at least one of these other objects, prompting the name of the mission to be changed to Pluto-Kuiper Express.

In 1998, Pluto-Kuiper Express was incorporated by NASA into its Origins program. This envisaged launching Europa Orbiter in 2003 to determine whether there really is an ocean beneath the icy crust of the eponymous Jovian moon, and Pluto-Kuiper Express in 2004. If necessary, Europa Orbiter would be postponed to permit Pluto-Kuiper Express to meet its launch window. It was hoped that a high degree of commonality and the use of the Delta II launcher would significantly reduce the cost of each mission. In September 2000, NASA halted work on Pluto-Kuiper Express to focus upon Europa Orbiter, but within a matter of months this was cancelled because the budget was judged unrealistic for such a complicated mission. The scientific community appealed to the US Congress, which compelled NASA to reinstate the Pluto mission.

Instead of resuming the previous development, in January 2001 the agency initiated a Discovery-style competition for a proposal to be led by a Principal Investigator. Two of five submissions were shortlisted for further study. On 19 November 2001, New Horizons by the Applied Physics Laboratory of Johns Hopkins University and the Southwest Research Institute was selected. The Principal Investigator was Alan Stern of SwRI, who had been involved in Pluto-Kuiper Express. NASA decided that this would be the first mission in its New Frontiers program of deep-space missions with budgets larger than a Discovery-class mission but considerably less than a Flagship-class mission such as Cassini-Huygens. The initial budget of $500 million was to cover the mission all the way to the Pluto encounter (as events transpired, this cost increased to $700 million). If there was to be an extension into the Kuiper Belt, then that would have to be funded separately when the time came. Initially, the launch was set for December 2004 with the Pluto flyby as early as 2012, but it wasn't long before the launch was slipped to 2006. At one point, the difficulty in obtaining sufficient plutonium for the radio-isotope thermal generator (RTG) that would power the spacecraft threatened to delay the launch even further (in the end, it was decided to buy some plutonium from Russia). If it was unable to launch in 2006, the Jovian gravity assist would be impracticable and the Pluto encounter would not occur until the 2020s, by which time a large portion of the surface would be in the deep darkness of winter.

NEW HORIZONS

New Horizons has an irregular 6-sided body that is 0.68 metres tall, 2.11 metres wide and 2.74 metres long, covered in blankets of insulation in order to retain as much heat as possible in the frozen environment so far from the Sun. The RTG projects out from one of the side faces. The spacecraft was to spin at 5 revolutions per minute for stability in the interplanetary cruise, but adopt 3-axis stability for its observations. During the Pluto flyby it would store science data in an

Figure 3. The New Horizons spacecraft undergoing ground tests at the Applied Physics Laboratory of Johns Hopkins University on 1 February 2005. (Image courtesy of NASA/Johns Hopkins University Applied Physics Laboratory/Southwest Research Institute.)

8 gigabyte solid-state memory, which would be downloaded to Earth via the 2.1-metre-diameter high-gain antenna dish in a fixed mounting on top of the body. This would maximize the time available for making observations by minimizing the time spent with the vehicle oriented to aim its antenna at Earth. Unlike the Voyagers, which had a scan platform, New Horizons will have to manoeuvre to aim its body-mounted instruments at a target.

At 478 kg, New Horizons is much heavier than was intended for the Pluto-Kuiper Express, and its 30 kg of payload comprises five instruments. A multispectral remote-sensing package that has a trio of monochrome and colour cameras, an infrared detector and an ultraviolet spectrometer will provide medium-angle mapping and spectroscopy of the surface and atmosphere of Pluto. A narrow-angle camera using a telescope with an aperture of 21 centimetres will supply high-resolution monochrome images and mapping in the weeks leading up to the flyby. A two-sensor particles package will investigate the solar wind in the outer solar system as well as gases which escape from Pluto and are then ionized by solar ultraviolet and 'picked up' by the magnetic field of the solar wind. A dust counter will characterize for the first time the dust environment at heliocentric distances exceeding 18 AU (where the dust sensor of Pioneer-Jupiter failed). In addition, the high-gain antenna will be able to function as a passive radiometer in the microwave range to study the nature of icy surfaces. The only typical 'first encounter' instrument which has been omitted is a magnetometer, but scientists hope to be able to infer the presence of a planetary magnetic field (if one exists) from the data supplied by the particles package. The trajectory will provide an occultation of Earth by Pluto, but instead of having the spacecraft transmit a signal to be monitored by the Deep Space Network, this time the DSN will transmit a signal for the spacecraft to monitor as a means of probing the atmosphere of Pluto.

Support for the mission came from the Space Sciences Board of the US National Research Council, which assessed projects for the decade of 2003–13 and, in its report *New Frontiers in the Solar System – An Integrated Exploration Strategy*, ranked a mission to Pluto and the Kuiper Belt very highly.

In July 2003, it was decided to launch New Horizons on an Atlas V rocket and employ a STAR 48B solid-fuelled 'kick stage' to give the spacecraft sufficient speed to escape the solar system at Earth departure, unlike the Voyagers, which used a Jovian gravity assist to achieve this state.

PLUTO AS 'KING OF THE KUIPER BELT'

In June 2005, Alan Stern and his colleagues discovered two additional satellites when analysing images taken by the Hubble Space Telescope the previous month. Named Nix and Hydra, they are respectively about 45 and 130 km in size. Nix has an almost circular orbit with a radius of 49,000 km about the barycentre of the system and period of 24.8 days. Hydra is at 65,000 km with a period of 38.2 days. The three satellites orbit in more or less the same plane, with the two smaller ones being beyond the orbit of Charon. Whereas Nix is slightly 'redder' than Pluto, Hydra more closely resembles Charon.

Figure 4. In 2005, the Hubble Space Telescope discovered the small moons Nix and Hydra in the Pluto system. (Image courtesy of NASA, ESA, H. Weaver of Johns Hopkins University Applied Physics Laboratory, A. Stern of Southwest Research Institute, and the HST Pluto Companion Search Team.)

Two other small satellites were discovered by a team led by Mark Showalter using long-exposure Hubble Space Telescope images to search for any evidence of faint rings around Pluto that might pose a threat to the New Horizons spacecraft. First, on 28 June 2011, was Kerberos. It has an estimated size of about 25 km and traces a circular orbit with a radius of 59,000 km (in between Nix and Hydra), with a period of 32.1 days. A year later, Styx was discovered. It is slightly smaller at about 18 km, and its orbit has a radius of 42,000 km (just inside that of Nix) with a period of 20.2 days.

Some of the Kuiper Belt objects proved to be surprisingly large and many had satellites, in some cases several satellites. In January 2005, a team led by Professor Mike Brown of the California Institute of Technology discovered (136199) Eris, which is similar in size to Pluto. Its orbit is steeply inclined (44 degrees) to the ecliptic and has a perihelion at about 38 AU, an aphelion at almost 100 AU and an orbital period of 560 years. It possesses a satellite several hundred kilometres in size named Dysnomia that travels in a circular orbit with a radius of

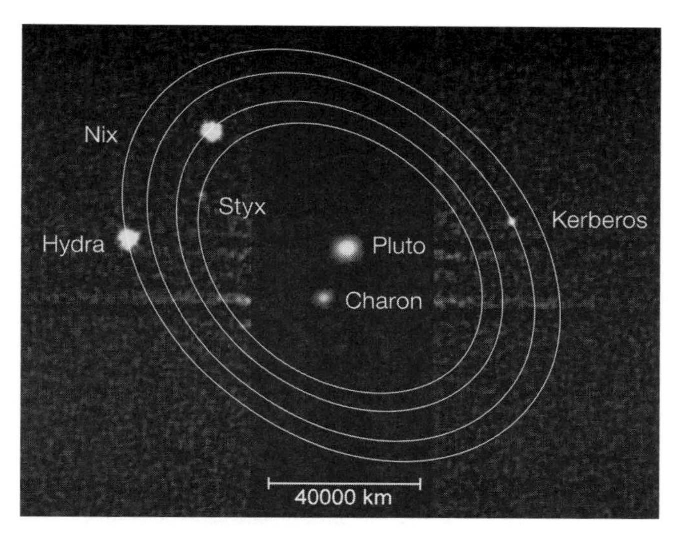

Figure 5. The Hubble Space Telescope discovered the small moons Kerberos and Styx in 2011 and 2012 respectively. (Image courtesy of NASA, ESA, and L. Frattare of Space Telescope Science Institute.)

about 37,350 km and a period of 15.8 days. The presence of the satellite enabled the mass of Eris to be calculated, and it proved to be 27 per cent *greater* than that of Pluto.

On 24 August 2006, the International Astronomical Union took note of the fact that Pluto is really one of the largest members of the Kuiper Belt and therefore demoted its official status from the ninth planet Pluto to dwarf planet (134,340) Pluto. Two other large objects in that region of space soon fell into this category. One was (136,108) Haumea, discovered by Mike Brown's team in 2004. Its orbit is inclined at 28 degrees to the ecliptic, has a perihelion at 35 AU, an aphelion at 43 AU, and a period of 283 years. It appears to be distinctly elongated, about 2,000 km by 1,000 km, and to spin very rapidly (3.9 hours). Its mass is one-third that of Pluto and it has a pair of small satellites. Dwarf planet (136,472) Makemake was found by the same team in 2005. Its orbit is 29 degrees to the ecliptic, has a perihelion at 38 AU, an aphelion at 53 AU, and a period of 310 years. As it does not appear to possess a satellite, its mass is undetermined, but a stellar occultation in 2011 implied a diameter of about 1,480 km, making it slightly larger than Haumea and therefore the third largest after Eris and Pluto.

NEW HORIZONS UNDERWAY

The New Horizons spacecraft was delivered to Cape Canaveral in September 2005 for mating with its RTG and final checkout. The launch window opened on 11 January 2006 and ran to 14 February. A launch later than 27 January would be unable to achieve a Jovian gravity assist, and would require extending the interplanetary cruise by five years. After several delays, it finally lifted off on 19 January. The Centaur upper stage released its payload into an orbit that had its aphelion in the asteroid belt between Mars and Jupiter, then the kick stage fired to give the spacecraft the greatest Earth-departure speed ever attained. The trajectory was so accurate that only minor mid-course refinements would be required. This provided a welcome reserve of propellant for targeting objects in the Kuiper Belt during an extended mission after the Pluto flyby.

On 7 April 2006, the New Horizons spacecraft passed the orbit of Mars. The telescopic camera obtained its first view of Pluto with a sequence of images taken on 21 and 24 September 2006, showing it as a speck moving against the background of stars, some 4.2 billion km away.

The trajectory through the Jovian system had its closest point of approach on 28 February 2007 at a range of 2.3 million km (some 32 Rj, well outside the orbits of the Galilean satellites), at a relative speed of 21.2 km/s. The atmosphere of the planet was observed at a resolution comparable to the best provided by the Voyagers during their much deeper penetrations of the system. New Horizons also inspected the intricate icy surface of Europa and monitored the vast plumes of the volcanoes on Io. With its imaging systems optimized for the low light levels at Pluto, New Horizons was able to provide the best-ever views of Jupiter's dark rings, showing details of their structure. Two 'movies' for a total of more than 100 frames were obtained on consecutive days, each spanning a full rotation period of the rings. One of the movies was inbound and viewed the rings backscattering sunlight, and the other was outbound with sunlight being forward scattered. Beyond Jupiter, New Horizons continued to monitor particles as it flew down the

Figure 6. On 21 September 2006, New Horizons spotted Pluto against the stars of the constellation of Sagittarius. (Image courtesy of NASA/Johns Hopkins University Applied Physics Laboratory/Southwest Research Institute.)

planet's magnetic tail for a distance of over 2,500 Rj (some 1.2 AU) before breaking out into the solar wind in late June.

The Jovian flyby increased New Horizons' heliocentric velocity by 4 km/s and deflected its trajectory in the direction of Pluto. On 25 September, the thrusters were fired for 15 minutes 37 seconds to achieve a 2.37 m/s correction to refine the trajectory for Pluto. The spacecraft then entered a hibernation from which it would awaken for twice-annual checks of its systems. (The dust counter remained active, storing its data.) In June 2008, the vehicle crossed the orbit of Saturn, although the planet was nowhere near. Another trajectory refinement was made on 30 June 2010.

Figure 7. Jupiter's high-altitude clouds imaged using a narrow filter centred on a methane absorption band near 890 nanometres taken by New Horizons on 28 February 2007 from a range of 2.3 million km. (Image courtesy of NASA/Johns Hopkins University Applied Physics Laboratory/Southwest Research Institute.)

During the summer awakening in 2012, the spacecraft rehearsed the most intense twenty-two hours of the Pluto flyby. In 2013, it performed an even more comprehensive rehearsal, this having been brought forward by a year in order to gain more time in which to deal with any problems that the test might reveal. It was revived in June 2014 for a trajectory refinement and another rehearsal, and it took images over the period of a complete orbit by Charon. In late August, it crossed Neptune's orbit, twenty-five years after the Voyager 2 flyby.

ONGOING STUDIES OF PLUTO

As New Horizons was cruising toward Pluto, scientific knowledge of its target increased, notably with the discovery of fresh ice and ammonia hydrate deposits on the surface of Charon that might be evidence of the existence of water geysers. A re-examination of telescopic pictures taken between the discovery of Pluto in 1930 through the early 1950s

Figure 8. The 300-km-tall volcanic plume of the Tvashtar caldera on Io, as imaged by New Horizons a few hours after closest approach on 28 February 2007. The sunlit hemisphere is deliberately overexposed to reveal the night hemisphere illuminated by 'Jupiter-shine'. (Image courtesy of NASA/Johns Hopkins University Applied Physics Laboratory/Southwest Research Institute.)

clearly showed that frost deposits migrate around the surface with the changing seasons that result from the ellipticity of its orbit and the obliquity of its axis. Other evidence showed fine structures in the atmosphere of Pluto and monitored the mixing ratio of methane, which declined, very possibly as the atmosphere cooled after the perihelion of 1989. The atmosphere was warmer than the surface, in part because the surface was chilled by the process of evaporation (being about 10 K colder than it would otherwise be). The atmosphere was expected to contract following perihelion, but in fact its pressure was observed to increase during the 2000s, and within a decade it expanded dramatically to some 3,000 km from the surface, which is 25 per cent of the distance to Charon. Carbon monoxide had been tentatively detected in 2000 and its concentration was shown to have significantly increased, probably as patches of carbon monoxide ice became exposed to sunlight. This implied a strong coupling between the surface and the atmosphere, with carbon monoxide regulating the temperature of

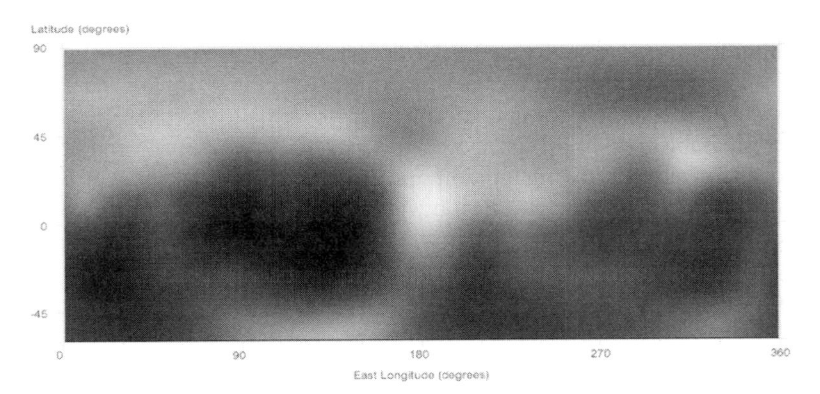

Figure 9. This map of Pluto compiled from data obtained by the Hubble Space Telescope in 2002–3 will remain the best until New Horizons approaches the system closely enough to obtain better resolution. In addition to methane ice, which seems to be everywhere, the lighter areas indicate nitrogen frost and the dark areas may be dirty water-ice. The bright spot near the centre seems to be carbon monoxide. (Image courtesy of Space Telescope Science Institute/Marc Buie of Lowell Observatory/NASA/ESA.)

the atmosphere as Pluto withdrew from the Sun. Interestingly, carbon monoxide spectra suggest this might be streaming away from Pluto in a comet-like tail. Nitrogen was expected to be the most common gas, but it would be undetectable to spectroscopes on Earth.

Pluto and Charon were observed employing the Hubble Space Telescope's Advanced Camera for Surveys in 2002 and 2003 and new, more accurate colour maps than those obtained for Pluto in 1994 were constructed. A striking result was that the northern polar regions seemed to have rapidly brightened. Several spots appeared to have moved, or to have changed their appearance. Amongst the features that had *not* changed (and had evidently been there since at least the 1950s) was a bright region rich in carbon monoxide. This will be prominently visible to New Horizons around closest approach. Hubble also discovered a strong ultraviolet absorber on the surface of Pluto, suggesting complex hydrocarbons or nitriles created by the interaction of ice with energetic solar particles and cosmic rays. The surface of Charon, on the other hand, remained almost featureless, but this could in large part be due to its small size. If the interior of Pluto possesses enough decaying potassium isotopes, there could even be an ocean of liquid water beneath the external icy crust.

The possibility has not been ruled out that Pluto has a faint ring system. In fact, the escape speeds from the smaller moons must be comparable to their orbital speeds, meaning that debris could end up in orbits of almost any inclination and form a torus or a cloud of dust instead of a ring. Even a collision with a particle 1 mm across would be catastrophic at a relative speed of 14 km/s. Therefore as a precaution against a threat being found, nine alternative trajectories of increasing distance from Pluto were considered. New Horizons will be targeted for the baseline trajectory, but will conduct a deep search for additional satellites, debris and rings from sixty-four to eighteen days out during the approach, and be able to execute a bail-out as late as ten days out with little effect on the fuel budget. The reasoning was that Charon ought to have cleared all the dangerous dust out of the targeted region. If there is a threat, the spacecraft will perform the most critical portion of the flyby with its high-gain antenna facing forward, as the Cassini spacecraft does for ring plane crossings at Saturn. An alternative strategy would be to fly within 3,000 km of Pluto in the expectation that the atmosphere has cleared away the dust.

PLUTO ENCOUNTER

The Pluto encounter will span about a year. After awakening on 7 December 2014, the distant approach phase will start in January 2015 and consist mostly of imaging for optical navigation. In fact, due to the uncertainty of the target's ephemeris, which amounts to several thousand kilometres, New Horizons will be more reliant upon optical navigation than any previous mission. In the second approach phase, lasting from 4 April to 23 June, New Horizons' observations will also image the smaller satellites and refine their orbits. It is expected that by about seventy-five days from the encounter the telescopic camera will start to supply images of Pluto that surpass those of the Hubble Space Telescope. The third approach phase will last until several days prior to closest approach. During this phase, the telescopic camera will document with a resolution of about 40 km the hemisphere that will be in darkness at the time of closest approach, thereby enabling a single spacecraft to achieve some of the scientific objectives expected of the

Figure 10. An artist's impression of New Horizons making its flyby of Pluto with Charon in the background. (Image courtesy of Johns Hopkins University Applied Physics Laboratory/ Southwest Research Institute.)

two-probe Pluto Fast Flyby. Optical navigation and imaging of the satellites will continue until twenty-four hours before closest approach.

New Horizons will reach its minimum distance from Pluto at 11h 50m UTC on 14 July 2015. Owing to the tilt of the orbital plane of the satellites, the flyby will resemble the Voyager 2 encounter with Uranus. During the approach, the spacecraft will observe the system nearly face-on, like a bull's-eye. The trajectory is required to create solar occultations for both Pluto and Charon to facilitate ultraviolet atmospheric studies, and an Earth occultation for Pluto in order to sound its atmosphere using the radio signal. The plan is to pass 12,500 km above the surface of Pluto at a relative speed of 13.8 km/s. The narrow-angle camera will provide high-resolution regional images while the multispectral imager takes global low-resolution pictures, maps the surface composition and temperature field, and provides ultraviolet measurements to study the structure of the upper atmosphere and the rate at which gas molecules leak away to space. Ionized particles that are 'picked up' by the solar wind will be analysed by the plasma sensors. New Horizons will also look for an exchange of material between Pluto and Charon, because it is possible that atmospheric gas molecules could reach the Lagrangian point between Pluto and Charon, some 15,000 km from the surface of the former, and thereafter spill into Charon's gravity sphere.

At the time of closest approach, Charon and Hydra will be on the same side of Pluto as the spacecraft (although Hydra will be much further away) and Nix will be on the opposite side. Although the distances from these objects will limit imagery to a resolution of a few kilometres per pixel, this should be sufficient to determine their shapes and general morphologies. About an hour after closest approach, the wide-angle camera should be able to obtain pictures of the night hemisphere and polar region of Pluto illuminated by 'Charon-shine', which will be 10,000 times fainter than sunlight at that heliocentric distance. The same observations will be repeated hours later by the long-range camera, and again several days later on the night-side of Charon itself.

As New Horizons downloads its data after the encounter, it will test using both redundant transmitters simultaneously to double the data rate and return the entire dataset in less than the nine months that it would require using a single transmitter. Depending on how this experiment turns out, the encounter will be wrapped up some time in early 2016.

KUIPER BELT MISSION

The final portion of the New Horizons mission will involve at least one flyby of a Kuiper Belt object. The trajectory through the Pluto system is obliged to produce eclipses and occultations, and owing to the extremely small mass of Pluto the encounter will produce only a minimal gravitational deflection. Hence most (if not all) of the targeting for the Kuiper Belt phase of the mission will have to be achieved by propulsive manoeuvres. The targeting burn could be carried out as soon as sixty to ninety days after the Pluto flyby, or be put off to as late as 2017. About 34 kg of fuel is expected to remain on board for this part of the mission. Unfortunately, none of the large Kuiper Belt objects mentioned above will be accessible. Whereas Pluto-Kuiper Express was to have searched for a target by aiming its camera in the direction of travel and awaiting an opportunity, New Horizons must rely on terrestrial telescopes.

Some of the largest telescopes on Earth and the Hubble Space Telescope are already surveying the cone of space accessible to New Horizons in search of objects as far out as 55 AU. Algorithms and techniques for an automated search of this type were tested in the context of the Pluto-Kuiper Express proposal and resulted in the discovery of two new Kuiper Belt objects. The ground-based telescopic survey relies on the help of a 'citizen science' project called Ice Hunters in which members of the public sift through pictures of the area of sky to suggest possible Kuiper Belt targets. If a good candidate is found, New Horizons will take navigation pictures to improve the ephemeris as early as possible, in order to make the best possible use of the propellant remaining. Statistical analyses suggest that the spacecraft ought to be able to reach several objects exceeding 35 km in size. In addition, distant flybys (of the order of tens of millions of kilometres) will be made of other Kuiper Belt objects that will be imaged in order to search for satellites.

New Horizons will leave the solar system in the direction of the constellation Aquila. Plans are being made for a deep heliospheric mission using the solar wind and energetic particle instruments, as well as the dust detector, lasting as long as the RTG will allow the vehicle to operate.

Poignantly, the New Horizons spacecraft carries a vial holding several grams of the ashes of Clyde Tombaugh, who died in 1997, aged ninety years.

Edmond Halley: Astronomer, Geophysicist, Meteorologist and Royal Navy Captain

ALLAN CHAPMAN

On 22 April 1715, southern and central England and south Wales witnessed the first total eclipse of the Sun seen in those regions since 20 March 1140. No one did more to promote the hoped-for widespread observation of this 'people's eclipse' than Dr Edmond Halley, Savilian Professor of Geometry at Oxford University. It was also Halley, evidence suggests, who not only alerted the English and Welsh to the forthcoming spectacle – 'The Novelty of the thing being likely to excite a general Curiosity' – but also seems to have calculated the date of the previous, 1140, eclipse.

The 1715 eclipse was, I suspect, the first call to the general population to take part in the 'mass observation' of an astronomical event. No less than twenty-eight individuals sent their results to Halley, including a tax collector and various country gentlemen, along with the Archbishop of Dublin and the Bishop of Exeter. We know all this because Halley listed and acknowledged them in an article which he published soon after the eclipse in the *Philosophical Transactions* of the Royal Society.

So, what kind of man was Edmond Halley, and how did he, amongst many other achievements, come to have a comet named after him? And in addition to his work in astronomy, how did he come to be one of the great early pioneers of scientific geomagnetism and global meteorology and a student of the aurora borealis?

In many respects, one might say, Edmond Halley 'had it all'. Born into a comfortably-off City of London business family, he was something of a child prodigy whose brilliance saw him through eighty-six years. He had a happy marriage that lasted an incredible – by seventeenth-century standards – fifty-four years, robust health of body

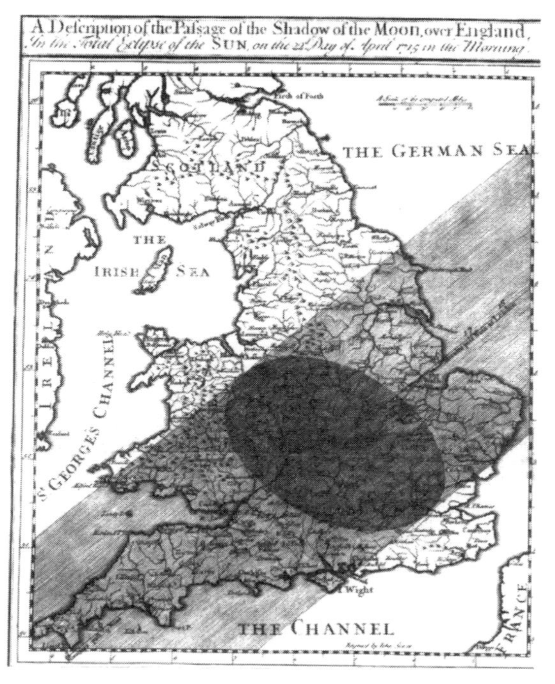

Figure 1. Halley's calculated map of the moon's shadow passing across England and Wales for the total solar eclipse of 3 May 1715 (22 April 1715, according to the Old Style Julian calendar, then in operation in Great Britain). Printed on single broadsheet. (Courtesy of the Royal Society Library.)

and mind, a 'merry' temperament and a gift for friendship, enterprise and initiative. He was elected Fellow of the Royal Society when he was twenty-two, and enjoyed a career that saw him variously as a scientist, international traveller, diplomat, Royal Navy captain, businessman, professor and Astronomer Royal. It was also Halley who cajoled Sir Isaac Newton to write and publish *Principia*. And it was all crowned by a quick and painless death, probably brought on by drinking a secret 'bumper' (or very large glass) of the wine of which he was fond, against his doctor's advice!

Halley was also astute at getting through to people via the 'media' of his day – the printed word – which is what made it possible for him to alert people to, and collate information from, the 1715 eclipse.

Figure 2. Edmond Halley, by Thomas Murray, c.1687. (Image courtesy of the Royal Society, London/Bridgeman Images.)

We are not quite sure where Halley was born; whether it was in the family's house at Winchester Street, in the City of London, or at one of their nearby 'country' homes in what were the then nearby villages of Shoreditch or Islington, where Edmond Halley Senior owned property. Either way, little Edmond entered the world on 8 November 1656 (New Style Calendar). Sadly, the Great Fire of London of 1666, when he was nine, destroyed many ancient City records and registers.

We know for a fact, however, that he was educated at St Paul's School, in the shadow of the Gothic ruins of Old St Paul's Cathedral, which had been devastated in the Fire, and which his fellow Oxford Astronomy Professor, Sir Christopher Wren, would later rebuild in its familiar 'modern' form, complete with its magnificent dome. Right from the start, Edmond was a conspicuously gifted pupil, rapidly mastering the Latin and Greek languages and classical and Christian culture which constituted the curriculum of the day. His fluent grasp of Latin and Greek stayed with him for the rest of his life, for Latin was the lingua franca of learned men across seventeenth-century Europe, and when in the years ahead he was in Italy, Germany, France and Austria, his Latinity and fluent French made it easy for him to communicate when dealing with learned people. At St Paul's, moreover, he just about came under the tutelage of the young Revd Dr Thomas Gale, appointed Head – or 'High' – Master of the School in 1672. Dr Gale was also fascinated by the new science and, in 1677, would put the letters 'F.R.S.' after his own name, following his election to the Fellowship of the Royal Society. Indeed, Gale was one of many clerical scientists and friends of science, and would later become Dean of York Minster.

Even as a schoolboy, Halley was involved in original scientific research, as he would later recall. He already possessed several astronomical instruments, but more significant were his pioneering researches into what appeared to be the mystery of the Earth's magnetic field. Professors at Gresham College, on Bishopsgate in the City (mercifully surviving the Great Fire) and at Limehouse, for example, had been making accurate measurements of 'magnetic' north in London for almost a century, recording the magnetic compass's exact 'variation' from true astronomical north, along with changes in the field's vertical 'dip'. For across a century or so, it had become clear that magnetic north, which was well to the *east* of true astronomical north in 1500, had slowly moved westwards, had been 'true' north in 1657, and since then had continued to move westwards. Could this be caused by the Earth's axial rotation, in accordance with the Copernican *hypothesis*, as it still was in 1660?

Schoolboy Halley not only began to make his own meticulous measurements of magnetic and astronomical north positions, but in 1672, when he was sixteen, he was also interested in trying to quantify the westwards drift by comparing his own observations with dated historical ones. For, as Halley would put it in a paper to the Royal Society in 1692, 'I myself observed it at 2° 30' to the West; and this present year 1692, I found it 6° 00' West.' Indeed, this could still be an interesting GCSE project today, but for a Latin- and Greek-trained schoolboy of 1672 it was amazing. His magnetism project would prefigure another enduring trait in Halley's life: his fascination with historical records, and how 'modern' research scientists could use old observations to provide firm points of reference for contemporary researches.

In 1673, when he was seventeen, Halley left St Paul's and 'went up' to The Queen's College, Oxford, to read for his classically oriented B.A. and M.A. degrees. Yet fewer scientifically minded undergraduates have gone to university with a private set of instruments that were as good as, if not better than, those of the Astronomy professor!

ST HELENA AND THE MAKING OF AN ASTRONOMER

Unlike their present-day counterparts, the astronomers of Halley's day did not spend most of their time looking *through* telescopes; they used them to make meticulous angular measurements of the Sun, Moon, planets and stars with relationship to the polar and equatorial co-ordinates. This celestial geometry was aimed at trying to find firm, *physical* evidence for the Earth's motion in space, such as the elusive stellar parallax. Likewise, it could yield more exact dimensions for the solar system, provide conclusive proof for the elliptical shape of lunar and planetary orbits and, hopefully, get to the bottom of the strange behaviour of comets. And if Europe's astronomers could accumulate a sufficiently large mass of accurate orbital data, this might supply a key to understanding the invisible attractive force acting between astronomical objects. All of this, moreover, depended not so much on more powerful telescopes as upon more accurately graduated angular scales, with their ancillary fittings.

Between 1640 and 1670, three key, transformative inventions had enabled the astronomer to measure to a wholly new level of accuracy. These were (1) the screw micrometer, which facilitated the accurate subdivision of tiny angles; (2) the telescopic sight, with cross-wires, enabling the astronomer to direct his instrument much more accurately than by the naked eye (both of these having been made by the Yorkshire amateur astronomer William Gascoigne); and (3) the pendulum clock, pioneered by Christiaan Huygens in Holland and then Robert Hooke in London, enabling, at last, the daily rotation of the stars to be measured to *seconds* of time. Precision indeed. Halley's private 'kit' of instruments in Oxford included all these devices, along with at least one good telescope – all thanks to Dad!

Following the establishment of the Royal Observatory at Greenwich in 1675, with the Revd John Flamsteed as first Astronomer Royal, the process was firmly in hand to measure and map the Northern Hemisphere skies with the new instruments. But what an achievement it would be to be the first man to map the Southern Hemisphere skies to the new level of accuracy!

Halley, still an undergraduate, leapt at the chance, and in 1676 he and a friend sailed to the remote south Atlantic island of St Helena – in the middle of nowhere, between the African and South American continents. Helped, no doubt, by Edmond Halley Senior, with his City of London connections, young Edmond secured a passage on the India-bound vessel *Unity*: the great long-haul sailing ship that brought the silks and spices of the East to the London warehouses. And as soon as he was on the island, he began to observe: first, from a temporary observatory on the dockside, then from the world's first 'high-altitude' observatory up the mountain.

And things fairly began to leap out of the sky. Why, for example, did the fine pendulum clock which he adjusted to keep exact time at sea level suddenly run slow when relocated in the mountain observatory? Could the Earth's mysterious pull, or 'gravitating principle' as they called it, become weaker the higher one rose above the Earth's surface? And if so, why? And in what mathematical ratio? Halley's meticulous pendulum observations would be discussed by Robert Hooke in his own gravitational researches, and then by Sir Isaac Newton.

Also, why did Halley's telescope lenses mist over when high in the mountain observatory, even under a clear night sky? (Halley, 1690–91, p. 471; Hooke, *De Potentia*, 1678, p. 39.) Perhaps this was the genesis of Edmond's lifelong fascination with air, refraction, condensation, 'dew-point' and meteorology. In addition to these unexpected discoveries, he achieved what he had set out to do – namely, to make a new map of the southern skies, containing 341 key star positions, which would stand for the next 80 years. On St Helena, he also began to exercise another talent which he would later use to effect as a Navy captain, diplomat, business-man and 'shrewd operator': that of knowing how to get his own way. For, following some kind of dispute with the island's Governor Gregory Field, Halley, along with some others, got him dismissed.

Back in London, the twenty-two-year-old Halley found he was a scientific celebrity. He was elected F.R.S., and everyone wanted to meet him – even King Charles II, especially as Halley had wisely named a new southern constellation 'Robur Carolinum', or 'Charles's Oak', in his newly published catalogue and atlas of the southern constellations (Halley, 1679). His Majesty even ordered the Vice-Chancellor of Oxford University to confer Halley's M.A. degree upon him without examination. Then, in 1679, when not yet twenty-three, Halley was officially sent by the Royal Society to sort out an acrimonious contro-versy between the forty-four-year-old Robert Hooke, F.R.S., and that great patriarch of European astronomy, the sixty-eight-year-old Johannes Hevelius, F.R.S., in Danzig, Poland.

DANZIG, FRANCE, AND ITALY

The controversy hinged upon whether the new telescopic sights did or did not distort the pinpoint images of stars. Hooke (who, so it would later transpire, was correct) said that telescopes did *not* distort star images; Hevelius, the great 'naked-eye' sight angle-measurer, said that they did. Indeed, Halley tells us that he and Hevelius began observing together, at the Polish astronomer's great private observatory, on his very night of arrival. Halley admitted that while he preferred measuring instruments with telescopic sights, Hevelius's naked-eye sight pieces were of stunning accuracy.

Halley's natural charm carried all before him. He and the great Hevelius became friends, while other prominent Danzig citizens were captivated by the handsome young Englishman. A malicious rumour, circulating around Oxford in 1713, and recorded by the diarist Thomas Hearne, even said that Halley had seduced the elderly astronomer's thirty-two-year-old wife, though Halley dismissed the accusation. (Hearne, *Reminiscences and Recollections*, 13 November 1713, cited in Cook, p. 362.) On the other hand, Elisabetha Catherina was clearly an accomplished lady: in the twentieth century, she was to win fame as the first woman to be depicted in an engraving as an astronomical observer in her own right – she acted as her husband's assistant – and after Johannes' death in 1687, she would complete the publication of his works and correspond, in Latin, with the Royal Society.

After his return home in August 1679, Edmond's successful diplomatic mission caused even more laurels to descend upon his head. Then towards the end of 1680, he went off to Paris, to win the applause of the Académiciens and to be hosted by Giovanni Domenico Cassini, Director of King Louis XIV's Royal Observatory. The 'Tycho of the South', or cartographer of the southern constellations, now occupied a secure place amongst the *savants* of Europe. Next, he travelled to Italy, to see the great sights of antiquity, and to be entertained by the great and the good. Fascinated as he would always be with accurate measurements, Halley made exact feet and inches measurements of some of the ancient public yardsticks set in stone in the still largely unexcavated classical city of Rome.

Back home in England, however, family tragedy struck. His father Edmond Senior, a prominent London Freeman, disappeared. His decomposed body was found some months later, in a ditch near Rochester. What he had been doing in Rochester was a mystery. Nor had Edmond Senior been robbed, for his watch and purse were still with the body, and his son was only able to identify his father by the new shoes which he had been wearing when he had left the house that morning. No perpetrator was ever found.

In addition to the obvious months of anxiety and their grisly termination on the discovery of Halley's widowed father's death, other family problems soon emerged. For Halley Senior had recently married a second wife, and the new Mrs Halley was not much older than Edmond. But the widow Joane (soon to re-marry and become Joane Cleeter) could claim a substantial chunk of her murdered husband's estate, leaving Edmond Junior done out of his inheritance and relatively hard up. Yet the legal wrangling involved resulted in useful clues for the historian, as they detailed Edmond Halley Senior's property and assets, the ownership of which Edmond and his 'child' stepmother were contesting – such as the *Dog* tavern in Billingsgate, and other properties in the City, Islington and Shoreditch, as well as various business assets.

Of course, the family fight still left Edmond an undisputed gentleman, although one who now faced the prospect of having to earn money! He now became salaried Clerk to the Royal Society, and began to cultivate his own commercial interests. And in 1682, he married Miss Mary Tooke, who herself was descended from a City of London family. His career was about to enter a new phase of brilliant creativity.

TRADE WINDS AND WEATHER

In 1686, Edmond Halley published his seminal 'An Historical Account of the Trade Winds and Monsoons' in the *Philosophical Transactions* of the Royal Society, scientific meteorology then being in its infancy. In the 1660s, Robert Hooke had suggested that data obtained from the newly invented mercury barometer, thermometer, hygrometer and wind-gauge should be recorded in public 'Registers' or records, so that a coherent 'History' (or 'rational account') of the weather could be compiled. In this way, the ancient mystery of the causes of the weather might be revealed, and the approach of storms or freak conditions hopefully be made as predictable as the phases of the Moon. Such knowledge would benefit farmers, sailors and travellers, and might perhaps even help to explain why – in an age still ignorant of germs – epidemic diseases mysteriously came and went.

Halley's 'Trade Winds' work was destined to become the foundation stone for a series of detailed papers in which he would establish the basic principles of global meteorology (Halley, 1686, pp. 164–8). On the one hand, these papers were visionary in their scope, yet they were based on meticulously marshalled data collected from sea captains, from Halley's own St Helena expedition and from carefully conducted experiments made onshore. For Halley was a relentless marshaller of measured physical evidences of all kinds: details of storms, astronomical observations, the behaviour of the Earth's magnetic field, and even the appearance of the elusive aurora borealis. And in addition to his astronomy, Halley was one of the first scientists to explore the interrelationships and consequences of heat, cold, barometric pressure, tides, winds and the moisture exchanges taking place between the land, oceans and atmosphere on a planet both rotating on its axis *and* whirling around the Sun at constantly changing speeds – in other words, to explore what today we call a dynamic 'eco-system'.

Distilling the contents of his 'Trade Winds' and related 'eco-system' papers into a coherent model, one might suggest that Halley saw something along the following lines taking place. First, the local solar warmth of the atmosphere was crucial in generating an equatorial updraught of heat directly above the Equator, as the Sun's local noontide position moved around the planet every twenty-four hours. This produced a westward 'Sun-chasing' wind, drawing in air from both hemispheres, which ascended above the Equator, cooled with altitude, and gradually descended to the northern and southern polar regions, only to get sucked back to the Equator to produce a matching pair of constant air-circulation systems in both hemispheres.

This climatic system generated the Atlantic Trade Winds, called thus because they blew the great sailing ships across the oceans. Yet Halley realized, from sea captains and other long-haul travellers, that places like the Malay archipelago, the China Seas, the Caribbean and other oceans displayed variant wind patterns resulting in monsoons and hurricanes. Could these be generated by climatic events caused by 'local' geographical conditions, such as chains of islands in shallow seas, or the proximity of great land masses, such as the African, Asian and American continents?

Indeed, it is amazing how many times Halley's inspired interpretations were more or less right, for land and sea masses do respond differently to solar heating, and thus have varying effects upon the 'local' air masses passing above them.

But where does the rain come from? Halley began with a simple experiment, as he told the Royal Society in 1687. He took an eight-inch-diameter pan and filled it to the brim with water, weighed it, then left it outside on a moderately warm day. Then, twenty-four hours later, he re-weighed the pan, to establish exactly how much water vapour had risen into the atmosphere by evaporation, not to mention that 'lick'd up' by the wind. From this Halley calculated that one square degree (or 69 by 69 miles) of ocean (about 4,571 square miles), in a similar latitude, would yield some 33 million tons of water vapour per day. And in the case of the British Isles, these accumulated billions of tons of water vapour would be forced to rise as they approached mountains and other land masses, thereby depositing their burden as rain (Halley, 1687, pp. 367–9).

Halley realized that one could never establish a precise equivalent for the amount of rain that fell and the volume of water in the rivers, for water was absorbed into the soil, but his mathematical approach and adherence to 'the Laws of Staticks [Statistics]' in scientific research spurred him to try. By surveying an exact depth-to-surface profile of the river at Kingston-upon-Thames bridge, he calculated the volume of water under the bridge at any one time. Then, multiplying the same by the measured rate of flow, Halley computed that some 2,030,000 tons of water flowed through Kingston each day (Halley, 1687, p. 369).

He even wrestled with the problem of how rivers seemed, from what was then known of the interiors of Asia and Africa, to flow out of arid deserts: did the cold of high mountains act as 'alembicks' or 'basons' which distilled water vapour out of the air (Halley, 1690–1, pp. 470–71)? This was an idea which may have had its origins in the puzzlement he had experienced at his St Helena mountain-top observatory, where not only did his lenses keep misting over on a perfectly clear night, but also the ambient moisture even made paper and cardboard soggy. Halley's global climatology researches next led him back in time, to ask why certain vast inland 'seas', such as Lake Titicaca in Peru and the Aral Sea in Kazakhstan, contained sweet water, whereas the rivers flowing into the oceans had somehow made them too salty to drink. Were the oceans much, much older than the present continents which accommodated the vast land-locked lakes? For while all rivers contain tiny saline deposits, if they have only been flowing for a relatively short period of geological time, the inland seas will not yet have acquired sufficient salt to make it noticeable to the taste. Could the Earth not only be immensely ancient, but also undergoing constant change, so that the present-day land masses, oceans, poles and latitudes were not the same, perhaps, as they had been in pre-Biblical times? And by carefully measuring the increase in oceanic salinity over a century or so, could one calculate the age of the Earth (Halley, 1715b, p. 298)?

In this respect, Halley drew on the 1660s and 1670s Royal Society 'Earthquake Discourses' of Robert Hooke. Had vast aeons of time rolled along between God's creation of the world from nothing 'In the Beginning' and the re-forming of the new, present-day world of Adam, Eve and their descendants from the old 'Chaos' (or debris) mentioned in Genesis 1? As Halley saw God as working out His Divine Plan through *natural* processes, governed by God-created natural laws, could these archaic global disturbances have even been brought about by collisions with astronomical bodies, such as comets? Could Noah's flood, also described in the book of Genesis, and the last recorded global disaster, have come about by 'natural' causes; albeit divinely created natural causes?

William Whiston, some years later, even suggested that the Flood had come about by the Earth passing through the vapour-tail of a comet. Indeed, in 1694, Halley applied his passion for quantification and statistics to Noah's Flood itself. On the basis of current measured tropical deluge data, Halley computed that forty days and nights of such downpour would only cover the globe with water to a depth of 132 feet. Of course, this might have inundated billiard-table-flat Mesopotamia, but it would certainly not have drowned the whole world. (Halley, Royal Society Ms. 1694, p. 41; for full citation, see below, Halley, 1724, pp. 405–6.)

And while it is true that several members of what Halley styled 'the Sacred Order', senior churchmen such as Archbishop Stillingfleet, were not happy with some of his conclusions, this was not a universal view, even amongst clergymen. For in his application of scientific techniques to the understanding of earth-forming processes, Halley was not in any way trying to advance a kind of atheist agenda. Mainstream Christian theology, moreover, had never been blindly fundamentalist in its interpretation of Scripture, especially if the Bible passages in question related to naturalistic events rather than to moral or spiritual teaching. For what had God given us brains for, if not to wrestle with fascinating puzzles and new evidences?

HALLEY GOES TO SEA

It strikes us as incredible today that a scientist and an astronomer – even an eminent one who had been to sea in a private capacity (St Helena) and who was intimately involved with geographical and navigational affairs – should be suddenly commissioned captain of a Royal Naval vessel. But this is exactly what happened between 1698 and 1701, when Dr Halley, F.R.S. was reborn as Captain Halley, R.N., Commanding Officer of *H.M.S. Paramore*, 'Pink'. A pink, in fact, was one of the smallest classes of ocean-going vessels in King William III's Navy, *H.M.S. Paramore*, 80 tons, being under 60 feet long and drawing 10 feet of water (Cook, 1998, p. 265). Just imagine going to sea to explore remote and iceberg-infested oceans in a vessel no bigger than a modern pleasure yacht! And to make matters worse, *Paramore* was a slow and poor sailer! Halley had come into his command owing to an

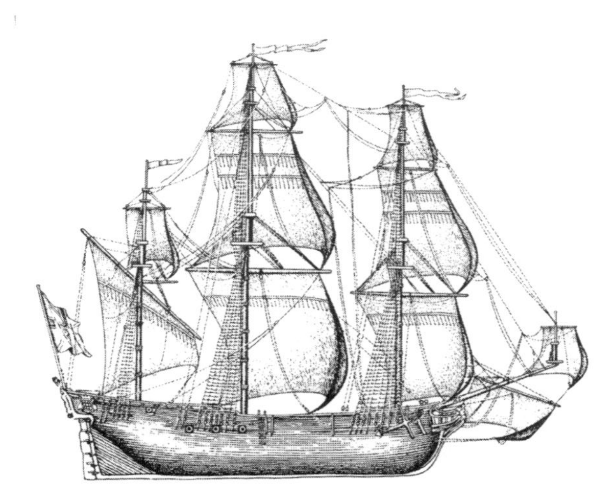

Figure 3. A 'pink', or small three-masted ocean-going ship, generally between fifty and sixty feet long. Similar to *H.M.S. Paramore*, in which Halley explored the iceberg-infested far south Atlantic Ocean in 1699–1700.

ingenious plan for finding the longitude at sea, which was still an enduring problem for sailors.

For while it was easy for a ship's master to find his latitude by using the Sun or stars, there were no astronomical guide points by which a ship's position could be found east or west, or in longitude. Various methods had been tried, such as clocks, the Moon's ecliptic position, and timing the moons of Jupiter, but none were reliable, and once a British navigator had sailed beyond Ireland, it was a matter of inspired guesswork exactly when he would hit the Grand Banks of Newfoundland, the Azores or Jamaica. Of course, the positions of all of these places had been mapped, using land-based techniques, to quite a high degree of accuracy by 1698, but once beyond the sight of land the navigator could only guess from prevailing winds, tides, seabirds and drift where he was on the map! And this is where Captain Halley comes in.

We saw above that Halley's fascination with geomagnetism went back to his schooldays, and between that time and his early forties he had accumulated a large body of geomagnetic data: from English and continental observations as well as from the log books of returning sea captains. What especially fascinated Halley, as it did many others, was both the westward 'drift' and the changing vertical 'dip' of the local compass needles in different parts of the globe.

And this was Halley's idea: if we can establish the exact 'variation' from astronomical north for specific locations at sea, then, by combining both horizontal and 'dip' compass readings with astronomical observations, we should be able to fix our longitude with pinpoint accuracy. So went the theory, and in consequence Halley, with his eminent reputation and persuasiveness, was not merely despatched as a passenger-scientist on a survey expedition, but was given full legal command of the vessel, sailing from Deptford on the Thames on 20 October 1698.

Not everyone was happy with this arrangement, however; in particular, Edward Harrison, Halley's career officer lieutenant and second in command, who proved obstructive once at sea. Not a man to be trifled with, Halley brought *Paramore* home in July 1699 and had his lieutenant court-martialled for disobedience, though he was annoyed that Harrison was only reprimanded by the Admiralty rather than receiving a more serious punishment. And then Halley set off on *Paramore*'s second voyage in August 1699.

Halley's three *Paramore* voyages were the first in a long and honourable series of great Royal Navy voyages of exploration and discovery, including the subsequent voyages of Captain James Cook in *H.M.S. Endeavour* and *H.M.S. Resolution* (1768–1779), and Captain Scott's 1910–12 expedition to the South Pole. Halley's strategy was to methodically cruise the Atlantic – the busiest ocean of the age – from the English Channel approaches down to the far southern latitude just below the Falkland Islands, between the European-African and American continents; some of the roughest and most forbidding waters on the planet in a fifty-two-foot wooden sailing ship! In Edmond

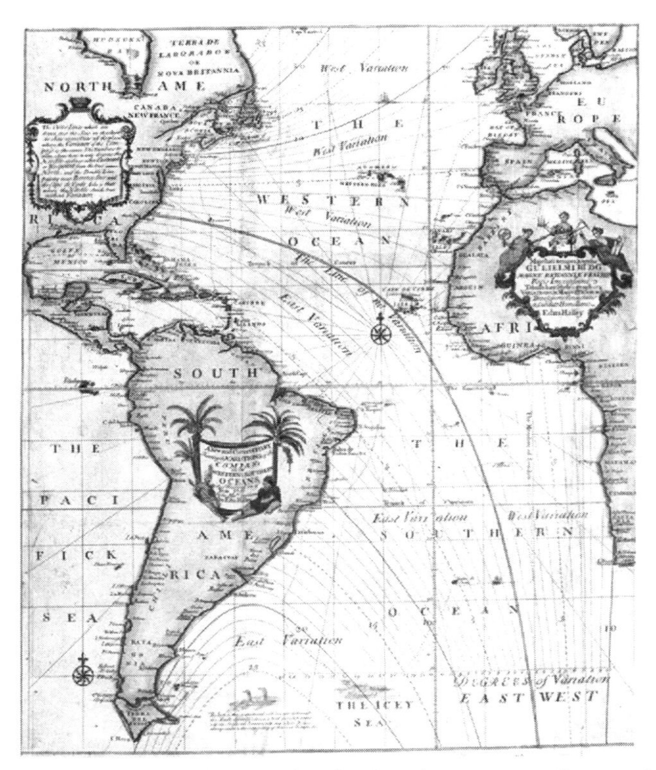

Figure 4. Halley's magnetic deviation chart for the Atlantic, 1701. Halley was the first cartographer to use a flowing 'isogonic' or contour line to connect places possessing the same mathematical properties. (Image courtesy of the Royal Geographical Society.)

Halley, scientific genius was combined with physical courage and daring. And in this way, he obtained hundreds of 'fixes' for magnetic variation angles, in conjunction with astronomical positions.

Halley's account of the three *Paramore* voyages, in 1698–1701, also contains some remarkable incidents, such as his observation of the indigenous Caribbean sponge divers, and his amazement at the length of time that they could stay underwater on one breath. What truly astonished Halley, however, was the vast size of the floating Ice Mountains encountered in the high southern latitudes, at 52° 24 south – icebergs so enormous that they seemed like great white islands, with ice cliffs that towered above *Paramore*'s main mast, one of which resembled Beachy Head on the Kentish coast.

In October 1701, back in England after a total of four years in Commission, Halley faced the problem of how to pinpoint those locations in the Atlantic and beyond which shared the same variation, so that a future navigator could identify them. So what he did in his magnetic charts of 1701 and 1702 was to link all similar places on a flowing contour (or 'isogonic') line that in some cases snaked across thousands of miles of sea. Halley's use of flowing 'contour' lines to designate places of equal magnetic variation, and later height or depth, was to become standard practice in cartography.

What an odd, twisted thing the Earth's magnetic field was shown to be in his world chart of 1702! Indeed, he had been working on an observation-based physical model for that field since at least 1683, and in 1691–2, he suggested that the Earth's interior contained several concentric iron-rock shells that rotated at slightly different speeds around an iron core, the cumulative effect of which was to give the magnetic field its twist and westwards drift.

Sadly, Halley's hope of using the magnetic variation as a way of finding the longitude at sea turned out to be impracticable. For unknown to him, or to anyone else in 1698, the variation was itself subject to variations, thereby rendering his meticulously compiled Atlantic and world magnetic maps useless for long-term longitude-finding purposes. But this in no way undermines the courage with which he undertook his heroic enterprise, nor Edmond Halley's standing as a founding father of geophysics.

COMETS

Halley was a physical scientist of genius, and one of the first people to recognize the interconnectedness of global meteorology, geophysics, long-term land mass and oceanic changes and geomagnetism. And as we shall see shortly, he would apply the same mode of thinking to his studies of the aurora borealis.

But what were comets, and what was the strange, invisible force that bound the whole of Creation together? In the century following 1577, when Tycho Brahe's precision measurements of the comet of that year demonstrated it to be in planetary space rather than in the classically postulated terrestrial atmosphere, a fascination with comets had come to grip Europe's astronomers. And there had been some spectacular comets, such as those of 1618, 1664, 1677 and 1680–1, to name a handful, studied by a veritable galaxy of astronomical talent, including Thomas Harriot, Galileo, Robert Hooke, Johannes Hevelius, John Flamsteed, Sir Isaac Newton and Edmond Halley. By the 1680s, moreover, several cometary characteristics were agreed upon: the nucleus of a comet was a solid body, and the tail seemed to be generated by a sort of corrosion which was related to the comet's distance from the Sun; comets' velocities also seemed to be related to their current solar distance; the nucleus of a comet was self-luminous in its own right, though the spectacular tail reflected sunlight, while it moved in a curved orbit. But were comets spat out from the sun or sucked into the Sun? And did their orbits correspond to conic-section curves, such as circles, parabolas, or ellipses? Yet most baffling of all was the nature of the force that drove them. Was it the same 'gravitating force' that Robert Hooke had recognized by 1674 as the agent which moved the planets around the Sun, or was it a force unique to comets?

By the time of these bright comets of the early 1680s, however, a new, precision astronomical technology was fast developing, and nowhere more than at the new Royal Observatory in Greenwich, under the inspired directorship of the Revd John Flamsteed, the first Astronomer Royal. The telescopic sight, screw micrometer, and pendulum clock instruments that Halley had taken to St Helena in 1676 were now available on an even more developed scale of size and precision at Greenwich, as Flamsteed was able to measure comet-star-planet-Sun angles to a new level of precision by 1680.

The comet of November 1680 to March 1681 was observed globally, across Europe, Maryland in the then American colonies, the Philippines, China and elsewhere, with several Jesuit missionary astronomers reporting it. But it was the splendid comet of 1682 that aroused the particular attention of European astronomers across the world. Many excellent astrometric measurements were made of the comet's changing stellar-solar positions over August and September 1682. Halley himself made accurate observations between 5 and 19 September, while Flamsteed's great seven-foot-radius equatorial sextant with telescopic

Figure 5. Halley's Comet, as depicted in the Bayeux Tapestry, which was made in England during the 1070s. A worried counsellor tells King Harold of the comet's appearance – a bad omen – while others wonder at the star (*isti mirant stellam . . .*). This is the earliest known depiction of Comet Halley.

sights at Greenwich probably secured the best measures of all, and these would be of special interest to Isaac Newton, who would analyse the comet's orbit in terms of gravitational attraction in his *Principia Mathematica* (1687), Book III.

It was the accurate observation of the comet of 1682 which would help change astronomical thinking. Its orbit was not only analysed by Newton, but also by Edmond Halley in his *A Synopsis of the Astronomy of Comets* (1705), for the 1682 object would come to immortalize the astronomer as 'Halley's Comet'.

Figure 6. [Sir] Isaac Newton, aged 46–47, by Sir Godfrey Kneller, 1689. (Detail of 159042. Image courtesy of Académic des Sciences, Paris/Giraudan/Bridgeman Images.)

In the *Synopsis*, Halley presented a masterly historical outline of cometary knowledge before getting down to the business of interpreting our rapidly advancing knowledge of comets as *gravitationally* driven bodies. He printed a list of twenty-four comets that had been observed between 1337 and 1698, and argued that comets move in 'very Eccentrick Orbits' under the force of solar gravity: sometimes being driven out 'into the remotest parts of the Universe' but then making 'their returns after long Periods of time'. In short, comets orbited the Sun! (Halley, *Comets*, 1705, p. 20.)

Of course, 'Halley's Comet' was not the only one which he believed was periodic – he being wrong regarding all the others – but his careful combing of astronomical records, combined with mathematical analyses of the superior data from that of 1682, correctly led him to associate it with the previous comets of 1607, 1531 and 1456. Consequently: 'Hence, I venture to foretell, that it will return again in the Year 1758' (Halley, *Comets*, 1705, p. 22.) And he was right! The *Synopsis* also contains a variety of observations of astronomical bodies which Halley made from the village of Islington over 1682–4, where, as we have learned, the Halley family owned property.

GRAVITY AND COSMOLOGY

One never ceases to be in awe of Edmond Halley's sheer intellectual range, not only as an astronomer and pioneer Earth scientist, but also as a fluent Latin and Greek classical scholar. For he was equally at home calculating the Moon-produced tidal currents in the English Channel and interpreting Greek and Roman narratives to fix the exact spot of Julius Caesar's landing site in the invasion of Britain in 54 BC, as in extracting stellar 'proper motions' from ancient Greek observations, or computing the orbits of comets.

But we must be eternally grateful to Halley's charm, persuasiveness and ability to get on with difficult people in the saga of cajoling Newton to write his monumental *Principia* in the mid-1680s, not to mention his business sense in facilitating the publishing and marketing of the treatise. For genius as he was, Newton was not an easy person to deal with, being reclusive and suspicious, and quickly taking offence. Indeed, it speaks volumes for Halley's persuasiveness, patience, and, no doubt, his ability to shrug off censorious rebukes like water from a duck's back, that he got Newton to write *Principia* in under three years.

It was in 1684, when calling into Cambridge on what was probably a Halley family property business trip to East Anglia, that Edmond had a decisive meeting with Newton. And it was at this time that he became convinced that Newton's mathematical analysis of 'gravitational' forces was the golden key for which scientists had been searching for almost a century. But let us not forget that, Halley's badgering him to finish his manuscript apart, Newton had much for which to be grateful to Halley, for not only did Edmond see the monumental treatise through the press – acting as Newton's editor and publisher – but he also stood the cost of publication out of his own pocket! No doubt he utilized his contacts in the City of London publishing and business worlds. Yet ironically, by 1687, Newton himself was very comfortably off, having acquired family property in Lincolnshire following his mother's death, not to mention his salary as a Professorial Fellow of Trinity College, Cambridge.

And by the early eighteenth century, Halley was becoming intrigued by the possible action of gravity in space. In particular, he began to examine three classes of 'deep-space' phenomena which were truly cosmological: the 'nebulae', the long-term movement of the 'fixed' stars and the infinity of the 'Sphere of the Fix'd Stars' (Halley, 1720, p. 23).

Around 1700, six 'nebulae', including the misty regions in Orion's Sword and Andromeda, were known to astronomers, but Halley had discovered another in Centaurus when he was at St Helena in 1677, and one in Hercules in 1714, which he duly reported to the Royal Society. All of this work being done, of course, using the small-aperture, long-focus simple refracting telescopes of that time, for though reflecting telescopes were around in 1714, they were not yet technologically viable as serious research instruments.

The Orion Nebula in particular had long puzzled astronomers. Christiaan Huygens had published a drawing of it in *Systema Saturni* (1659) and speculated what it might be. Huygens's speculations would not only puzzle Halley in a strangely anonymous paper of his in 1715–16, but they were still puzzling William Herschel in the 1790s: namely, what were those glowing objects made of, what was their relation to the stars, and how far away were they?

In a pair of short papers of 1715–16 (Halley, 1715c and 1716a), which, while anonymous, openly announced the discovery of two 'new' nebulae by 'MR Edm. Halley', Halley speculates about the nature of the 'Medium' in space that could seemingly glow in its own right. But how far away could those nebulae be which appear as stars to the naked eye, yet may be several arc minutes across when viewed through a telescope? These nebulae, moreover, display no annual parallax, thus placing them at immense distances from the Earth. But as they are so large, in angular terms, they must be enormous, 'and perhaps not less than our whole solar system'. A staggering thought!

Then in 1720, Halley published a pair of signed papers, 'On the Infinity of the Sphere of the Fix'd Stars', which, in many ways, first addressed the question of deep-space stellar distribution. Knowing all about gravity by 1720, he was faced with the question of why the universe – as we might put it today – does not *implode* into its gravitational centre and collapse into one vast mass. The answer, Halley suggested, was because it must be *infinite*: the stars (and nebulae) being scattered throughout an infinity of space means that the universe has no geometrical centre, so that an infinite *matrix* of stars, as it were, holds the whole in gravitational equipoise.

This, in turn, led him to face that problem which would become known in the nineteenth century as 'Olber's Paradox' (Halley, 1720, p. 24). For if stars went on for ever and ever to infinity, should not every arc second of the sky contain a star, thus making the sky blaze with a permanent light? Yet, oddly enough, even Halley had not been the first astronomer to think of this one, for Leonard Digges, English disciple of Copernicus and a speculator about cosmological infinity, had first aired this possibility in the 1570s.

Halley, who had believed space to be wholly transparent and had no knowledge of light- and energy-absorbing cosmic dust, put forward an elegant geometrical solution. For if progressively remote stars displayed smaller and smaller angular surface areas, albeit unmeasurable in 1720, and hence appeared correspondingly dim, then each visible star must be surrounded by an annulus of black sky, no matter how small in extent. Hence, the sky will always seem to contain more black sky than glowing light. This may look like a fudge to us today, but it was a brilliant idea in 1720, especially from a man who, by that date, was Oxford University's Savilian Professor of Geometry. For while we have noted already that Halley was a scientist who saw physical truth as emerging through the making and marshalling of precision measurements, we must also remember that his creative instincts in astronomy centred upon geometry.

It was this instinct for geometry, combined with his concern for precision measurement and the analysis of historical data, that led him to question the ancient idea of the 'fixed' stars. And by 1718, two distinct sets of data were leading him to do so. The first of these was a growing body of data regarding the luminosity changes of individual stars. Of course, there had been the baffling 'supernovae' of 1572 and 1604, analysed respectively by Tycho Brahe and Johannes Kepler. Accurate geometrical measurements had shown these stars to have no discernible parallaxes, thereby demonstrating that, just like the comets, they were not in the Earth's upper atmosphere, but in astronomical space.

But in addition to supernovae that appeared out of nowhere, dimmed, and then disappeared for good, there were stars that displayed regular changes in brightness, or magnitude, such as Mira in Cetus, Algol in Perseus, and one or two other known 'variables'.

On the other hand, Halley had come to realize by 1718 that certain stars actually displayed a systematic motion of their own in a given direction – what would come to be called a 'proper motion'. He noted, for example, that Aldebaran and Arcturus seemed to occupy slightly different positions in 1718 than they had in the classical Greek catalogues, and while he noted that Sirius was two arc minutes more northerly than it had been in around 1590, the meticulous Halley was cautious. For in 1718, we had a much better knowledge of atmospheric refraction and its slight distorting effect on star positions than had Tycho Brahe. Even so, Halley concluded that over an 1800-year period certain stars *had* moved. Most of them were very bright stars, however, and 'in all probability the nearest to the Earth', though they may have possessed motions of their own as well as line-of-sight motions. Once again, one never ceases to be amazed at Halley's acute perceptiveness when it came to evaluating measurements and physical data, and drawing conclusions from them.

THE AURORA BOREALIS, METEORS AND LIGHTS IN THE SKY

The 6th of March 1716 (or 16 March in the New Style Calendar) had been a mild, calm, pleasant day in London. Then, between 9 and 10 p.m. in the evening, someone told Halley that a curious light was shining in the northern sky, and had been there since 7 p.m. or so. It was a flame-coloured yellow light, and especially conspicuous in the region of the constellation Cygnus. By 11 p.m., no distinctly new permutations of the aurora were occurring, beyond a circle of light, flashing streaks, and ray patterns similar to the insignia of the Knights of the Garter. By 3 a.m., after observing through the window as it was a cold night, Halley seems, after measuring the angular size of the light circle, to have gone to bed.

It is interesting to note, however, that though he was a relentless traveller by land and sea, not to mention being an assiduous celestial observer, Halley had never seen an aurora before his sixtieth year. He wanted to know why, and began to collect historical references to aurorae. He discovered that a few had been seen between Ireland and Berlin across 1707–8 (when, one presumes, it had been cloudy in London), yet the last notable aurora had been the one seen by the eminent French astronomer Pierre Gassendi on 2 September (Old Style Calendar) 1621. But no aurorae were reported in the 'Registers' of the Royal Society, founded in 1660, or its journal *Philosophical Transactions*.

Halley next set about the business of collecting observations of the 1716 aurora from scientists across Europe, and from returning sea captains. And yes, it had been observed across a fair expanse of the Northern Hemisphere, and even as far down as the Mediterranean. Yet whether one saw the lights in Aberdeen or Paris, they were always in the *northern* sky. Why?

With the auroral bit between his teeth, Halley next began to investigate how lights without flame could appear in the sky, and initiated pioneering researches into not only atmospheric optics but also atmospheric chemistry and physics. Indeed, one never ceases to be struck by the breadth and diversity of Halley's learning, and by his ability to see connections. Could the auroral light be like 'magnetical effluvia' circulating the North Pole, analogous to the circulation of the 'atoms' which produced the attractive force of the magnet? (Halley, 1716b, p. 421.) Indeed, laboratory parallels ran deep in Halley's astronomical thinking. This enabled him to imaginatively link aurorae with sparks emitted from one of the newly invented electrostatic machines, and even with the 'nitro-sulphurous' fires that produced the mysterious after-glow in the glass sphere of a vacuum pump after gunpowder had been ignited *in vacuo*. This had been an experiment he had seen performed in the darkened laboratory of Oxford's Science Museum (now the Museum of the History of Science on Broad Street) by his friend the Revd John Whiteside, the Chemistry Professor, around 1710 (Halley, 1716b, p. 420).

Electrical sparks, 'magnetical effluvia', and mysterious glows in the dark all took Halley towards a faltering sense of what we would now style atmospheric ionization. Why were there no notable aurorae between 1621 and 1707? Two centuries later, Edwin Maunder of the Royal Greenwich Observatory would also identify these decades as a time of hardly any sunspot activity, or what would later be called the 'Maunder Minimum'.

On the same subject of atmospheric phenomena, in 1714, Halley published a list of notable meteoritic displays. But, once again, what could the meteors be? He proposed that the showers may have been occasioned by 'some fortuitous Concourse of Atoms, and that the Earth met with as it past [*sic*] along in its Orb' (Halley, 1714, p. 162). Indeed, it was a remarkably good 'guess', for we now know that meteor showers are produced by 'clouds' of fine particles from space hitting the Earth's atmosphere and burning up.

MERCURY AND VENUS TRANSITS AND THE SOLAR DISTANCE

On 28 October (Old Style) 1677 (7 November New Style), Halley had observed a passage of the planet Mercury across the Sun's disk. Of course, the observation of such phenomena was not new by that date, and several had been seen since Gassendi made the first telescopic observation of a Mercury transit in 1631. But it gave Halley a brilliant idea: for if Mercury's apparent position on the solar disk, along with the exact time of its transit, could be accurately measured (as he actually did) on St Helena, and similar observations were made in Europe, then a parallax angle for Mercury could be calculated.

This angle could be derived from a computed Greenwich to St Helena base-line through the Earth, for the two locations are about 70 degrees apart on the globe. But Halley came to realize that Mercury's disk was much too small and its motion too fast, making it unsuitable for a parallax determination. Venus would be much better, though there would not be a Venus transit until 1761, when all currently living astronomers would be long dead.

Yet Halley was a scientist who thought in long terms, both when it came to using historical data and to alerting future generations. Consequently, he began to analyse the gravitational characteristics of the Sun-Venus system, and devised a method by which future astronomers might use the 1761 transit to extract a more accurate solar parallax value, publishing several preliminary papers before his definitive one in 1716 – a paper written in Latin, moreover, intended for an international readership. This paper, and subsequent ones by French and other astronomers, would establish the basic technical rationale used by the international expeditions sent across the globe to observe the 1761 and 1769 transits, including Captain James Cook's Pacific expedition of 1767–71.

In many respects, Newtonian gravitation after 1687 gave a new urgency to the definitive measurement of the Earth–Sun distance, for through an extension of the Inverse Square Law, and with the use of Kepler's Laws, the proportional distances of the rest of the planets could be established. Indeed, this mathematical knowledge made it possible to determine the orbital characteristics of the new planet Uranus, discovered by Sir William Herschel in 1781, while providing the rationale behind Adams's and Le Verrier's search for Neptune in 1845–6, not to mention defining numerous asteroidal and cometary orbits after 1801. Indeed, it made possible the testing of Newtonian theory against real objects in the universe. And central to this wider enterprise was Edmond Halley's concern with attempting to define the 'Astronomical Unit' as a base-line upon which subsequent solar-system gravitational investigations could be founded.

SAVILIAN PROFESSOR AND ASTRONOMER ROYAL

Few scientific careers, even in the wide-ranging seventeenth century, were as diverse as that of Halley: youthful prodigy, bold traveller, geophysicist, astronomer, Newton's publisher, Royal Navy Captain and even diplomat. For in the autumn of 1702, Edmond was commissioned by Queen Anne to present his diplomatic credentials at the Imperial Court in Vienna, and then to supervise the survey of the fortifications of the northern Adriatic sea, as part of the complex political and military minuet being danced by Great Britain's allies and enemies across Europe. Indeed, it was part of that same cycle of conflicts – the War of the Spanish Succession – which won Gibraltar for the British Empire in 1703.

In January 1704, Oxford University elected Edmond Savilian Professor of Geometry. But not everyone was happy, certainly not Astronomer Royal John Flamsteed, who, no friend of Halley's, had fulminated in December 1703 that he 'talks, swears and drinks brandy like a sea captain' (Flamsteed, *Correspondence* III, no. 922, p. 47; Cook 321). I suspect, however, that the socially astute Halley did *not* behave in such a way in an Oxford Senior Common Room, before Queen Anne, or at the court of the Emperor of Austria!

But as an Oxford Professor, the forty-eight-year-old ex-sea captain had a new role: namely, to be a learned sage. And, true to form, he fulfilled it admirably. For while an active advocate of the new instrument-based science, the 1619 statutes of his Savilian Professorship required Halley to lecture on classical geometry. But as he was an accomplished classical scholar, with a fluent command of Latin and Greek, this simply presented another avenue for intellectual exploration. In particular, he set about the task of complex classical Greek geometrical scholarship. One aspect of geometry which the third-century-BC Greeks had started to explore was that of the curious geometry of the cone, and how exact angled slices through the cone produced a set of elegant, perfect curves, such as the circle, ellipse, hyperbola and parabola, some of which were 'open' and others 'closed'. These curves were playing a crucial part in modern astronomy, such as the Keplerian ellipses of planetary and cometary orbits and studies of how lenses and mirrors with circular and parabolic surfaces refracted and focussed light. Book VIII of the *Conics* of Apollonius of Perga (262–190 BC) had survived only in scattered fragments of Greek text and Arabic translation, and the task which Professor Halley took on was the reconstruction of that geometrical treatise in a logical and coherent form. This was a formidable task, and part of a wider project of the Oxford mathematicians to publish the surviving works of all the classical Greek geometers. With Apollonius behind him, and his clear statement in the 'Praefatio' or Preface to this work that subsequent geometry and mathematics stood firmly on ancient Greek foundations, Halley's standing as a classical scholar became unassailable.

Figure 7. Flamsteed House and the courtyard of the Royal Observatory, Greenwich. Halley succeeded the Revd John Flamsteed as Astronomer Royal in 1720.

Edmond Halley was to hold the Savilian Professorship for the rest of his life, living, when in Oxford, in a fine house in New College Lane. The memorial plaque on the wall means that it is often photographed by visitors to Oxford today. But Halley was not only a fine and respected scholar and university teacher – much to John Flamsteed's chagrin – but also an assiduous observer of the heavens; he had a 'gazebo' platform built on the roof of the house from which he continued to make astronomical observations, for astronomy was always Halley's first love.

On the very last day of 1719, Halley's old rival (whom, it must be admitted, Halley had not treated fairly), the Revd John Flamsteed, Astronomer Royal, died at his house within the Royal Observatory, Greenwich. And much to the understandable fury of his widow Margaret, the well-connected and persuasive Halley succeeded him. Let us not forget that by 1720 Halley was at our modern-day retirement age, being in his sixty-fourth year, so why did he want the job, with its miserable salary of £100 per annum, and how did he juggle his commitments at Greenwich and his continuing Savilian Professorship in Oxford? I would suggest that Halley's still-undiminished energy and zest for life played a part, plus his eye for the 'main chance' of winning yet more fame and making more money.

The longitude remained as elusive as ever in 1720, and Halley's hoped-for use of the magnetic variation as a way of finding it had come to nothing. Following the tragic loss of Admiral Sir Cloudsley Shovell's triumphal home-coming fleet on the rocks of the erroneously calculated Scilly Isles in bad weather in 1707, Parliament had been moved to pass the Longitude Act of 1714, which offered a spectacular £20,000 reward to anyone who devised a fool-proof method for finding the longitude at sea.

As the Astronomer Royal was not formally excluded from the prize, it was, almost certainly, the prospect of claiming this reward, not to mention the accompanying prestige for finding the longitude, that spurred Halley to become Astronomer Royal. Nor was he dependent on the salary, which he simply added to his Oxford stipend, quite apart from any private income which he enjoyed. Halley even succeeded in claiming his 'half-pay' stipend as an ex-Royal Navy captain – a sort of RN retirement pension – which would, all together, have left him comfortably off, and with a free residence at Greenwich.

Yet, thorough as always, Halley proposed to undertake a rigorous set of observations of the Moon's angular position amongst the fixed stars. For in this 'lunars' method, it was hoped that a sailor in any part of the globe could fix his longitude from Greenwich using the Moon-star parallax discrepancy. To make this practicable, however, it was necessary to undertake a long run of Moon-star angles at Greenwich which, in conjunction with the new Newtonian gravitational Inverse Square Law calculations, would enable a table of definitive lunar positions for Greenwich to be constructed, published and carried on board ship. The problem, however, lay in the Moon's complex orbital cycle over a known period of 18.6 years, and to aspire to observe the Moon through such a long period was no mean feat for a sixty-four-year-old. The fact that he successfully accomplished this gruelling cycle of observations, which took him into his early eighties, says much for Edmond Halley – especially as in 1720 there were no instruments in the Royal Observatory to work with!

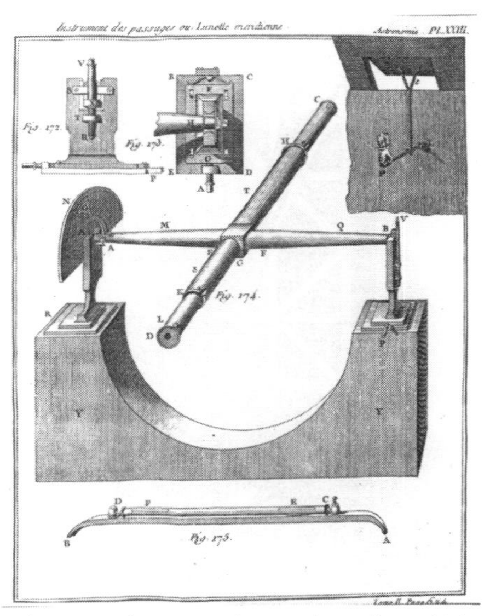

Figure 8. George Graham F.R.S.'s transit telescope for Halley, which came into use at Greenwich in 1721. The telescope, mounted on precision trunnions, delineated the meridian through 180 degrees. Halley used it in conjunction with a precision pendulum clock to time star transits across the meridian, which enabled him to measure precise Right Ascension angles. (J. J. L. de Lalande, *Astronomie* II, Plate XXIII (Paris, 1792).)

All John Flamsteed's instruments, between 1675 and 1719, were shown in a lawsuit to be Flamsteed's personal property, and his widow Margaret promptly sold them. Halley, however, had no intention of spending his own money on re-equipping Greenwich, as Flamsteed had done, and soon extracted a handsome official grant of £500 with which to commission new instruments. Central to his new 'kit' were a pair of large instruments build by the scientist and precision-engineer clock-maker George Graham, F.R.S.

The first to be completed was a five-foot 'transit telescope' which rotated on east-west-pointing pivots so that the telescope itself described a meridian arc from the southern to the northern horizons,

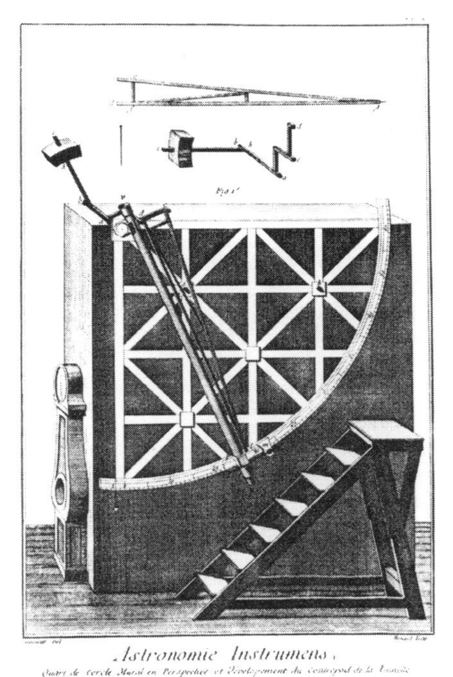

Figure 9. The great eight-foot-radius mural quadrant which Halley commissioned from the craftsman George Graham, F.R.S., and which was completed in 1725. In 1725, this was the most accurate angle-measuring instrument in the world, and a triumph of precision engineering. It was set up in the meridian plane to measure the declination angles of astronomical objects as they came to the meridian. (D. Diderot, *Encyclopédie* XIII, 1765.)

bisecting the north polar points. At a cost of £61 10/-, it was finished in 1721, and enabled Halley to start making exact right ascension (east-west) timings of the Moon and stars as they passed the meridian, in conjunction with a precision pendulum clock. Halley's transit became the ancestor of every subsequent observatory transit, down to the Carlsberg Meridian Circle after 1953 in Copenhagen and, subsequently, in the Canaries.

His second major instrument, also built by George Graham, was the great eight-foot-radius telescopic mural quadrant, secured to a stone wall in the meridian plane, which enabled him to take declination (vertical) angles to an unprecedented level of accuracy. Completed in 1725, at a cost of £362, and inspired by earlier mural instruments by Tycho Brahe, Hooke and Flamsteed, it in turn was destined to be the prototype of all large mural astronomical instruments.

Figure 10. Halley aged eighty, and still Astronomer Royal. Halley holds a drawing of his 1690s model of the Earth's interior. He proposed that different concentric shells within the Earth rotated at slightly different speeds, thereby occasioning the drift and changes within the Earth's magnetic field. Halley's theory was wrong, yet, nevertheless, it gives him a strong claim for being the 'father of geophysics'. (Michael Dahl, c.1736. Image courtesy of The Royal Society.)

Halley's declared ambition to observe the Moon in the meridian through its complete 18.6-year cycle – weather permitting – had finally been achieved by the octogenarian Astronomer Royal by 1740. This intention, however, already stood on two decades of detailed lunar position observing by Halley, which he had undertaken in his private capacity between 1704 and 1720.

Yet Edmond Halley never found the longitude. For, quite simply, even George Graham's superlative instruments could not measure down to the *single* arc second accuracies necessary to compile lunar tables of sufficient exactitude to enable a sailor to fix his position anywhere in the world. This would only become possible with the 'next generation' angle-measuring instruments of Graham's successor, John Bird, in the hands of Halley's protégés and Greenwich successors, the Revd Drs James Bradley and Nevil Maskelyne, and Tobias Mayer at Göttingen, all of whom worked with Bird's angle-measuring instruments. The 'lunars' method finally became viable with Maskelyne's *Nautical Almanac* tables (1767), which afforded a cheaper way of finding the longitude than did John Harrison's ingenious, exquisite, but formidably expensive early chronometers.

A FULL LIFE AND CAREER

Edmond Halley lived a truly remarkable life, being not only an astronomer of genius, but also a pioneer of geophysics, geodesy, meteorology, auroral studies and even 'deep-sea' diving (Halley, 1716c). Indeed, on several occasions around 1691, the dare-devil Halley descended to the sea bed to try out his pioneering diving bell, and his friend Newton was to report in his *Opticks* in 1704 (Book I, Part II, Proposition 10, Problem 5) that Halley had told him that as the bell sank to a depth of sixty feet, the sunlight became progressively redder. Furthermore, Halley's passion for both historical data collection and meticulous measurement epitomized the rationale behind the early Royal Society, to which he had been elected on 30 November 1678. As we saw above, Halley died suddenly and peacefully on 14 January 1742. He was buried alongside Mary, his wife of fifty-four years, who had pre-deceased him in 1736. They both lie in Lee churchyard, near the Royal Observatory, Greenwich.

A SELECTION OF EDMOND HALLEY'S ORIGINAL PUBLICATIONS

Nearly all of Halley's researches were published in the *Philosophical Transactions* of the Royal Society, though his St Helena (1679), cometary studies (1705), and some classical works were published as books. *Phil. Trans.*, and probably the books, are available online. A full list of Halley's publications is to be found in Cook, *Edmond Halley*, pp. 502–9. References in the main text inside brackets, e.g. (Halley, 1720), are to articles below, which are listed in date order.

Catalogus Stellarum Australium (*Catalogue of Southern Stars*) (London, 1679).

'An Historical Account of the Trade Winds, and Monsoons', *Phil. Trans.* Vol. 16, No. 183, pp. 153–68 (1686).

'An Estimate of the Quantity of Vapour Raised out of the Sea by . . . the Sun . . .', *Phil. Trans.* Vol. 16, No. 189, pp. 366–70 (1687).

'An Account of the Circulation of the Watry [*sic*] Vapours of the Sea, and Cause of Springs . . .', *Phil. Trans.* Vol. 16, No. 192, pp. 468–73 (1690/1).

'An Account of the Cause of the Change of the Variation of the Magnetical Needle . . .', *Phil. Trans.* Vol. 16, No. 195, pp. 563–78 (1692).

A Synopsis of the Astronomy of Comets . . . (London, 1705).

'An Account of Several Extraordinary Meteors or Lights in the Sky', *Phil. Trans.* Vol. 29, No. 341, pp. 159–64 (1714).

'Observations of the Late Total Eclipse of the Sun, 22 of April [1715]', *Phil. Trans.* Vol. 29, No. 343, pp. 245–62 (1715 [1715a]).

'A Short Account of the Course of the Saltness of the Ocean, and of Several Lakes that Emit no Rivers; with a Proposal to Help Thereof, to Discover the Age of the World . . .', *Phil. Trans.* Vol. 29, No. 343, pp. 296–300 (1715 [1715b]).

'A Short History of Several New-Stars . . .', *Phil. Trans.* Vol. 29, No. 346, pp. 354–6 (1715 [1715c]).

'An Account of Several Nebulae or Lucid Spots Like Clouds . . .', *Phil. Trans.* Vol. 29, No. 347, pp. 390–2 (1716 [1716a]).

'An Account of the late surprizing Appearance of Lights in the Air on the Sixth of March Last . . .', *Phil. Trans.* Vol. 29, No. 347, pp. 406–28 (1716 [1716b]).

'The Art of Living Under Water . . .', *Phil. Trans.* Vol. 29, No. 349, pp. 492–9 (1716 [1716c]).

'Consideration of the Changes of the Latitudes of Some of the Principal Fix'd Stars', *Phil. Trans.* Vol. 30, No. 355, pp. 736–8 (1717–18).

'An Account of the Phenomenon of a Very Extraordinary Aurora Borealis Seen in *London* on 10 November 1719 both Morning and Evening', *Phil. Trans.* Vol. 30, No. 363, pp. 1099–1100 (1719).

'Of the Infinity of the Sphere of the Fix'd Stars', *Phil. Trans.* Vol. 31, No. 364, p. 22–4 (1720).

'Some considerations about the cause of the universal deluge, laid before the Royal Society, on 12 December 1694', *Phil. Trans.* Vol. 33 No. 383, pp. 118–23 (1724). (The original MS 'A Discourse, Concerning an Hypothesis of the Manner of the Generall Deluge', 12 December 1694. Royal Society Register Book Copy, 9, pp. 40–4.)

FURTHER READING

Allan Chapman, *Dividing the Circle: The Development of Critical Angular Measurement in Astronomy, 1500–1800* (Praxis-Wiley, 1990, 1995).

Allan Chapman, 'Edmond Halley's Use of Historical Evidence in the Advancement of Science' [Royal Society Wilkins Lecture, 1994], *Notes and Records of the Royal Society of London*, No. 48, Vol. 2, pp. 167–91 (1994).

Allan Chapman, 'Edmond Halley', in *Oxford Figures: 800 Years of the Mathematical Sciences*, eds John Fauvel, Raymond Flood and Robin Wilson (OUP, 2000, 2013), pp. 17–36.

Sir Alan Cook, *Edmond Halley: Charting the Heavens on the Sea* (OUP, 1998).

John Flamsteed, *The 'Preface' to John Flamsteed's 'Historia Coelestis Britannica' [vol. III] 1725*, edited and introduced by Allan Chapman, based on a translation by Dione Johnson (Greenwich, National Maritime Museum Monograph No. 52, 1982).

The Correspondence of John Flamsteed, Astronomer Royal, Vol. III, 1703–1719, eds Eric Forbes, Lesley Murdin and Frances Willmoth (Institute of Physics, Bristol and Philadelphia, 2002).

Eric G. Forbes, *Greenwich Observatory. Vol. I, Origins and Early History (1675–1835)* (London, 1975).

Robert T. Gunther, *Early Science in Oxford*, Vol. XI (Oxford, 1937).
Robert Hooke, *De Potentia Restitutiva, or of Spring* . . . (London, 1678), pp. 39ff. for Halley's St Helena mountain observations.
Derek Howse, *Greenwich Observatory: Vol. III, The Buildings and Instruments* (London, 1975).
H. C. King, *A History of the Telescope* (London, 1955).
E. F. MacPike (ed.), *The Correspondence and Papers of Edmond Halley* (OUP, 1932).
E. F. MacPike, *Hevelius, Flamsteed and Halley* (London, 1937).
Colin Ronan, *Their Majesties' Astronomers: A Survey of Astronomy in Britain Between the Two Elizabeths* (London, 1967).
Colin Ronan, *Edmond Halley: Genius in Eclipse* (London, 1970).

Part Three

Miscellaneous

Some Interesting Variable Stars

JOHN ISLES

All variable stars are of potential interest, and hundreds of them can be observed with the slightest optical aid – even with a pair of binoculars. The stars in the list that follows include many that are popular with amateur observers, as well as some less well-known objects that are, nevertheless, suitable for study visually. The periods and ranges of many variables are not constant from one cycle to another, and some are completely irregular.

Finder charts are given after the list for those stars marked with an asterisk. These charts are adapted with permission from those issued by the Variable Star Section of the British Astronomical Association. Apart from the eclipsing variables and others in which the light changes are purely a geometrical effect, variable stars can be divided broadly into two classes: the pulsating stars, and the eruptive or cataclysmic variables.

Mira (Omicron Ceti) is the best-known member of the long-period subclass of pulsating red-giant stars. The chart is suitable for use in estimating the magnitude of Mira when it reaches naked-eye brightness – typically from about a month before the predicted date of maximum until two or three months after maximum. Predictions for Mira and other stars of its class follow the section of finder charts.

The semi-regular variables are less predictable, and generally have smaller ranges. V Canum Venaticorum is one of the more reliable ones, with steady oscillations in a six-month cycle. Z Ursae Majoris, easily found with binoculars near Delta, has a large range, and often shows double maxima owing to the presence of multiple periodicities in its light changes. The chart for Z is also suitable for observing another semi-regular star, RY Ursae Majoris. These semi-regular stars are mostly red giants or supergiants.

The RV Tauri stars are of earlier spectral class than the semi-regulars, and in a full cycle of variation they often show deep minima and double maxima that are separated by a secondary minimum. U Monocerotis is one of the brightest RV Tauri stars.

Among eruptive variable stars is the carbon-rich supergiant R Coronae Borealis. Its unpredictable eruptions cause it not to brighten, but to fade. This happens when one of the sooty clouds that the star throws out from time to time happens to come in our direction and blots out most of the star's light from our view. Much of the time R Coronae is bright enough to be seen in binoculars, and the chart can be used to estimate its magnitude. During the deepest minima, however, the star needs a telescope of 25-cm or larger aperture to be detected.

CH Cygni is a symbiotic star – that is, a close binary comprising a red giant and a hot dwarf star that interact physically, giving rise to outbursts. The system also shows semi-regular oscillations, and sudden fades and rises that may be connected with eclipses.

Observers can follow the changes of these variable stars by using the comparison stars whose magnitudes are given below each chart. Observations of variable stars by amateurs are of scientific value, provided they are collected and made available for analysis. This is done by several organizations, including the British Astronomical Association, the American Association of Variable Star Observers (49 Bay State Road, Cambridge, Massachusetts 02138, USA), and the Royal Astronomical Society of New Zealand (PO Box 3181, Wellington, New Zealand).

Star	RA		Declination		Range	Type	Period	Spectrum
	h	m	°	′			(days)	
R Andromedae	00	24.0	+38	35	5.8–14.9	Mira	409	S
W Andromedae	02	17.6	+44	18	6.7–14.6	Mira	396	S
U Antliae	10	35.2	–39	34	5–6	Irregular	—	C
Theta Apodis	14	05.3	–76	48	5–7	Semi-regular	119	M
R Aquarii	23	43.8	–15	17	5.8–12.4	Symbiotic	387	M+Pec
T Aquarii	20	49.9	–05	09	7.2–14.2	Mira	202	M
R Aquilae	19	06.4	+08	14	5.5–12.0	Mira	284	M
V Aquilae	19	04.4	–05	41	6.6–8.4	Semi-regular	353	C
Eta Aquilae	19	52.5	+01	00	3.5–4.4	Cepheid	7.2	F–G
U Arae	17	53.6	–51	41	7.7–14.1	Mira	225	M
R Arietis	02	16.1	+25	03	7.4–13.7	Mira	187	M
U Arietis	03	11.0	+14	48	7.2–15.2	Mira	371	M

Some Interesting Variable Stars

Star	RA		Declination		Range	Type	Period	Spectrum
	h	m	°	′			(days)	
R Aurigae	05	17.3	+53	35	6.7–13.9	Mira	458	M
Epsilon Aurigae	05	02.0	+43	49	2.9–3.8	Algol	9892	F+B
R Boötis	14	37.2	+26	44	6.2–13.1	Mira	223	M
X Camelopardalis	04	45.7	+75	06	7.4–14.2	Mira	144	K–M
R Cancri	08	16.6	+11	44	6.1–11.8	Mira	362	M
X Cancri	08	55.4	+17	14	5.6–7.5	Semi-regular	195?	C
R Canis Majoris	07	19.5	−16	24	5.7–6.3	Algol	1.1	F
VY Canis Majoris	07	23.0	−25	46	6.5–9.6	Unique	—	M
S Canis Minoris	07	32.7	+08	19	6.6–13.2	Mira	333	M
R Canum Ven.	13	49.0	+39	33	6.5–12.9	Mira	329	M
*V Canum Ven.	13	19.5	+45	32	6.5–8.6	Semi-regular	192	M
R Carinae	09	32.2	−62	47	3.9–10.5	Mira	309	M
S Carinae	10	09.4	−61	33	4.5–9.9	Mira	149	K–M
l Carinae	09	45.2	−62	30	3.3–4.2	Cepheid	35.5	F–K
Eta Carinae	10	45.1	−59	41	-0.8–7.9	Irregular	—	Pec
R Cassiopeiae	23	58.4	+51	24	4.7–13.5	Mira	430	M
S Cassiopeiae	01	19.7	+72	37	7.9–16.1	Mira	612	S
W Cassiopeiae	00	54.9	+58	34	7.8–12.5	Mira	406	C
Gamma Cas.	00	56.7	+60	43	1.6–3.0	Gamma Cas.	—	B
Rho Cassiopeiae	23	54.4	+57	30	4.1–6.2	Semi-regular	—	F–K
R Centauri	14	16.6	−59	55	5.3–11.8	Mira	546	M
S Centauri	12	24.6	−49	26	7–8	Semi-regular	65	C
T Centauri	13	41.8	−33	36	5.5–9.0	Semi-regular	90	K–M
S Cephei	21	35.2	+78	37	7.4–12.9	Mira	487	C
T Cephei	21	09.5	+68	29	5.2–11.3	Mira	388	M
Delta Cephei	22	29.2	+58	25	3.5–4.4	Cepheid	5.4	F–G
Mu Cephei	21	43.5	+58	47	3.4–5.1	Semi-regular	730	M
U Ceti	02	33.7	−13	09	6.8–13.4	Mira	235	M
W Ceti	00	02.1	−14	41	7.1–14.8	Mira	351	S
*Omicron Ceti	02	19.3	−02	59	2.0–10.1	Mira	332	M
R Chamaeleontis	08	21.8	−76	21	7.5–14.2	Mira	335	M
T Columbae	05	19.3	−33	42	6.6–12.7	Mira	226	M
R Comae Ber.	12	04.3	+18	47	7.1–14.6	Mira	363	M
*R Coronae Bor.	15	48.6	+28	09	5.7–14.8	R Coronae Bor.	—	C
S Coronae Bor.	15	21.4	+31	22	5.8–14.1	Mira	360	M
T Coronae Bor.	15	59.6	+25	55	2.0–10.8	Recurrent nova	—	M+Pec
V Coronae Bor.	15	49.5	+39	34	6.9–12.6	Mira	358	C
W Coronae Bor.	16	15.4	+37	48	7.8–14.3	Mira	238	M
R Corvi	12	19.6	−19	15	6.7–14.4	Mira	317	M
R Crucis	12	23.6	−61	38	6.4–7.2	Cepheid	5.8	F–G

Star	RA		Declination		Range	Type	Period	Spectrum
	h	m	°	′			(days)	
R Cygni	19	36.8	+50	12	6.1–14.4	Mira	426	S
U Cygni	20	19.6	+47	54	5.9–12.1	Mira	463	C
W Cygni	21	36.0	+45	22	5.0–7.6	Semi-regular	131	M
RT Cygni	19	43.6	+48	47	6.0–13.1	Mira	190	M
SS Cygni	21	42.7	+43	35	7.7–12.4	Dwarf nova	50±	K+Pec
*CH Cygni	19	24.5	+50	14	5.6–9.0	Symbiotic	—	M+B
Chi Cygni	19	50.6	+32	55	3.3–14.2	Mira	408	S
R Delphini	20	14.9	+09	05	7.6–13.8	Mira	285	M
U Delphini	20	45.5	+18	05	5.6–7.5	Semi-regular	110?	M
EU Delphini	20	37.9	+18	16	5.8–6.9	Semi-regular	60	M
Beta Doradûs	05	33.6	–62	29	3.5–4.1	Cepheid	9.8	F–G
R Draconis	16	32.7	+66	45	6.7–13.2	Mira	246	M
T Eridani	03	55.2	–24	02	7.2–13.2	Mira	252	M
R Fornacis	02	29.3	–26	06	7.5–13.0	Mira	389	C
R Geminorum	07	07.4	+22	42	6.0–14.0	Mira	370	S
U Geminorum	07	55.1	+22	00	8.2–14.9	Dwarf nova	105±	Pec+M
Zeta Geminorum	07	04.1	+20	34	3.6–4.2	Cepheid	10.2	F–G
Eta Geminorum	06	14.9	+22	30	3.2–3.9	Semi-regular	233	M
S Gruis	22	26.1	–48	26	6.0–15.0	Mira	402	M
S Herculis	16	51.9	+14	56	6.4–13.8	Mira	307	M
U Herculis	16	25.8	+18	54	6.4–13.4	Mira	406	M
Alpha Herculis	17	14.6	+14	23	2.7–4.0	Semi-regular	—	M
68, u Herculis	17	17.3	+33	06	4.7–5.4	Algol	2.1	B+B
R Horologii	02	53.9	–49	53	4.7–14.3	Mira	408	M
U Horologii	03	52.8	–45	50	6–14	Mira	348	M
R Hydrae	13	29.7	–23	17	3.5–10.9	Mira	389	M
U Hydrae	10	37.6	–13	23	4.3–6.5	Semi-regular	450?	C
VW Hydri	04	09.1	–71	18	8.4–14.4	Dwarf nova	27±	Pec
R Leonis	09	47.6	+11	26	4.4–11.3	Mira	310	M
R Leonis Minoris	09	45.6	+34	31	6.3–13.2	Mira	372	M
R Leporis	04	59.6	–14	48	5.5–11.7	Mira	427	C
Y Librae	15	11.7	–06	01	7.6–14.7	Mira	276	M
RS Librae	15	24.3	–22	55	7.0–13.0	Mira	218	M
Delta Librae	15	01.0	–08	31	4.9–5.9	Algol	2.3	A
R Lyncis	07	01.3	+55	20	7.2–14.3	Mira	379	S
R Lyrae	18	55.3	+43	57	3.9–5.0	Semi-regular	46?	M
RR Lyrae	19	25.5	+42	47	7.1–8.1	RR Lyrae	0.6	A–F
Beta Lyrae	18	50.1	+33	22	3.3–4.4	Eclipsing	12.9	B
U Microscopii	20	29.2	–40	25	7.0–14.4	Mira	334	M
*U Monocerotis	07	30.8	–09	47	5.9–7.8	RV Tauri	91	F–K

Some Interesting Variable Stars

Star	RA		Declination		Range	Type	Period	Spectrum
	h	m	°	′			(days)	
V Monocerotis	06	22.7	–02	12	6.0–13.9	Mira	340	M
R Normae	15	36.0	–49	30	6.5–13.9	Mira	508	M
T Normae	15	44.1	–54	59	6.2–13.6	Mira	241	M
R Octantis	05	26.1	–86	23	6.3–13.2	Mira	405	M
S Octantis	18	08.7	–86	48	7.2–14.0	Mira	259	M
V Ophiuchi	16	26.7	–12	26	7.3–11.6	Mira	297	C
X Ophiuchi	18	38.3	+08	50	5.9–9.2	Mira	329	M
RS Ophiuchi	17	50.2	–06	43	4.3–12.5	Recurrent nova	—	OB+M
U Orionis	05	55.8	+20	10	4.8–13.0	Mira	368	M
W Orionis	05	05.4	+01	11	5.9–7.7	Semi-regular	212	C
Alpha Orionis	05	55.2	+07	24	0.0–1.3	Semi-regular	2335	M
S Pavonis	19	55.2	–59	12	6.6–10.4	Semi-regular	381	M
Kappa Pavonis	18	56.9	–67	14	3.9–4.8	W Virginis	9.1	G
R Pegasi	23	06.8	+10	33	6.9–13.8	Mira	378	M
X Persei	03	55.4	+31	03	6.0–7.0	Gamma Cas.	—	O9.5
Beta Persei	03	08.2	+40	57	2.1–3.4	Algol	2.9	B
Zeta Phoenicis	01	08.4	–55	15	3.9–4.4	Algol	1.7	B+B
R Pictoris	04	46.2	–49	15	6.4–10.1	Semi-regular	171	M
RS Puppis	08	13.1	–34	35	6.5–7.7	Cepheid	41.4	F–G
L^2 Puppis	07	13.5	–44	39	2.6–6.2	Semi-regular	141	M
T Pyxidis	09	04.7	–32	23	6.5–15.3	Recurrent nova	7000±	Pec
U Sagittae	19	18.8	+19	37	6.5–9.3	Algol	3.4	B+G
WZ Sagittae	20	07.6	+17	42	7.0–15.5	Dwarf nova	1900±	A
R Sagittarii	19	16.7	–19	18	6.7–12.8	Mira	270	M
RR Sagittarii	19	55.9	–29	11	5.4–14.0	Mira	336	M
RT Sagittarii	20	17.7	–39	07	6.0–14.1	Mira	306	M
RU Sagittarii	19	58.7	–41	51	6.0–13.8	Mira	240	M
RY Sagittarii	19	16.5	–33	31	5.8–14.0	R Coronae Bor.	—	G
RR Scorpii	16	56.6	–30	35	5.0–12.4	Mira	281	M
RS Scorpii	16	55.6	–45	06	6.2–13.0	Mira	320	M
RT Scorpii	17	03.5	–36	55	7.0–15.2	Mira	449	S
Delta Scorpii	16	00.3	–22	37	1.6–2.3	Irregular	—	B
S Sculptoris	00	15.4	–32	03	5.5–13.6	Mira	363	M
R Scuti	18	47.5	–05	42	4.2–8.6	RV Tauri	146	G–K
R Serpentis	15	50.7	+15	08	5.2–14.4	Mira	356	M
S Serpentis	15	21.7	+14	19	7.0–14.1	Mira	372	M
T Tauri	04	22.0	+19	32	9.3–13.5	T Tauri	—	F–K
SU Tauri	05	49.1	+19	04	9.1–16.9	R Coronae Bor.	—	G
Lambda Tauri	04	00.7	+12	29	3.4–3.9	Algol	4.0	B+A
R Trianguli	02	37.0	+34	16	5.4–12.6	Mira	267	M

Star	RA		Declination		Range	Type	Period	Spectrum
	h	m	°	′			(days)	
R Ursae Majoris	10	44.6	+68	47	6.5–13.7	Mira	302	M
T Ursae Majoris	12	36.4	+59	29	6.6–13.5	Mira	257	M
*Z Ursae Majoris	11	56.5	+57	52	6.2–9.4	Semi-regular	196	M
*RY Ursae Majoris	12	20.5	+61	19	6.7–8.3	Semi-regular	310?	M
U Ursae Minoris	14	17.3	+66	48	7.1–13.0	Mira	331	M
R Virginis	12	38.5	+06	59	6.1–12.1	Mira	146	M
S Virginis	13	33.0	–07	12	6.3–13.2	Mira	375	M
SS Virginis	12	25.3	+00	48	6.0–9.6	Semi-regular	364	C
R Vulpeculae	21	04.4	+23	49	7.0–14.3	Mira	137	M
Z Vulpeculae	19	21.7	+25	34	7.3–8.9	Algol	2.5	B+A

V CANUM VENATICORUM 13h 19.5m +45° 32′ (2000)

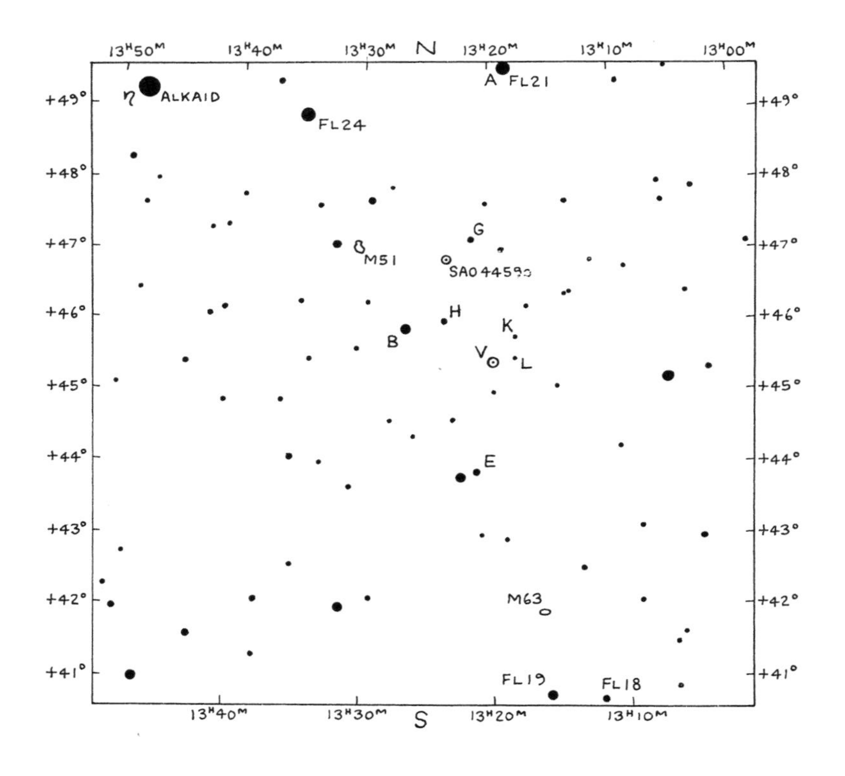

A 5.1 H 7.8
B 5.9 K 8.4
E 6.5 L 8.6
G 7.1

(MIRA) CETI 02h 19.3m −02° 59′ (2000)

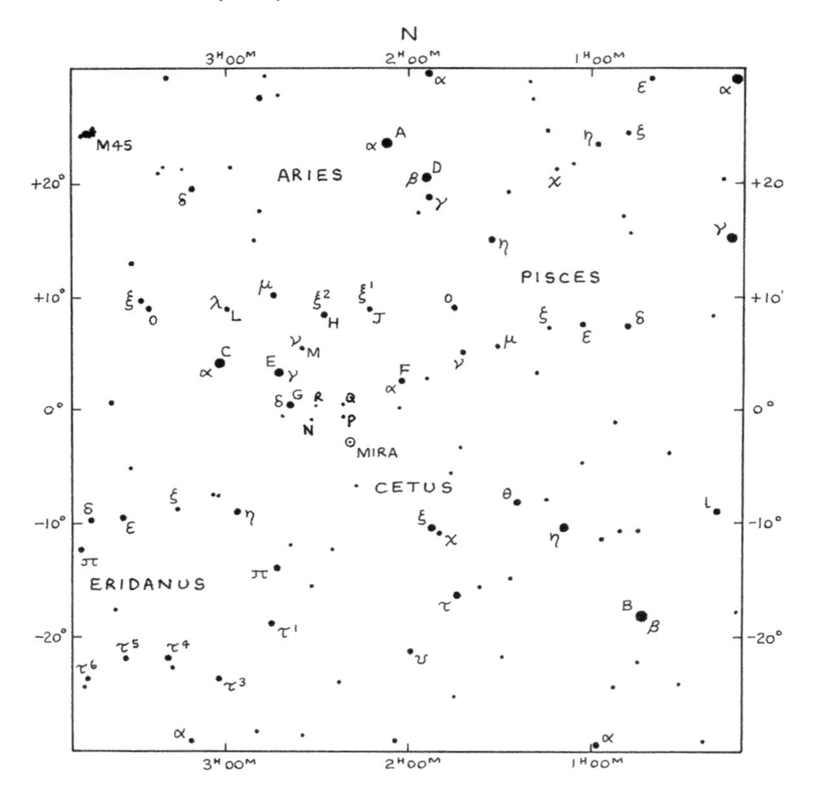

A 2.2	J 4.4
B 2.4	L 4.9
C 2.7	M 5.1
D 3.0	N 5.4
E 3.6	P 5.5
F 3.8	Q 5.7
G 4.1	R 6.1
H 4.3	

R CORONAE BOREALIS 15h 48.6m +28° 09′ (2000)

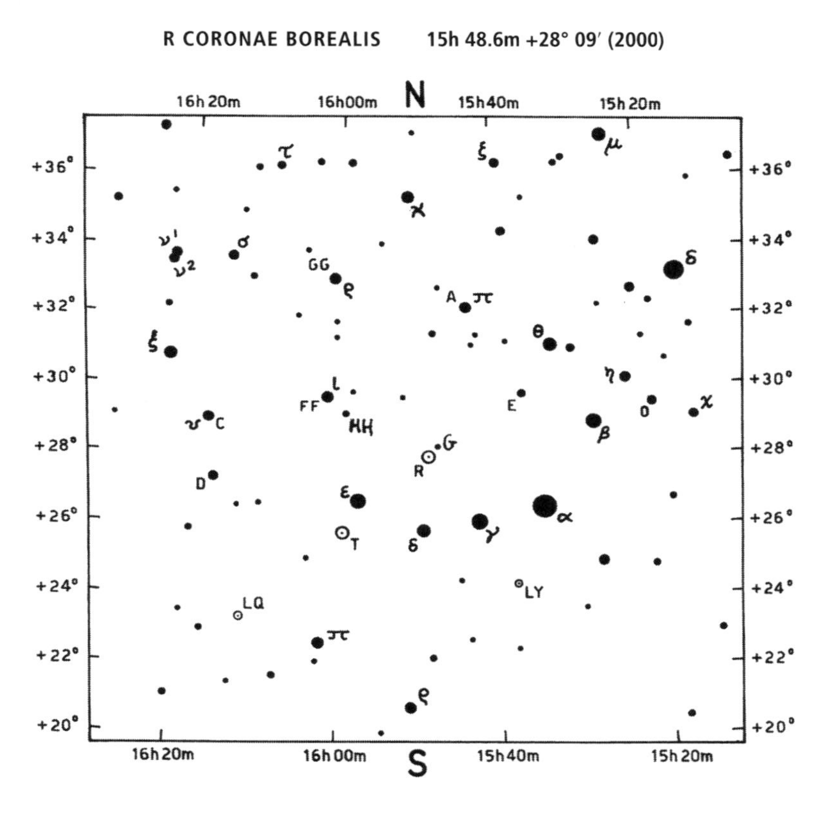

FF	5.0	C	5.8
GG	5.4	D	6.2
A	5.6	E	6.5
		HH	7.1
		G	7.4

CH CYGNI 19h 24.5m +50° 14′ (2000)

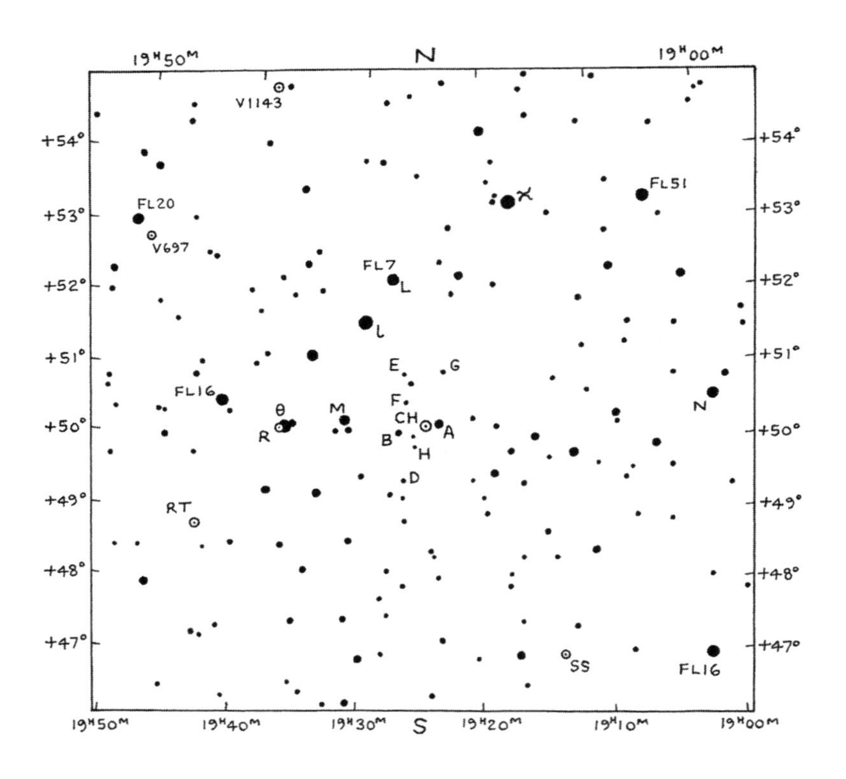

N 5.4	D 8.0
M 5.5	E 8.1
L 5.8	F 8.5
A 6.5	G 8.5
B 7.4	H 9.2

U MONOCEROTIS 07h 30.8m −09° 47′ (2000)

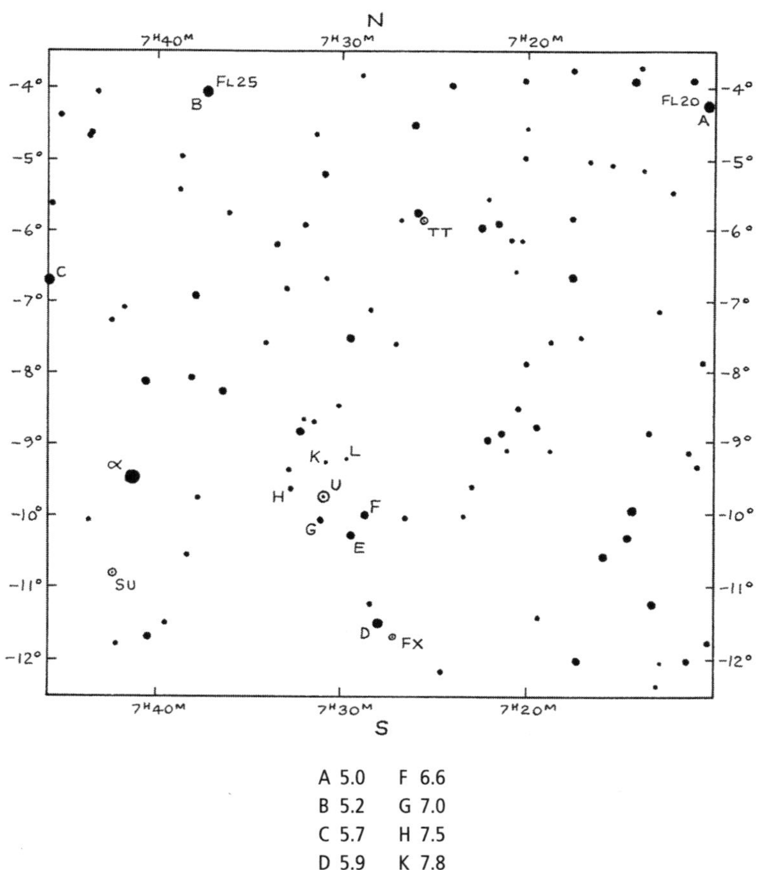

A 5.0	F 6.6
B 5.2	G 7.0
C 5.7	H 7.5
D 5.9	K 7.8
E 6.0	L 8.0

RY URSAE MAJORIS 12h 20.5m +61° 19′ (2000)
Z URSAE MAJORIS 11h 56.5m +57° 52′ (2000)

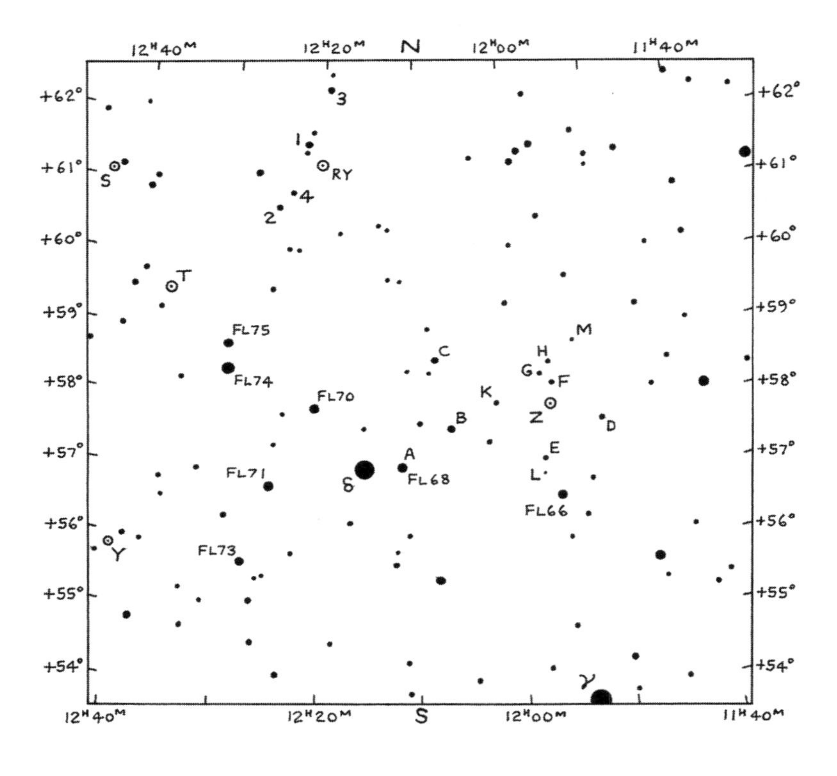

A	6.5	F	8.6	M	9.1
B	7.2	G	8.7	1	6.9
C	7.6	H	8.8	2	7.4
D	8.0	K	8.9	3	7.7
E	8.3	L	9.0	4	7.8

Mira Stars: Maxima, 2015

JOHN ISLES

Below are the predicted dates of maxima for Mira stars that reach magnitude 7.5 or brighter at an average maximum. Individual maxima can in some cases be brighter or fainter than average by a magnitude or more, and all dates are only approximate. The positions, extreme ranges and mean periods of these stars can be found in the preceding list of interesting variable stars.

Star	**Mean magnitude at maximum**	**Dates of maxima**
R Andromedae	6.9	17 Mar
R Aquarii	6.5	14 Apr
R Aquilae	6.1	8 Sept
R Boötis	7.2	6 Feb, 17 Sept
R Cancri	6.8	30 July
S Canis Minoris	7.5	12 June
R Carinae	4.6	2 Feb, 7 Dec
S Carinae	5.7	27 Feb, 26 Jul, 23 Dec
R Centauri	5.8	22 Feb
T Cephei	6.0	30 Apr
U Ceti	7.5	25 May
Omicron Ceti	3.4	3 May
T Columbae	7.5	14 May, 26 Dec
S Coronae Borealis	7.3	27 Aug
V Coronae Borealis	7.5	5 Nov
R Corvi	7.5	2 Mar
U Cygni	7.2	22 Apr
RT Cygni	7.3	12 Apr, 19 Oct
Chi Cygni	5.2	7 Aug
R Geminorum	7.1	28 Feb

Star	Mean magnitude at maximum	Dates of maxima
U Herculis	7.5	3 May
R Horologii	6.0	25 Aug
R Hydrae	4.5	2 Dec
R Leonis	5.8	22 Aug
R Leonis Minoris	7.1	21 Mar
RS Librae	7.5	3 July
V Monocerotis	7.0	11 Oct
R Normae	7.2	9 Sept
T Normae	7.4	9 Aug
V Ophiuchi	7.5	1 Sept
X Ophiuchi	6.8	20 June
U Orionis	6.3	30 Mar
R Sagittarii	7.3	18 Jan, 14 Oct
RR Sagittarii	6.8	21 Aug
RT Sagittarii	7.0	20 May
RU Sagittarii	7.2	1 Feb, 29 Sept
RR Scorpii	5.9	11 May
RS Scorpii	7.0	18 Apr
S Sculptoris	6.7	24 Dec
R Serpentis	6.9	28 July
R Trianguli	6.2	19 June
R Ursae Majoris	7.5	25 Oct
R Virginis	6.9	1 Mar, 25 July, 17 Dec
S Virginis	7.0	17 Mar

Some Interesting Double Stars

BOB ARGYLE

The positions, angles and separations given below correspond to epoch 2015.0.

No.	RA	Declin–ation	Star	Magni–tudes	Separa–tion	PA	Cata–logue	Comments
	h m	° ′			arcsec	°		
1	00 31.5	–62 58	β Tuc	4.4,4.8	27.1	169	LCL 119	Both stars again difficult doubles
2	00 49.1	+57 49	η Cas	3.4,7.5	13.3	324	STF60	Easy. Creamy, bluish. P = 480 years.
3	00 55.0	+23 38	36 And	6.0,6.4	1.1	329	Σ73	P = 168 years. Both yellow. Slowly opening.
4	01 13.7	+07 35	ζ Psc	5.6,6.5	23.1	63	Σ100	Yellow, reddish-white.
5	01 39.8	–56 12	p Eri	5.8,5.8	11.7	187	Δ5	Period = 484 years.
6	01 53.5	+19 18	γ Ari	4.8,4.8	7.5	1	Σ180	Very easy. Both white.
7	02 02.0	+02 46	α Psc	4.2,5.1	1.8	261	Σ202	Binary, period = 933 years.
8	02 03.9	+42 20	γ And	2.3,5.0	9.6	63	Σ205	Yellow, blue. Relatively fixed. BC now beyond
			γ2 And	5.1,6.3	0.0	42	OΣ38	range of amateur instruments (0″.02!)
9	02 9.1	+67 24	ι Cas AB	4.9,6.9	2.6	228	Σ262	AB is long-period binary. P = 620 years.
			ι Cas AC	4.9,8.4	7.1	117		
10	02 33.8	–28 14	ω For	5.0,7.7	10.8	245	HJ 3506	Common proper motion.
11	02 43.3	+03 14	γ Cet	3.5,7.3	2.3	298	Σ299	Not too easy.
12	02 58.3	–40 18	θ Eri	3.4,4.5	8.3	90	PZ 2	Both white.

No.	RA	Declin-ation	Star	Magni-tudes	Separa-tion	PA	Cata-logue	Comments
	h m	° ′			arcsec	°		
13	02 59.2	+21 20	ε Ari	5.2,5.5	1.3	210	Σ333	Closing very slowly. P=1216 years? Both white.
14	03 12.1	–28 59	α For	4.0,7.0	5.4	300	HJ 3555	P = 269 years. B variable?
15	03 48.6	–37 37	f Eri	4.8,5.3	8.2	215	Δ16	Pale yellow. Fixed.
16	03 54.3	–02 57	32 Eri	4.8,6.1	6.9	348	Σ470	Fixed. Deep yellow and white.
17	04 32.0	+53 55	1 Cam	5.7,6.8	10.3	308	Σ550	Fixed.
18	04 50.9	–53 28	ι Pic	5.6,6.4	12.4	58	Δ18	Good object for small apertures. Fixed.
19	05 13.2	–12 56	κ Lep	4.5,7.4	2.0	357	Σ661	Visible in 7.5 cm. Slowly closing.
20	05 14.5	–08 12	β Ori	0.1,6.8	9.5	204	Σ668	Companion once thought to be close double.
21	05 21.8	–24 46	41 Lep	5.4,6.6	3.4	93	HJ 3752	Deep yellow pair in a rich field.
22	05 24.5	–02 24	ε Ori	3.8,4.8	1.8	77	DA 5	Slow-moving binary.
23	05 35.1	+09 56	λ Ori	3.6,5.5	4.3	44	Σ738	Fixed. Both stars white.
24	05 35.3	–05 23	θ Ori AB	6.7,7.9	8.6	32	Σ748	Trapezium in M42.
			θ Ori CD	5.1,6.7	13.4	61		
25	05 40.7	–01 57	ζ Ori	1.9,4.0	2.2	167	Σ774	Can be split in 7.5 cm. Long-period binary.
26	06 14.9	+22 30	ε Gem	var,6.5	1.6	253	β1008	Well seen with 20 cm. Primary orange.
27	06 46.2	+59 27	12 Lyn AB	5.4,6.0,	1.9	67	Σ948	AB is binary, P = 908 years.
			12 Lyn AC	5.4,7.3	8.8	308		
28	07 08.7	–70 30	γ Vol	3.9,5.8	14.1	298	Δ42	Very slow binary.
29	07 16.6	–23 19	h3945 CMa	4.8,6.8	26.8	51	-	Contrasting colours. Yellow and blue.
30	07 20.1	+21 59	δ Gem	3.5,8.2	5.5	228	Σ1066	Not too easy. Yellow, pale blue.

Some Interesting Double Stars

No.	RA	Declin-ation	Star	Magni-tudes	Separa-tion	PA	Cata-logue	Comments
	h m	° ′			arcsec	°		
31	07 34.6	+31 53	α Gem	1.9,2.9	5.1	55	Σ1110	Widening. Easy with 7.5 cm.
32	07 38.8	–26 48	κ Pup	4.5,4.7	9.8	318	H III 27	Both white.
33	08 09.5	–47 20	γ Vel	1.8,4.1	40.3	229	Δ 65	Spectacular pair in lovely field.
34	08 12.2	+17 39	ζ Cnc AB	5.6,6.0	1.2	20	Σ1196	Period (AB)= 59.6 years. Near maximum separation.
			ζ Cnc AB–C	5.0,6.2	5.9	66	Σ1196	Period (AB-C) = 1115 years.
35	08 46.8	+06 25	ε Hyd	3.3,6.8	2.8	307	Σ1273	PA slowly increasing. A is a very close pair.
36	09 18.8	+36 48	38 Lyn	3.9,6.6	2.6	226	Σ1334	Almost fixed.
37	09 47.1	–65 04	υ Car	3.1,6.1	5.0	129	RMK 11	Fixed. Fine in small telescopes.
38	10 20.0	+19 50	γ Leo	2.2,3.5	4.6	126	Σ1424	Binary, period = 510 years. Both orange.
39	10 32.0	–45 04	s Vel	6.2,6.5	13.5	218	PZ 3	Fixed. Both white.
40	10 46.8	–49 25	μ Vel	2.7,6.4	2.4	56	R 155	P = 149 years. Near widest separation.
41	10 55.6	+24 45	54 Leo	4.5,6.3	6.6	111	Σ1487	Slowly widening. Pale yellow and white.
42	11 18.2	+31 32	xi UMa	4.3,4.8	1.8	177	Σ1523	Binary, 59.9 years. Needs 7.5 cm.
43	11 23.9	+10 32	ι Leo	4.0,6.7	2.1	96	Σ1536	Binary, period = 186 years.
44	11 32.3	–29 16	N Hya	5.8,5.9	9.4	210	H III 96	Both yellow. Long-period binary.
45	12 14.0	–45 43	D Cen	5.6,6.8	2.8	243	RMK 14	Orange and white. Closing.
46	12 26.6	–63 06	α Cru	1.4,1.9	4.0	114	Δ 252	Glorious pair. Third star in a low power field.
47	12 41.5	–48 58	γ Cen	2.9,2.9	0.2	169	HJ 4539	Period = 83.6 years. Nearing periastron. Both yellow.

No.	RA	Declin-ation	Star	Magni-tudes	Separa-tion	PA	Cata-logue	Comments
	h m	° ′			arcsec	°		
48	12 41.7	−01 27	γ Vir	3.5,3.5	2.3	6	Σ1670	Now widening quickly. Beautiful pair for 10 cm.
49	12 46.3	−68 06	β Mus	3.7,4.0	0.9	59	R 207	Both white. Closing slowly. P = 194 years.
50	12 54.6	−57 11	μ Cru	4.3,5.3	34.9	17	Δ126	Fixed. Both white.
51	12 56.0	+38 19	α CVn	2.9,5.5	19.3	229	Σ1692	Easy. Yellow, bluish.
52	13 22.6	−60 59	J Cen	4.6,6.5	60.0	343	Δ133	Fixed. A is a close pair.
53	13 24.0	+54 56	ζ UMa	2.3,4.0	14.4	152	Σ1744	Very easy. Naked-eye pair with Alcor.
54	13 51.8	−33 00	3 Cen	4.5,6.0	7.7	102	H III 101	Both white. Closing slowly.
55	14 39.6	−60 50	α Cen	0.0,1.2	4.1	288	RHD 1	Finest pair in the sky. P = 80 years. Closing.
56	14 41.1	+13 44	ζ Boo	4.5,4.6	0.4	290	Σ1865	Both white. Closing. Needs at least 30 cm.
57	14 45.0	+27 04	ε Boo	2.5,4.9	2.9	344	Σ1877	Yellow, blue. Fine pair.
58	14 46.0	−25 27	54 Hya	5.1,7.1	8.3	122	H III 97	Closing slowly. Yellow and reddish.
59	14 49.3	−14 09	μ Lib	5.8,6.7	1.8	6	β106	Becoming wider. Fine in 7.5 cm.
60	14 51.4	+19 06	ξ Boo	4.7,7.0	5.6	303	Σ1888	Fine contrast. Easy. P = 151.6 years.
61	15 03.8	+47 39	44 Boo	5.3,6.2	1.0	67	Σ1909	Period = 210 years. Closing quickly.
62	15 05.1	−47 03	π Lup	4.6,4.7	1.7	65	HJ 4728	Widening. Both pale yellow.
63	15 18.5	−47 53	μ Lup AB	5.1,5.2	0.9	300	HJ 4753	AB closing. Under-observed.
			μ Lup AC	4.4,7.2	22.7	127	Δ180	AC almost fixed.
64	15 23.4	−59 19	γ Cir	5.1,5.5	0.8	357	HJ 4757	Closing. Needs 20 cm. Long-period binary.

Some Interesting Double Stars

No.	RA	Declin-ation	Star	Magni-tudes	Separa-tion	PA	Cata-logue	Comments
	h m	° ′			arcsec	°		
65	15 34.8	+10 33	δ Ser	4.2,5.2	4.0	172	Σ1954	Long-period binary.
66	15 35.1	–41 10	γ Lup	3.5,3.6	0.8	277	HJ 4786	Binary. Period = 190 years. Needs 20 cm.
67	15 56.9	–33 58	ξ Lup	5.3,5.8	10.2	49	PZ 4	Fixed. Both pale yellow?
68	16 14.7	+33 52	σ CrB	5.6,6.6	7.2	238	Σ2032	Long period binary. Both white.
69	16 29.4	–26 26	α Sco	1.2,5.4	2.6	277	GNT 1	Red, green. Difficult from mid-northern latitudes.
70	16 30.9	+01 59	λ Oph	4.2,5.2	1.4	42	Σ2055	P = 129 years. Fairly difficult in small apertures.
71	16 41.3	+31 36	ζ Her	2.9,5.5	1.2	138	Σ2084	Period 34.5 years. Now widening. Needs 20 cm.
72	17 05.3	+54 28	μ Dra	5.7,5.7	2.5	3	Σ2130	Period 812 years. Both stars white.
73	17 14.6	+14 24	α Her	var,5.4	4.6	103	Σ2140	Red, green. Long-period binary.
74	17 15.3	–26 35	36 Oph	5.1,5.1	5.0	141	SHJ 243	Period = 471 years.
75	17 23.7	+37 08	ρ Her	4.6,5.6	4.1	319	Σ2161	Slowly widening.
76	17 26.9	–45 51	HJ 4949 AB	5.6,6.5	2.1	251	–	Beautiful coarse triple. All white.
			Δ216 AC	,7.1	105.0	310		
77	18 01.5	+21 36	95 Her	5.0,5.1	6.5	257	Σ2264	Colours thought variable in C19.
78	18 05.5	+02 30	70 Oph	4.2,6.0	6.3	126	Σ2272	Opening. Easy in 7.5 cm. P = 88.4 years.
79	18 06.8	–43 25	h5014 CrA	5.7,5.7	1.7	0	–	Period = 450 years. Needs 10 cm.
80	18 25.4	–20 33	21 Sgr	5.0,7.4	1.7	279	JC 6	Slowly closing binary, orange and green.
81	18 35.9	+16 58	OΣ358 Her	6.8,7.0	1.5	147	—	Period = 380 years.

No.	RA	Declin-ation	Star	Magni-tudes	Separa-tion	PA	Cata-logue	Comments
	h m	° ′			arcsec	°		
82	18 44.3	+39 40	ε1 Lyr	5.0,6.1	2.3	346	Σ2382	Quadruple system with epsilon2. Both pairs.
83	18 44.3	+39 40	ε2 Lyr	5.2,5.5	2.4	76	Σ2383	Visible in 7.5 cm.
84	18 56.2	+04 12	θ Ser	4.5,5.4	22.4	104	Σ2417	Fixed. Very easy. Both stars white.
85	19 06.4	−37 04	γ CrA	4.8,5.1	1.4	349	HJ 5084	Beautiful pair. Period = 122 years.
86	19 30.7	+27 58	β Cyg AB	3.1,5.1	34.3	54	Σ I 43	Glorious. Yellow, blue-greenish.
			β Cyg Aa	3.1,5.2	0.4	84	MCA 55	Aa. Needs 40 cm. Period = 214 years.
87	19 45.0	+45 08	δ Cyg	2.9,6.3	2.7	217	Σ2579	Slowly widening. Period = 780 years.
88	19 48.2	+70 16	ε Dra	3.8,7.4	3.0	19	Σ2603	Slow binary. Yellow and blue.
89	19 54.6	−08 14	57 Aql	5.7,6.4	36.0	170	Σ2594	Easy pair. Contrasting colours.
90	20 46.7	+16 07	γ Del	4.5,5.5	9.0	265	Σ2727	Easy. Yellowish. Long-period binary.
91	20 59.1	+04 18	ε Equ AB	6.0,6.3	0.2	282	Σ2737	Fine triple. AB a test for 40 cm. P = 101.5 years.
			ε Equ AC	6.0,7.1	10.3	66		
92	21 06.9	+38 45	61 Cyg	5.2,6.0	31.6	152	Σ2758	Nearby binary. Both orange. Period = 678 years.
93	21 19.9	−53 27	θ Ind	4.5,7.0	7.0	271	HJ 5258	Pale yellow and reddish. Long-period binary.
94	21 44.1	+28 45	μ Cyg	4.8,6.1	1.6	322	Σ2822	Period = 789 years.
95	22 03.8	+64 37	ξ Cep	4.4,6.5	8.4	274	Σ2863	White and blue. Long-period binary.
96	22 14.3	−21 04	41 Aqr	5.6,6.7	5.1	113	H N 56	Yellowish and purple?
97	22 26.6	−16 45	53 Aqr	6.4,6.6	1.3	61	SHJ 345	Long-period binary; periastron in 2023.

Some Interesting Double Stars

No.	RA		Declin-ation	Star	Magni-tudes	Separa-tion	PA	Cata-logue	Comments
	h	m	° ′			arcsec	°		
98	22	28.8	–00 01	ζ Aqr	4.3,4.5	2.2	165	Σ2909	Period = 487 years. Slowly widening.
99	23	19.1	–13 28	94 Aqr	5.3,7.0	12.3	351	Σ2988	Yellow and orange. Probable binary.
100	23	59.5	+33 43	Σ3050 And	6.6,6.6	2.4	339	–	Period = 717 years. Visible in 7.5 cm.

Some Interesting Nebulae, Clusters and Galaxies

Object	RA		Declination		Remarks
	h	m	°	′	
M31 Andromedae	00	40.7	+41	05	Andromeda Galaxy, visible to naked eye.
H VIII 78 Cassiopeiae	00	41.3	+61	36	Fine cluster, between Gamma and Kappa Cassiopeiae.
M33 Trianguli	01	31.8	+30	28	Spiral. Difficult with small apertures.
H VI 33–4 Persei, C14	02	18.3	+56	59	Double cluster; Sword-handle.
Δ142 Doradûs	05	39.1	–69	09	Looped nebula round 30 Doradûs. Naked eye. In Large Magellanic Cloud.
M1 Tauri	05	32.3	+22	00	Crab Nebula, near Zeta Tauri.
M42 Orionis	05	33.4	–05	24	Orion Nebula. Contains the famous Trapezium, Theta Orionis.
M35 Geminorum	06	06.5	+24	21	Open cluster near Eta Geminorum.
H VII 2 Monocerotis, C50	06	30.7	+04	53	Open cluster, just visible to naked eye.
M41 Canis Majoris	06	45.5	–20	42	Open cluster, just visible to naked eye.
M47 Puppis	07	34.3	–14	22	Mag. 5.2. Loose cluster.
H IV 64 Puppis	07	39.6	–18	05	Bright planetary in rich neighbourhood.
M46 Puppis	07	39.5	–14	42	Open cluster.
M44 Cancri	08	38	+20	07	Praesepe. Open cluster near Delta Cancri. Visible to naked eye.
M97 Ursae Majoris	11	12.6	+55	13	Owl Nebula, diameter 3′. Planetary.
Kappa Crucis, C94	12	50.7	–60	05	'Jewel Box'; open cluster, with stars of contrasting colours.
M3 Can. Ven.	13	40.6	+28	34	Bright globular.
Omega Centauri, C80	13	23.7	–47	03	Finest of all globulars. Easy with naked eye.
M80 Scorpii	16	14.9	–22	53	Globular, between Antares and Beta Scorpii.
M4 Scorpii	16	21.5	–26	26	Open cluster close to Antares.

M13 Herculis	16	40	+36	31	Globular. Just visible to naked eye.
M92 Herculis	16	16.1	+43	11	Globular. Between Iota and Eta Herculis.
M6 Scorpii	17	36.8	−32	11	Open cluster; naked eye.
M7 Scorpii	17	50.6	−34	48	Very bright open cluster; naked eye.
M23 Sagittarii	17	54.8	−19	01	Open cluster nearly 50′ in diameter.
H IV 37 Draconis, C6	17	58.6	+66	38	Bright planetary.
M8 Sagittarii	18	01.4	−24	23	Lagoon Nebula. Gaseous. Just visible with naked eye.
NGC 6572 Ophiuchi	18	10.9	+06	50	Bright planetary, between Beta Ophiuchi and Zeta Aquilae.
M17 Sagittarii	18	18.8	−16	12	Omega Nebula. Gaseous. Large and bright.
M11 Scuti	18	49.0	−06	19	Wild Duck. Bright open cluster.
M57 Lyrae	18	52.6	+32	59	Ring Nebula. Brightest of planetaries.
M27 Vulpeculae	19	58.1	+22	37	Dumb-bell Nebula, near Gamma Sagittae.
H IV 1 Aquarii, C55	21	02.1	−11	31	Bright planetary, near Nu Aquarii.
M15 Pegasi	21	28.3	+12	01	Bright globular, near Epsilon Pegasi.
M39 Cygni	21	31.0	+48	17	Open cluster between Deneb and Alpha Lacertae. Well seen with low powers.

(M = Messier number; NGC = New General Catalogue number; C = Caldwell number.)

Our Contributors

Pete Lawrence has been a keen amateur astronomer since the mid-1960s. He qualified with a degree in Physics with Astrophysics from the University of Leicester in the early 1980s and subsequently became interested in digital imaging and processing techniques. His images now regularly appear all over the world on the Internet, in magazines and in books. Many examples of Pete's work appear on his website, www.digitalsky.org.uk. He currently writes for the BBC *Sky at Night* Magazine as their resident imaging expert, observing guide and equipment reviewer. He is also a regular contributor to the *Sky at Night* television programme.

Martin Mobberley is one of the UK's most active imagers of comets, planets, asteroids, variable stars, novae and supernovae and served as President of the British Astronomical Association from 1997 to 1999. In 2000, he was awarded the Association's Walter Goodacre Award. He is the sole author of seven popular astronomy books published by Springer as well as three children's 'Space Exploration' books published by Top That Publishing. In addition he has authored hundreds of articles in *Astronomy Now* and numerous other astronomical publications.

Dr Stephen Webb is the author of several books, including *Where Is Everybody?* (winner of the Contact in Context award, and shortlisted for the 2003 Aventis Prize for science books) and the more recently published *New Eyes on the Universe: Twelve Cosmic Mysteries and the Tools We Need To Solve Them* (which gives an overview of the many observatories that are planned or are already under construction). He blogs on these and other topics in science at stephenwebb.info.

Dr John Mason is the Editor of the *Yearbook of Astronomy*. He is a past President of the British Astronomical Association and Director of the

BAA's Meteor Section. He is currently Principal Lecturer at the South Downs Planetarium and Science Centre in Chichester. He appeared many times with Sir Patrick Moore on BBC TV's *The Sky at Night*. For over thirty years he has been leading overseas expeditions to observe and record annular and total solar eclipses, the polar aurora and major meteor showers. He was made an MBE in the 2009 New Year's Honours List for his services to science education.

Richard Myer Baum is a former Director of the Mercury and Venus Section of the British Astronomical Association, an amateur astronomer and an independent scholar. He is author of *The Planets: Some Myths and Realities* (1973), (with W. Sheehan) *In Search of Planet Vulcan: The Ghost in Newton's Clockwork Universe* (1997) and *The Haunted Observatory* (2007). He has contributed to the *Journal of the British Astronomical Association*, *Journal for the History of Astronomy*, *Sky & Telescope* and many other publications including *The Dictionary of Nineteenth-Century British Scientists* (2004), and *The Biographical Encyclopedia of Astronomers* (2007).

Dr David M. Harland gained his B.Sc. in astronomy in 1977 and a doctorate in computational science. Subsequently, he has lectured in computer science, worked in industry and managed academic research. In 1995, he 'retired' and has since published many books on space themes.

Dr Allan Chapman, of Wadham College, Oxford, is probably Britain's leading authority on the history of astronomy. He has published many research papers and several books, as well as numerous popular accounts. He is a frequent contributor to the *Yearbook*.

Note from the Editor

Previous editions of the Yearbook have included a list of Astronomical Societies in the British Isles. It has become apparent that much of this information is rather out-of-date and so it has been decided to omit this section from the Yearbook this time.